COPING WITH INCREASING COMPLEXITY

COPING WITH INCREASING COMPLEXITY

Implications of General Semantics and General Systems Theory

Edited by
Donald E. Washburn
North Adams State College
and
Dennis R. Smith
Temple University

GORDON AND BREACH SCIENCE PUBLISHERS
New York London Paris

Gordon and Breach, Science Publishers, Inc.
1 Park Avenue
New York, N.Y. 10016

Editorial office for the United Kingdom
Gordon and Breach, Science Publishers Ltd.
42 William IV Street
London W.C.2.

Editorial office for France
Gordon & Breach
7-9 rue Emile Dubois
Paris 75014

Library of Congress catalog card number 72-80160. ISBN 0 677 14940 9. Printed in Great Britain.

Preface

Change, growth, complexity are the hallmarks of the modern age. Science and technology have transformed life in our century in ways that are both qualitative and quantitative, proceeding at an exponential rate. We find ourselves in the midst of countless explosions: the population explosion, the information explosion, the arms race, the accelerating consumption of energy and natural resources, the increase in wastes and pollutants, which threaten to make our planet less inhabitable. Everything seems to be getting bigger, more rapid, harder to comprehend. And so we struggle to find new ways to cope with what is happening, to understand and control forces that threaten to overwhelm us.

In the world of information the problem has been worsened by the tendency of scholarship and research to become increasingly more fragmented and specialized. As more and more knowledge has accumulated in an ever widening circle of investigation, it has become more difficult for one person to encompass even a portion of a growing field of study. Moreover, there are fewer thinkers who are able to transcend the limits of their specialty or who have the time to master the special techniques and vocabularies of fields other than their own. The result is a kind of academic tower of Babel which is contributing to the problem of increasing complexity rather than helping to solve it.

Fortunately there have been some efforts in our time to develop a basis for integrating knowledge and for making people aware of the ways in which information is abstracted and communicated. One of these is the Non-Aristotelian system of Alfred Korzybski, the application of which he called general semantics. The aim of general semantics is to enhance human sanity by making people conscious of the way the nervous system (including symbolizing processes of all kinds) works. In this way an allowance can be made for the inevitable distortions that result when people talk to each other about what they think is happening. General semantics provides a basis for unifying our conception of man's relation to the illimitable flux of events that make up the content of all knowledge.

Another effort in this direction has become known as general systems theory. This has brought together people who are interested in what systems of all sorts, models from every field of knowledge, have in common. It is a search for isomorphism, patterns of similarity that seem fundamental no matter what magnitudes or dimensions of the universe are involved. And it is carried out with the hope that such an inquiry will provide a basis for a synthesis and a restructuring of knowledge so as to simplify the assimilation of complex bodies of information. Although it is a relatively new concern, general systems theory seems to be the intellectual equivalent of the planetary consciousness which is gradually emerging in our century, and, indeed, must prevail if we are to survive as a race.

Two organizations are now active in promoting these ideals of personal and epistemological integration. The first is the Institute of General Semantics, founded in 1938 by Count Korzybski, and dedicated to "linguistic, epistemologic, scientific research and education". The Institute has sponsored annual lectures, seminars, and the General Semantics Bulletin. It has also in recent years promoted a series of conferences bringing together people who have made a contribution to the development of general semantics and others working in related disciplines. The second organization is the Society for General Systems Research, with offices in Washington D.C. It is the largest international professional society devoted to the task of organizing and unifying knowledge. The Society publishes General Systems (Yearbook for the Society for General Systems Research) and fosters interdisciplinary communication through meetings, conferences, reports, and special standing committees and task forces.

In keeping with the ideal of promoting interdisciplinary dialogue, the two organizations collaborated in May of 1970 in sponsoring a joint conference at the Phipps Conference Center, University of Denver. The theme of the conference was "Coping with Increasing Complexity". Approximately thirty-five scholars from general semantics and general systems theory presented papers and discussed the problem of how the two disciplines could be related in a mutually productive way. There was also much interest in the ecological crisis, student unrest, information overload, alienation and depersonalization, and other symptoms of complexity. Revisions of papers presented at the conference and others subsequently solicited from scholars not present make up the content of this volume. They represent an attempt to formulate in a preliminary way some basis for helping twentieth century man to contend with a world that is rapidly becoming incomprehensible and unmanageable.

The collection represents a wide range of disciplines and concerns. The first article, by Dr. Elwood Murray, former director of the

Institute of General Semantics, is a kind of overture, announcing the theme of the volume and indicating ways in which general semantics and general systems theory might contribute to solving the problem of increasing complexity. The articles by Kenneth G. Johnson and Charlotte Read will be useful to the reader who needs to know something more about general semantics itself. T. Samuel Bois' article also deals with some fundamental issues. The logical basis of general semantics is explored by C. A. Hilgartner in an article which shows that ordinary set theory is not adequate to express the premises of a non-Aristotelian system. He proceeds to make the necessary revisions.

For the reader interested in what general semantics and general systems theory have in common, there is Robert P. Pula's article, which makes a point-by-point comparison. Other articles show how the two disciplines might be applied to specific problems. The Bourlands deal with improving clarity and exactness in language and with keeping track of information, using E-prime and the extensional devices of general semantics. Information retrieval is also the subject of an article by Glynn Harmon. Subsequent articles by Messeurs Broadus, Flagg, Silverman, and Banathy discuss how the two disciplines might be taught or might influence the nature of the teaching process itself.

There are also several articles that apply to research efforts and model building. Alvin A. Goldberg reviews the status of research in general semantics. Other contributions by James C. McFarland; Dennis R. Smith and Bob R. Hunt; and Russell W. Jennings, Joe Vinovich, and Thomas J. Pace explore the relevance of general semantics and general systems theory to a comprehensive theory of communication. Donald Washburn's article discusses how a systems orientation might be useful in integrating many of the traditional concepts of literary criticism and analysis.

A final group of articles deals with broad epistemological, social, and ecological issues. Anthony Wilden, Lee Thayer, and Alton Barbour analyze the mentality that lies behind the current technological and ecological crisis. E. S. Stafford looks ahead to the challenges of the communication revolution and the information explosion. Jere W. Clark and Edward Haskell have written articles the aim of which is to move us closer to the integration of knowledge and to the educational revisions that will be necessary if we are to cope with increasing complexity. Perhaps an important instrument in the integration of knowledge will be Stuart C. Dodd's Epicosm

Model of the Universe, which represents an effort to develop a testable theory of the cosmos based on a small number of repeating, combining, permuting, and interacting patterns of relationship. It is a somewhat abstract contribution, but very rewarding for the reader who is willing to spend time working through the formulations.

The problem of dealing with complexity is obviously itself complicated. If we are to cope with increasing complexity in our modern world, we must strive, insofar as possible, for wisdom. And wisdom, as Alfred North Whitehead wrote in 'Science and the Modern World', "is the fruit of a balanced development". People can no longer be mere specialists with fixed duties, competent in only a very limited way. We must attain to a kind of global awareness. And in the academic world, the traditional boundaries must be somehow transcended. It is to be hoped that the cross-fertilization of disciplines will help to foster the emergence of more balanced and far-sighted men who can solve the complex problems that are pressing upon us from all sides and who can help Man evolve in harmony with his universe.

D.E.W.
D.R.S.

Table of Contents

General Semantics and General Systems Theory: The Foundations for an 'Ecology of Knowledges'

ELWOOD MURRAY

The pertinence of the theme "Coping With Increasing Complexities", has been well stated by Roy F. Larson.

> Whether we like it or not, life in the Seventies is unprecedently complicated. What seems simple is complex; what seems obvious is subtle. In fact, anything that seems to be, probably is not.[1]

In a sort of brilliant stupidity, Mankind continues to labor under widespread illusions, if not delusions, in a tendency to confirm and defend various self-deceptions.[2]

With the continuing failure of education to inculcate foresight, proper evaluation, and decent human communication, Man's technology seems to be having exponential consequences – more for the worse than for the better. New dangers threaten our survival on several fronts at the very time in which science has made almost everything desired possible for human well-being.

While many old problems are now no longer serious, we are faced with new crises such as environmental degradation, deprivation, ignorance of what is necessary to carry on our civilization, alienation and breakdown of interpersonal and organizational communication, vast problems in the economy, war, a multiplication of mental health disorders, along with increases in crime and injustice.

Most people who are able to comprehend our difficulties and who recognize the challenge of change are not able to communicate this to their leaders. The leaders who have the necessary perceptions are beset with the old thought-action dichotomy in which they fail to get through to their followers. This apathy occurs in spite of the mass and multi-media today. At the same time these media contribute to an overload of information, with a saturation of trivia. As our seemingly smallest problems often escalate into global predicaments, there have arisen complications apparently hopeless of resolution.

In regard to this situation, R. Buckminster Fuller says of the scholarly communities:

> Because the experts had been sifted out to do highly specialized work and because they couldn't integrate their work and realize the potential for society of their individual capabilities, they just began to talk baseball to one another. It had to be left to someone else to harvest the potential to society.
>
> Having deliberately sifted the brights from the dulls and made all the brights specialists, it was left to the dulls to put things together again.[3]

Thus far there have been only weak movements to break through the narrow academic interests and the verbal departmental barriers. The proliferating jargons and terminologies and the one-sided emphasis on specialization defeat the educational process and confirm the pedagogical institutions in their failures. This is accentuated in the exclusiveness of professions profiting from their monopolies of knowledges. The interferences in human evaluating-communicating are part and parcel of every problem above mentioned. Until this mis-education is alleviated, there will be little prevention and improvement, let alone resolution of our difficulties.

Because of the absorption with parts and pieces, the blindness to wholes and relationships, and the incomprehension of underlying systems, decision-making comes too late or too soon and with too little effect or too much. These mis-evaluations extend into practically all systems, their sub-systems and their super-systems, including social relationships. Interpersonal and organizational communication is lamentably weak in the necessary deep listening – a failure of the social feedbacks.[4] There is far too little proper balancing of freedoms with restraints (and vice versa) with pernicious effects on organizations and institutions.

Our social systems often interfere with each other to the point of destructive competition and mutual death. An outgrowth of the issue of what is the most viable form of human organization is the confrontation with communism. But although totalitarianism buttressed by space-age science challenges hard-won democracies, the communists have similar and perhaps more virulent problems.

As we perceive general systems, the most important formulation is suggested by the word INTERACTION.[5] Interaction is a chief object of investigation of systems at all levels and dimensions in the general systems discipline. Ludwig von Bertalanffy defined systems as complex interacting elements:

> Interaction means that elements, P, stand in relations, R, so that the behavior of an element P in R is different from its behavior

> in another R, R^1. If the behaviors of R and R^1 are not different, there is no interaction and the elements behave independently with respect to R and R^1.[6]

Interaction is explicated in great depth with concepts such as linkages, connections, transformations, functionings, correlations, isomorphisms, parallels, relations, order, disorder, organization, organized complexity, feedback, adaption, states, purpose, equilibrium, transactions, entropy, etc. etc. A distinguishing feature of general systems is the emphasis upon RELATIONS from the entire gamut of sciences.

Walter Buckley perceives a system as complex of components:

> directly or indirectly related to a casual network, such that at least some of the components are related to some others in a more or less stable way AT ANY ONE TIME. The interrelations may be mutual or unidirectional, linear, non-linear or intermittent, and varying in degrees of casual efficacy or priority. The particular kinds of more or less stable relationships of components that become established at any time constitute the particular structure of the system at the time.[7]

A similar stress on relationships has been a major theme of general semantics from its inception in 1933. The whole methodology of this discipline is designed to bring about a relational orientation whereby practically everything and every piece is perceived as existing only in its whole complex of relationships instead of as separate, static, piecemeal objects.

Derived from the widest of interdisciplinary sources, Alfred Korzybski's Science and Sanity (1933) remains the fountainhead for General Semantics.[8] Korzybski is to General Semantics as Freud is to psychiatry in some respects. An applied epistemology, General Semantics is relevant wherever there is concern for improving the knowing and communication processes. The sub-title of Korzybski's book is 'An Introduction to Non-Aristotelian Systems.'

In the chapters concerning order, function, structure, relations, Korzybski may be considered as a forerunner of general systems. The whole discipline-methodology asks that the Universe be viewed as an interaction of relational structures. Students of General Semantics often tend to regard the literatures of general systems as supplementary and extensions of General Semantics (and vice-versa). The findings of general systems, more recent than General Semantics, may often be directly incorporated into General Semantics.

While general systems theory appears to supplement General

Semantics, in a similar sense General Semantics provides a powerful supplement to general systems. This occurs at the very points wherein general systems is lagging in application at points vital in the coherence and unity of general systems itself. General Semantics is particularly necessary, we think, in the resolution of problems now stymied by the vast terminologies encompassed. Terminologies are often so specialized that communication across the boundary lines in the same general area (mathematics, as an example) keeps scholars isolated, compounding confusions and misunderstandings.

In the realm of General Semantics there are some 3,000 extant language systems which have emerged from countless socio-cultures. According to the anthropologist Benjamin Wharf, "all higher levels of thinking are dependent on language. Also, the structure of the language one habitually uses influences the manner in which he understands his environments. The picture of the Universe shifts from tongue to tongue".[10] The 3,000 languages carry with them 3,000 sets of doctrines, ways of observing, and orientations. In this sense, Korzybski says:

> A (language) system represents a complex whole of coordinated doctrines resulting in methodological rules and principles of procedure (grammar, etc.) which affect the orientation by which we act and live. Any system involves an enormous number of assumptions, pre-suppositions, etc. which in the main are not obvious but operate unconsciously.[11]

For the relating of Man to his various environments, general systems must necessarily be much concerned with languages as systems – and this includes linguistic codes, grammar, logic and their rules. Perhaps the relatively small references to language in the general system literature arises from a frustration in the teaching of these subjects. Traditionally, and in large measure today, the teaching professions hold tightly to only one dimension or set of relationships: word to word, statement to statement, paragraph to paragraph. Organization and composition proceeds inefficiently without adaption to the personality, group dynamics, and isomorphic factors. Many teachers of English and Speech remain naive to the deeper-level central matters of the communication process and then wonder why so many of their efforts are ignored by the "effective" communicators.

Concerning symbols (and language systems), Heinz Von Foerster points out that

> . . . it becomes impossible to separate a symbol from its

symbolizer, his sensory motor and mental capabilities and constraints. And further, it becomes impossible to separate symbol and symbolizer from his environment which we have to populate with other symbolizers in order that symbolization makes any sense at all.[12]

The structure for interaction of Man with his environment may be described as a series of isomorphic transformations. Beginning with sensory abstractings from object-events, there is a progression of abstractings into symbolizing, language structures, and other behaviors.

Isomorphism, that is, structural similarity, has been derived from morphology, the science of structure. A. D. Fagan and R. E. Fagan define isomorphism as "a one-to-one correspondence between objects which preserve the relationships".[14] For sanity, fidelity of statements, and credibility, language must have correspondence with the "fact" structures of which the language structures are "maps". In a changing world, there must be constant correction of assumptions (that is, neural semantic patterns) into a workable isomorphism with "reality".[15] Adequacy of human evaluations basically depends upon this matching of structures.

The manner in which language outputs are brought into isomorphism with the "inputs" from the "fact" structures is outlined in Korzybski's STRUCTURAL DIFFERENTIAL. This diagram indicates the circularity which occurs between the sensory inputs from the environments into the behavioral outputs, then back into the environments in which the organism lives.

It may be said that Man lives in two environments simultaneously — with both of which he must come to terms. He must not permit the potent array of sensory signals inside his body (including those from memory) to become confused with or to overwhelm the sensory signals from the outside.

Insofar as neurological studies of the brain and nervous system have made the information available, General Semantics would describe the central phenomena of symbolizing and language structure. Often General Semantics has been called "neuro-semantics" and "neuro-linguistics" with the semantic reaction and pattern as a unit for meaning. This all makes available a feasible way in which to view Man-as-a-whole in his environment-as-a-whole.

Joined into the structural differential are some fifteen formulations which comprise the General Semantics "model". Of primary importance for an optimum ability to cope with change and complexity is what is called the extensional orientation. This attitude, which becomes built-in with maturity, powers an intensive searching into the levels of operations and relationships as it is "turned on" into the environments – verbal and otherwise – and the inner intensionality of the person himself and others. It is an outlook ever searching for order, function, transformations, structures.

The extensional orientation becomes reinforced through CONSCIOUSNESS OF ABSTRACTING; by abstracting is meant the characteristics of an event to which there are responses – that which is noticed. Consciousness of abstracting makes possible in an applied epistemology the knowing about knowing. Abstracting of abstracting proceeds into higher orders indefinitely from the sensory inputs through the language and other behavioral outputs. Abstracting implies that all knowing is partial; that no one knows all about anything (including General Semantics).

The pattern and shape of assumptions emerges from abstracting. Their correction into isomorphism with the event-structures requires DELAY OF REACTION – an engaging of the cortex and experience in longer nerve circuits. Correction of assumptions to their referents involves a review and checking of inferences for correspondence to events being evaluated. This occurs in the above mentioned circularity (a feedback) from the events to the various orders ("levels") of abstracting. These acts of reviewing tend to overcome impulsive and signal responses. Anything short of this deeper cerebral functioning defeats predictability and evaluation. This certainly has no place in science or in any complex evaluation.

Most assumptions by which we live remain unconscious. The checking for correspondence to "fact" structures tends to bring them into consciousness to make their correction and control possible. Delay permits linking of knowledges and experiences. This is required for the guided spontaneity of intuitive and creative insights, for rational-critical thought. There is no meaning and symbolizing without the joining of two or more perceptions. There is no relating without a semantic connection of at least two things.

Unconscious allness attitudes fuel dogmatism, prejudice, and rigidities, which often build into conflict. They tend to screen out what is not doctrinally acceptable.

No process in the Universe is identical with another process or with itself. This statement must apply to persons and what they say. If this is so, no language structure is identical with its "fact" structure. Description, summaries, higher order abstractions must be distinguished. But this unconscious confusion of orders of abstraction is the most common semantic blockage. Important to General Semantics is the principle of NON-IDENTIFICATION. "Maps are NOT their territories; words are NOT their facts." Young scientists and others waste great efforts until they become aware of the semantic traps and static implications in various forms of the verb "to be" and in numerous words implying only two possibilities.

For greater precision in bringing language into correspondence with the dynamic existences are the five extensional devices. These are usually applied implicitly – that is, mentally as the sentences are being put together and before they are spoken or written. This "thinking before you speak" reduces the use of the more static and antomizing words and permits more accurate details closer to the reality being expressed. These devices, which have important implications, can only be mentioned here.

More specific thinking comes from use of the semantic INDEX. The affixing of a number mentally brings the event into circumstances in space. For instance, crime indexed to a number becomes $crime_{81}$. Further toward operational perception is the use of the mental date, such as $crime_{81}$ January 30, 1971. These devices break through dead level abstracting by means of a circularity in working back and forth from the higher to lower orders of abstracting. This flexibility becomes increasingly necessary for new inductions and deductions as the events being evaluated become more global and complex.[16]

Other extensional devices include the use of mental QUOTES as a warning that words are only approximations; the ETCETERA, a reminder that the statement does not say all about the object-event; and the mental HYPHEN which overcomes dichotomous thinking about unsplitable objectives, systems, and situations. For example, space separated from time exists as space-time; mind separated from body as mind-body-soul; intellect separated from emotion as intellect-feeling; your automobile is the result of a unified process of capital-labor-management. These are examples of verbal splitting of what must be unitary processes. Upon this verbal fragmenting are based our most important institutions. The whole human being is treated by inefficient psychologies (which in themselves are anatomized) by the medical professions and the religious institutions. Your automobile precariously arrives from the assembly line in spite of frequent conflicts of three institutions.

Indeed, the mental hyphen may have especial importance in

overcoming the piece-meal thinking and fragmentation holding back general systems and the learning institutions. As Sanford has pointed out, we have separated – and then pretty well institutionalized the separation – action-research: the analysis of the problem, conceptualization, data gathering, planning, execution, training – all of the many activities that make up Kurt Lewin's model.[17]

A most important mechanism of language systems which we can only refer to here is called MULTI-ORDINALITY. An understanding of m-o will help resolve many semantic and epistemological problems. M-o refers to the important words in science and the general vocabulary which take their meanings from their contexts and hence have as many meanings as there are contexts for them. For specific meanings, they are functions of other words. Example: $cause_1$ (cause of an explosion); $cause_2$ (cause of John dropping out); $cause_3$ (cause of Mary's measles). There are hundreds of such words as yes, no, true, false, order, function, cause, effect, structure, system, meaning, evaluation, relations, etc.

General Semantics makes possible an ECOLOGY of KNOWLEDGES. Along with the formulations above mentioned, there is the ORDERING OF ABSTRACTIONS, the ability to separate reports and descriptions from summaries, summaries of summaries, and higher order abstractions. The guiding principle for this Korzybski called the NATURAL ORDER OF EVALUATION. M-o makes the objects and the sensory abstractions of them of primary importance, while summaries and higher order abstractings are kept secondary. Persons who can put into effect this ordering of abstractions as they approach the vast deposits of abstractions in the library will be less confused in respect to the first order events and structure. Events are not there operationally and relationally, and what is said about any event is not all. In this there is an application of epistemology, a knowing about one's knowing.

Formation of symbols and the assembling of words into statements occur through what is called SELF-REFLEXIVENESS of semantic reaction to s-r and word to word. In the warming-up of anger, fear, love, etc., one s-r often triggers another s-r etc. in sequences, indefinitely. In most all critical exchanges, words are used to talk about words, as when maps may be made of maps of higher order abstraction. A statement (an order of abstraction) about another statement (another order of abstraction) must keep in isomorphism with the fact-territory of the first statement.

Persons not conscious of this mechanism tend to fall into verbal traps, misunderstanding, impasses, and even conflicts. Unchecked self-reflexiveness at each order of abstraction may escalate into various emotional disturbances. For example, fear (a second order reaction) of fear (a first order of reaction), and love of love, etc. lead into hopeless impasses characteristic of not a few mental disorders. Consciousness of abstracting enables the speaker or writer to talk sensibly about the evaluation-communication process (the interactions) with the other person(s). They become better able to control signal reactions and to break through the unconscious identifications, confusions, and allnesses.

Self-reflexiveness, in short, is central in idea formation and language behaviors. Guidance into goals and purposes emerge from the many feedback levels of the brain and nervous system. General semantics is especially concerned with FEED-FORWARD as powered by POSITIVE FEEDBACK into which NEGATIVE FEEDBACK is joined.[18]

Feedback refers to the response to a signal as inferred by the sender. Positive feedback occurs as the response seems understanding and accepting. Negative feedback restrains or stops the sender until he changes his message. In ways not known at present, the critical, selecting action of the cortex inhibits positive freewheeling action irrelevant to goals and purposes. It also corrects non-isomorphism by asking factual questions.

A basic "unit" of language structure has been devised by Bert Decker. It is called the VERIFIABLE FUNCTION. If the order of abstraction is so "high" and vague, if it cannot be verified, it remains in limbo as useless as well as meaningless. The extensional devices raise countless factual questions in such statements as "civilization advances", "man is a mistake of Nature", etc. But "He turned on the switch ten times yesterday", and "He left his home at ten years of age" become verifiable and functional in space-time. The verifiable function, which Bert Decker makes the countable unit of language structure, consists of the joining of a noun with a verb, both of which can be demonstrated.[16] Thus far we know of no way of measuring the underlying unconscious semantic reactions not attached to words. This precludes many of the uses of "to be" and "is".

The basis for creativity and innovation depends upon the amount of fuel fed into positive feedback, that is, how much the person knows, the depth of his experiences, and whether his knowledges are relationally connected at operational levels. The chief enemy for this comes from the semantic disorders of the INTENSIONAL, verbal, and definition-ORIENTED person.

But the fire of creativity must be kept under control. There must be sufficient negative feedback to keep the ideas relevant to purpose. The creativity of most persons has been early snuffed out by too much criticism from his environment. They soon learn not to "stick their necks out". This is a serious problem in the proper balancing of feedbacks. Too much emphasis upon the free-wheeling of positive feedback can result in emotional disturbances; in groups it can result in panics and riots. Too much negative feedback may inhibit all action. And tangential behaviors from feedforward may occur as goals, and purposes are not clear.

The need of a way to talk about words for all contents – that is, for a META-LANGUAGE – is acute. This is as necessary in training the so-called generalists as it is for others. There are a few persons with enormous memories for details from many areas but whose abilities to derive the necessary perspectives to assess wholes is lacking. Others seemingly caught in dead levels of higher order abstractions have very thin factual knowledges.

But a meta-language such as General Semantics is subject to the same gamut of semantic disorders and hazards of isomorphism as the words, technical and otherwise, used about any other content. What meta-language can be used to talk about itself? That is, in this case, about the PROCESS STRUCTURE of communication-evaluating? Inspection of the formulations above outlined in the General Semantics model is as applicable to General Semantics as process as it is applicable to other languages concerned with content. For instance, when statements evoke allness or signal reactions to a content, we would analyze these General Semantic words IN THEIR CONTENTS AND SITUATIONS for allness or for signal reaction. The labeling of allness or signal reaction to a statement is amenable to the application of these formulations to the labels themselves. However, we must hasten to qualify this "claim" of General Semantics as a META-META language for all languages by "as nearly as we know".

Hopefully, this small beginning synthesis will reveal the structural bare-bones of the evaluating-communicating process. We think it applies to everything accessible to human investigation. It applies directly to research within and upon both of these general disciplines. In impasses of definition, analysis, safeguarding of inferences, avoidance of confusions, and unraveling of paradoxes,

General Semantics has been particularly useful if not indispensible.

Presumably applicable to everything researched are the so-called information sciences. General Semantics, obviously, should help humanize these increasingly dehumanizing activities. Also, General Semantics approaches ethics, morals, religion, and history in a way that leads to provocative questions and hypotheses. Its neuro-semantic, neuro-linguistic structure also provides a core for the organization of the various psychologies, sociologies, and psychiatries.

Heavy resistance to general systems and General Semantics is typical of the educational establishments. But the greatest difficulty is in the Humanities and Political sciences, which should take – along with the whole pedagogical establishment – more responsibility for the critical problems of today.

The dramatic achievements of space-age communication should remove all doubts as to the physical unity of the Universe. The vast manifold of magnetic, radiational, induction, gravitational fields by which everything is influenced leads us to infer that there are no gaps and no vacuums. Perhaps this belief will renew the dreams for a Unified Science which a few scholars have long talked about in a sort of muted way and to which general systems seem to be a most powerful force.

Alfred Korzybski in his seminars was less timid. He emphatically pointed out that there could be no unified science without a unified evaluating-communicating structure whereby man could relate properly to the vast complex of structure-systems. This structure of General Semantics, which is called timebinding, may eventually exert that NEGATION OF ENTROPY to defeat the global human death which some pessimists say is pending. With Man as the center, perhaps a Science of Man will arrive in the indefinite future.

As this and further syntheses are continued, if the world doesn't fall apart in the meantime, perhaps the efforts initiated in the 1970 Denver Conference might become another "small step" of the sort we saw and heard from the moon.

REFERENCES

1. "The Thinking Man's Medium," Saturday Review, Nov. 14, 1970, p. 72.
2. S. I. Hayakawa. Language in Thought and Action. Harcourt, Brace and Company, New York, 1949, pp. 301-303.
3. University of Iowa Spectrum, November 6, 1970, p. 6.

4. Elwood Murray, Gerald Phillips, David Truby. Speech: Science-Art. Bobbs-Merrill, Indianapolis, Ind. 1969, Chapter 7.
5. Interaction is often called TRANSACTION in general semantics. It refers to the criss-crossing of abstractings. Abstracting is necessarily derived from assumptions as two or more people relate.
6. General System Theory. George Braziller, New York, 1968, pp. 55-56.
7. Modern Systems Research for the Behavioral Scientist. Aldine Publishing Company, Chicago, Illinois, 1968, p. 493.
8. Distributed by Institute of General Semantics, Lakeville, Connecticut, 06039.
9. Structure, a bundle of relations. When two relations are similar, we say they have similar structure, which is defined as a class of all relations similar to a given relation. Science and Sanity, p. 57.
10. Language, Thought, and Reality. Edited by John B. Carroll, John Wiley & Son: N.J., 1956, p. vi.
11. Science and Sanity, p. 148.
12. "From Stimulus to Symbol: The Economy of Biological Computation," in Modern Systems Research for the Behavioral Scientist. Aldine Publishing Co., Chicago, 1968, p. 120.
13. Speech: Science-Art. Chapter 4.
14. General Systems Yearbook. Mental Health Institute, Ann Arbor, 1956, p. 24. (Of course, this cannot be absolute no matter how "finely milled" the interfaces might be.)
15. F. S. C. Northrup. Philosophical Anthropology and Practical Politics. Macmillan, New York, 1960, Chapter 3. (His EPISTEMIC NEUROLOGICAL CORRELATIONS have many similarities to language isomorphisms.)
16. Inductions are powered by ability to abstract SIMILARITIES in the differences: deduction occurs in perception of DIFFERENCES in the similarities. Competency increases with ability to perceive more and more of both.
17. "What Has Happened to Action Research?" Kurt Lewin Memorial Aware Presentation, Journal of Social Issues, vol. 26, number 4, 1970, p. 8.
18. See Bess Sondel. Power Steering by Words. Follette Publishing Co., Chicago, 1964.

Non-Aristotelian Premises and Isomorphism

KENNETH G. JOHNSON

Both general semantics and general systems are concerned with "similarity of structure" or "isomorphism" – the general semanticist with the structural similarities of language and the world it attempts to "map", the general systems theorist with the structural similarities in different fields of knowledge. To use the map-territory analogy of Alfred Korzybski, the general semanticist emphasizes map-territory isomorphism, the general systems theorist examines map-map isomorphism, comparing maps (models, theories, equations) from different fields of knowledge. Both fields are, of course, much more complex than these simple dichotomies would suggest.

I believe a good case could be made for considering Korzybski, founder of General Semantics, an early general systems theorist. He read widely in many fields of human knowledge, always seeking underlying structural similarities, looking for the broadest generalizations that emerged from the converging inferences of these diverse fields. He maintained that ". . . we should leave no stone unturned to discover beneath the varying phenomena more and more general and invariant foundations on which human understanding and agreement may be based."[1]

But to begin at the beginning we must go back to the early work of Korzybski and his concerns about the nature of Man. Looking back at the destruction and brutality of World War I, he asked the age-old question: "What is Man?" He felt Man must be different from other forms of life, but how? Because he had been educated as an engineer, he sought OBSERVABLE differences and classified forms of life in terms of what they DO.

Plants, he said, can transform energy from the sun and the soil into organic chemical energy, so he called them the "chemistry-binding class of life." Animals bind chemicals but can also move about from one place to another, so he called them the "space-binding class of life." Man binds both chemicals and space, but, in addition, has the unique capacity to "freeze" experience and the products of his creativity in the form of language, then "unfreeze" them later. In this way he can use nervous systems miles and centuries away. He can use the accumulated achievements of past generations, build upon them, and pass them on to the next generation. Korzbyski called this the "time-binding" capacity and Man the "time-binding class of life."[2]

Once he had arrived at this definition of Man, he became interested in the MECHANISMS of time-binding. What characteristics of Man make time-binding possible? If these are the characteristics that distinguish Man from animals, can we find ways to improve or make more extensive use of these characteristics – to help Man to be more human?

Clearly Man's symbol-making and symbol-manipulating potentialities play a major role in time-binding, so Korzybski focussed on those aspects of behavior. He studies what he considered the effective symbolic behavior of scientists in their scientific work and the ineffective symbolic patterns and behavior of mental patients: out of these studies emerged the discipline of general semantics.

The three non-artistotelian Premises that underly and summarize the field of general semantics do not appear grouped together as such in Science and Sanity, but they were spelled out in his seminars and a paper presented at the Second American Congress on General Semantics in 1941.[3]

PREMISE OF NON-IDENTITY

The first premise can be stated, "A word is not the fact, feeling, situation, relation, etc.", or, to use the map-territory analogy, "The map is not the territory."

Bourland, restating this premise in what he calls E-prime (standard English with the exclusion of all the inflectional forms of the verb 'to be'), formulates it this way:

> No two structures can exhibit absolute sameness in all respects; most particularly, a structure on the symbolic level (such as a 'word') has characteristics which differ significantly from those of other structures associated with it on the object level or event level (such as the fact, feeling, situation, relation, etc., labeled by the given 'word')[4]

Beginning with this premise, that words are not the objects which they represent, Korzybski argued that structure, and structure alone, becomes the only link which connects our verbal processes with events. "To achieve adjustment and sanity and the conditions which follow from them, we must study structural characteristics of this world FIRST, and, then only, build languages of similar structure, instead of habitually ascribing to the world the primitive structure of our language."[5]

Although a growing number of scientists, mathematicians, and philosophers, are concerned with the role of language and problems of structural similarity, language and the structural implications of language remain insufficiently examined.

> Here we come across a tremendous fact; namely, that a language, any language, has at its bottom certain metaphysics, which ascribe, consciously or unconsciously, some sort of structure to this world. . . .
>
> Neither Aristotle nor his immediate followers realized or could realize what has been said here. They took the structure of the primitive-made language for granted, and went ahead formulating a philosophical grammar of this primitive language, which grammar – to our great semantic detriment – they called "logic", defining it as the "laws of thought." Because of this formulation in a general theory, we are accustomed even today to inflict this "philosophical grammar" of primitive language upon our children, and so from childhood up imprison them unconsciously by the structure of the language and the so-called "logic", in an anthropomorphic, structurally primitive universe. . . .
>
> We do not realize what tremendous power the structure of an habitual language has. It is not an exaggeration to say that it enslaves us through the mechanism of semantic reactions and that the structure which a language exhibits, and impresses upon us unconsciously, is automatically projected upon the world around us. . . .[6]

Since "STRUCTURE" is central to general semantics and general system theory, it might be well to examine this notion in some detail.[7] Korzybski suggested that structure can be considered as a complex of RELATIONS, and ultimately as MULTIDIMENSIONAL ORDER.

> In this vocabulary, 'structure' is the highest abstraction, as it involves a whole, taken as-a-whole, made up of INTER-RELATED PARTS, the relations of which can be defined in still simpler terms of order. "Knowing", in its broadest as well

> as its narrow human sense, is conditioned by structure, and so consists of STRUCTURAL knowings. All empirical structures involve relations, and the last depend on multi-dimensional order. A language of ORDER, therefore, is the simplest form of language, yet in structure it is similar to the structure of the world and ourselves. Such a language is bound to be useful for adaptation and, therefore, sanity; it results in the understanding of the structural, relational, multi-dimensional order in the environment on all levels.[8]

The relations of greatest interest to scientists (and, perhaps, all men) are not the simple ones of "larger than", "above", "between", etc., but those that express FUNCTIONAL RELATIONSHIPS. Stated simply, a FUNCTION labels something that happens if something else happens: if so, then so. This could be considered the paradigm of a scientific assertion. If two variables are related in such a way that if you change one of them you change the other, the variables are said to be FUNCTIONS of each other. One variable may be a function of several others or several may be a function of one.

> The concept of function is of the greatest importance, not only in pure mathematics but also in practical applications. Physical laws are nothing but statements concerning the way in which certain quantities depend on others when some of these are permitted to vary. Thus the pitch of the note emitted by a plucked string depends on the length, weight, and tension of the string, the pressure of the atmosphere depends on the altitude, and the energy of a bullet depends on its mass and velocity. The task of the physicist is to determine the exact or approximate nature of this functional dependence.[9]

Based on the space-time notions of Einstein and Minkowski, Korzybski assumed a dynamic world of unique events: no two "objects" are identical, nor is any one "object" twice the same. He visualized a restless universe of constant change. Within such a world, how is it possible to make "maps" that fit "territories"? How is it possible to make predictions? To deal with these problems, Korzybski borrowed the notions of FUNCTION and INVARIANCE UNDER TRANSFORMATION from mathematics and generalized them into a method of evaluation that can be applied to all areas of human concern.

Suppose a mathematical transformation is given by the function $Y = 2x$. A set of integers 1, 2, 3, etc. would be transformed into a set of even integers 2, 4, 6, etc. While the VALUES of the integers are changed, the RELATIONSHIP remains INVARIANT.

This example deals with the "rules of the game" of a self-contained system. But if we can establish correspondence between mathematical quantities and physical quantities (measurements of the external world), we can establish a SIMILARITY OF STRUCTURE between verbal and non-verbal levels, based on INVARIANCE of relations which are found or discovered in both.

A physicist, for example, may make a number of measurements of the time it takes objects to fall certain distances. He notices certain RELATIONS between time and distance; that is, distance appears to be a FUNCTION of time. If he then establishes a correspondence between the physical measurements and mathematical quantities, he finds that the SAME RELATIONS hold for the physical quantities as for the corresponding mathematical quantities. He has established a SIMILARITY OF STRUCTURE between the nonverbal and the verbal (in this case, mathematical) world. The map and the territory are found to be ISOMORPHIC. The functional relationship (first established by Galileo) may be expressed $D = 1/2\ gt^2$ where g is a constant. This relationship proved to be invariant within certain limits. The usefulness of statements of invariant relationships lies in their prediction value. (We must be careful here not to interpret "invariance" as "permanence", for we are often talking about the RELATIVE invariance we perceive in the world around us.)

The search for invariance, for order amid seeming chaos, is not confined to science or scientists. We all have a need for stability, for predictability, for invariance in a changing world. When we ask "Will drug x cure disease y?" we are seeking an invariant functional relationship. When we ask "Should a child be taught to read before he goes to school?" we are seeking an invariant relationship between certain actions on our part (teaching) and certain long-range benefits. (In this case the "benefits" would have to be defined operationally before we could begin to answer the question.)

> The structure of the world is, in principle, unknown; and the only aim of knowledge and science is to discover this structure. The structure of languages is potentially known, if we pay attention to it. Our only possible procedure in advancing our knowledge is to match our verbal structures, often called theories, with empirical structures, and see if our verbal predictions are fulfilled empirically or not, thus indicating that the two structures are either similar or dissimilar.[10]

The notion might be summarized as $P = f(I)$; that is, predictability is a function of isomorphism. Korzybski used the term "similarity of structure" rather than "isomorphism", though I believe they could be used interchangeably. For him, similarity of structure included notions of order, function, relation, invariance under transformation, and predictability. And predictability, he argued, is necessary for science, for sanity, for survival.

Korzybski suggested a number of techniques (he called them "extensional devices") for increasing the structural similarity of everyday language and the external world. One of these is called "indexing":

> In general terms, the structure of the external world is such that we deal always on the objective levels with absolute differences. The structure of the human nervous system is such that it abstracts, or generalizes, or integrates, etc., in higher orders, and so finds similarities, discovering often invariant (sometimes relatively invariant) relations. To have 'similar structure', a language should comply with both structural exigencies, and this characteristic is found in the mathematical notation of x_i, which can be enlarged to the daily language as 'Smith$_i$', 'Fido$_i$', where i = 1, 2, 3, . . . n.[11]

To solve the problem of representing a process world with static language, he suggested a time index or date: Smith$_{1970}$, Fido$_{6 \text{ p.m. MAY } 1, 1970}$, etc.

PREMISE OF NON-ALLNESS

The second premise can be stated, "A word covers not all the characteristics of an object, situation, fact, feeling, etc." Or, to use Bourland's E-prime version and the map-territory analogy: "A 'map' does not contain all the structural characteristics of the 'territory' it represents."

Until relatively recently in human history, little was known about the role of the nervous system in the process of perception. Statements were made (and ARE made) with little awareness of the processes of abstraction, selective perception, integration, and projection.

W. W. Sawyer in Mathematician's Delight points out that "Without abstraction, thought is impossible. We should spend all our lives collecting information if we tried to make a perfect picture of even a simple event." And I believe it was J. Robert Oppenheimer who said, "All human knowledge is purchased at the price of ignoring something else."

The functional relationships discussed earlier are of great value because they enable us to make predictions, but we must not lose sight of the fact that they represent higher order abstractions – that much of the detail, the uniqueness of each event, is lost. A scientific generalization represents an abstraction of an invariant relationship in a welter of transformations. It makes no attempt to say "all". The invariant relationship can be isolated only at the price of ignoring much of what is going on. (Note that this appears to be an inherent limitation of science, or of human knowledge generally. We cannot hope to know "all".)

In discussing this diagram, Korzybski wrote:

> Most of us IDENTIFY IN VALUE levels I, II, III, and IV and react AS IF our verbalizations ABOUT the first three levels were 'it'. Whatever we may SAY something 'is' obviously IS NOT the 'something' on the silent levels. Indeed, as Wittgenstein wrote, 'What CAN be shown, CANNOT be said.' . . . Unfortunately, people in general, including many scientists, DISREGARD LEVELS II AND III COMPLETELY, and react as if unconscious that IV 'is not' I. In other words, we do not take into account the mechanisms of the human nervous system or 'think electro-colloidally' about our reactions. Such disregard leads to misunderstandings, heated two-valued ('either-or') debates, hostilities, prejudice, bitterness, etc. . . An awareness of the processes of abstracting clarifies the STRUCTURE of a great many of our inter-personal, professional, etc., difficulties which may become trivial or non-existent if we become conscious of the identifications involved.[12]

Note that this diagram concerns the "organism-as-whole". Korzybski used the term "semantic reactions" to stand for the total evaluational reaction of the organism, including "thoughts", "feelings", "electro-colloidal changes", etc. He contended that the aristotelian structure of language tends to be elementalistic and implies, through structure, a split or separation of what in actuality cannot be separated.

> For instance, we can verbally split 'body' and 'mind', 'emotion' and 'intellect', 'space' and 'time', etc., which as a matter of fact cannot be separated empirically, and can be split only verbally. These elementalistic, splitting, structural characteristics of

language have been firmly rooted in us through the aristotelian training. It built for us a FICTITIOUS, ANIMISTIC WORLD not much more advanced than that of the primitives, a world in which under present conditions an optimum adjustment is in principle impossible.[13]

PREMISE OF SELF-REFLEXIVENESS

The third premise can be stated, "Language is self-reflexive in the sense that in a language we can speak about language". Or, in Bourland's E-prime elaboration: "We can use language self-reflexively, in the sense that in a language we can speak about the language, but we must recognize that the resulting symbolic structures will belong to higher orders of abstraction than those we ascribe to the symbolic structures of the initial language".[14]

The notion of self-reflexiveness has a multitude of ramifications that have yet to be explored. In order to give a "complete" definition, we would have to define every term in the definition, then define every word used to define, etc., etc. We can speak words about words, words about words about words, etc. In behavioral terms, this means we can react to our reactions, react to our reactions to our reactions, etc.

The self-reflexiveness of languages introduces serious complexities which Korzybski attempted to solve by introducing the notion of MULTI-ORDINALITY. He pointed out that many of our most important terms can be used at many orders of abstraction and can therefore be called "multiordinal". Furthermore, multi-ordinal terms can be used self-reflexively: that is, they can be applied to themselves. For example, one may learn how to learn, think about thinking, have feelings about feelings, suffer fear of fear, etc. Note that $feeling_2$ about $feeling_1$ is not the same as $feeling_1$. Since multiordinal terms 'mean' different things at different levels of abstraction, confusion results if the speaker or the context does not specify the level of abstraction.

All three premises deal with the processes of abstraction – from event to sensation to perception to silent reactions to labels to descriptions to inferences to generalizations to statements about statements, etc. Korzybski stressed the need to develop CONSCIOUSNESS OF ABSTRACTING, an awareness of this complex,

Figure 1 An extensional analysis of the process of abstracting from an electro-colloidal non-Aristotelian point of view.

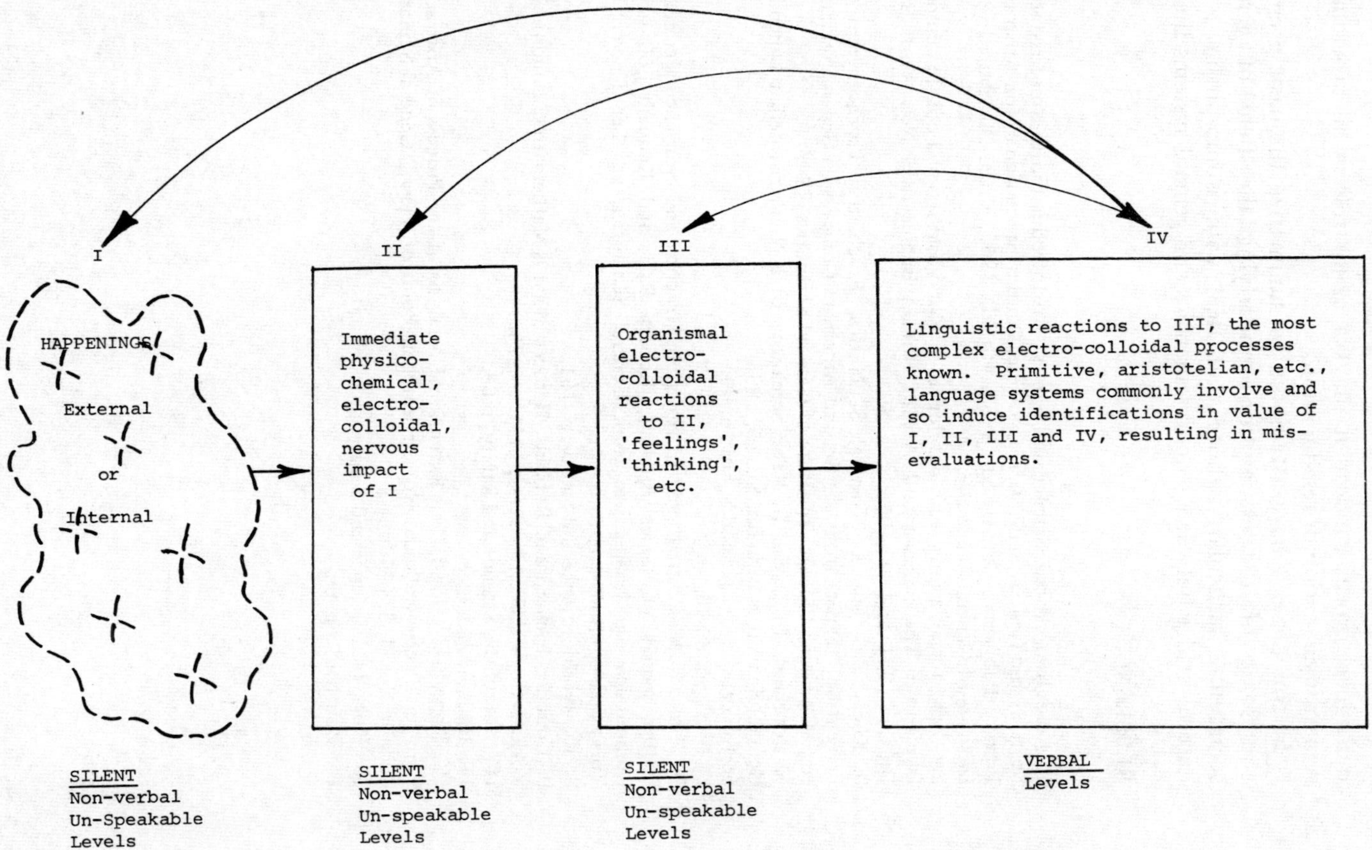

multi-level process; indeed it plays a major role in his program for semantic education or re-education.

In this paper I have attempted to summarize the basic premises and some of the assumptions underlying the field of general semantics, particularly those that deal with isomorphism – the natural bridge between general semantics and general systems theory.

REFERENCES

1. Korzybski, Alfred. Science and Sanity: An Introduction to Non-Aristotelian Systems and General Semantics. Lakeville, Conn.: International Non-Aristotelian Library Publishing Co., 1st ed., 1933, 4th ed., 1958, p. 23. Distributed by the Institute of General Semantics.
2. These formulations were developed in Korzybski's first book Manhood of Humanity: The Science and Art of Human Engineering. New York: E. P. Dutton, 1921, 2nd ed., 1950.
3. Korzybski, Alfred. "General Semantics, Psychiatry, Psychotherapy and Prevention," Papers from the Second American Congress on General Semantics, ed., M. Kendig. Chicago: Institute of General Semantics, 1943.
4. Bourland, D. David, Jr. "The Semantics of a Non-Aristotelian Language," General Semantics Bulletin, No. 35. 1968.
5. Korzybski, Science and Sanity, p. 59.
6. Ibid., pp. 89-90.
7. A more detailed treatment of function, structure and invariance is contained in the papers of James Broadus, Robert Pula, and Kenneth Johnson in Communication: Perspectives from General Semantics, ed. Lee Thayer. New York: Spartan Books, 1970.
8. Korzybski, Science and Sanity, p. 161.
9. Courant, Richard and Robbins, Herbert. What is Mathematics? London, New York: Oxford University Press, 1941. p. 275.
10. Korzybski, Science and Sanity, p.63.
11. Ibid., p. 263.
12. Korzybski, Alfred. "An Extensional Analysis of the Process of Abstracting from an Electro-Colloidal Non-Aristotelian Point of View," General Semantics Bulletin Nos. 4, 5, 1950.
13. Korzybski, Science and Sanity, p. 1.
14. Bourland, op. cit.

Basic Korzybskian Orientations in the Organizing of Experience

CHARLOTTE READ

The study of general semantics deals with man and his world considered through the understanding, use and development of a particular system. Since Korzybski's non-aristotelian system was made public in 1933 it has been worked with, applied, taught, developed, etc., although the basic postulates have not been altered. This paper focusses on some of its central concerns, in the theory and in experiencing it. Can the general orientations be taught? Are they related to general systems theory, and also to some Eastern philosophic outlooks?

The study of general systems theory deals with "the formulation and derivation of those principles which hold for systems in general."[1] From the point of view of "systems in general," general semantics would be more limited than general systems theory. The particular non-aristotelian system, based on many scientific disciplines methodologically and dealing with how we humans evaluate, seems at this date to be the most general one of its kind, and would encompass general systems theory as among existing theories and happenings.

In considering these two studies, our aims and main concerns appear different and complementary. There are many similarities in outlook, since both have grown out of western scientific formulations and attitudes. The following list gives some of the more obvious similarities in orientations:

1. Their holistic, non-elementalistic, non-split, outlook, dealing with our internal-external environment as-a-whole.
2. Their basis on a contemporary scientific world view.
3. Their concern with inter-relatedness and integration on many levels.
4. Their concern with hierarchical orders.
5. Their assumption of a dynamic process world rather than static.
6. Their expression of a need for more generalized formulations of scientific knowledge.

In general semantics, epistemological in character and linking science with factors of sanity, we mainly give attention to the study and improvement of human evaluating processes, and we draw upon the findings in many fields to try to understand their bearing on how we function and relate ourselves to the world. We ask: "How can the generalizations derived from these findings be translated into methodology and into action?"

With the basic orientations mentioned above, the distinguishing characteristic of Korzybski's formulation as a system consists of its construction in terms of orders of abstractions (with its modus operandi called "general semantics"). The overall goals of realizing our human potentialities, achieving cultural evolutions peacefully, etc., are to be worked toward through "consciousness of abstracting."

Let us consider for a moment what this term stands for, and why it was chosen. Quoting some of the reasons given in 'Science and Sanity':

> By now we know how important it is for a non-aristotelian system to abandon the older implications and adopt an actional, behavioristic, operational, or functional language. On the neurological level, what the nervous system DOES is abstracting [selecting, omitting, separating, summarizing, etc.], of which the summarization, integration, etc., are only special aspects . . . . The structure of the nervous system is in ordered levels, and all levels go through the process of abstracting from the other levels.
>
> The term implies a general activity, not only of the nervous system as-a-whole, but even of all living protoplasm. The characteristic activities of the nervous system, such as summarizing, integrating, etc., are also included by implication . . . . The term "first order abstractions" or "abstractions of lower order" does not distinguish between "body" and "mind". PRACTICALLY, it corresponds roughly to "senses" or immediate feelings, except that by implication it DOES NOT ELIMINATE "MIND". Neither does the term "abstractions of higher orders" eliminate "body" or "senses", although it corresponds roughly to "mental" processes . . . .the term "abstracting" implies "abstracting from something" and so involves the environment as an implication. [2]

We see that instead of dividing orders of abstractions into "physical," "mental," etc., they are called "lower" and "higher" orders, to remain flexibly within a non-elementalistic, non-split framework. While we abstract "from" the environment, we also project "onto" the environment (including ourselves) in continual interaction, maintaining what is termed the "circularity of human knowledge." This makes possible a continual feedback on many levels. Since we are a part of each situation, our evaluations affect and are affected by, the whole.

What does the formulation "consciousness of abstracting" imply about our relationship to our world? We find that it involves the many aspects of our organism. We can try to "feel it through." Whatever activity we may be engaged in, whether we are walking down the street, working on a mathematical problem, looking at the television, teaching, etc., the processes of abstracting are happening in us, including our organismic reactions, our feelings, talking, etc. As M. Feldenkrais expressed it in Body and Mature Behaviour:

> The most abstract thought has emotional-vegetative and sensory-motor components. Abstract thinking is possible only in conjunction with a special configuration or pattern or state of the body. The whole nervous system, therefore, participates in every act; whether it is easily observable or not is only a matter of knowing what and how to observe.[3]

To learn to become conscious of abstracting, we must function on a higher level of awareness. We "know" that we have selected, omitted, summarized. We are "aware" that we approach our daily living with our undefined terms, our prejudices, our learned interpretations. A "general consciousness of abstracting" involves a more or less continuing state, an inner condition of open readiness with which to meet the world. It includes a generalized state of awakeness, a sense of rapport with where we are, a feeling of being connected with our environment, with what we are engaged in.

How can we learn this ourselves, and help others to experience it? "Intellectual" understanding is only a beginning. What is needed is a readiness to learn, the capacity to be "touched."

Each person who teaches this work – if "teaching" would be the appropriate word – or writes about it, does so in his own style, according to his understanding, emphasizing those aspects which seem to him especially important. The methods may be simple, but practice is needed from as many facets as possible according to the ingenuity of the teacher.

The procedure is not as simple as that in the story of the young monk who went to a master to learn the secret of Zen. The young man asked again and again for some guidance, some answer, to his question, but no help was given. Finally the master said, "All right, tomorrow morning you will go with me up the mountainside and I will show you." When they walked up the mountainside the next day, they came to a place where trees were blossoming, filling the air with fragrance. The master stopped and asked, "Do you smell them?" "Yes," the monk replied. "Well, you see, I am hiding nothing from you."

The aims of general semantics are formulated differently from those in Zen. There are some points of similarity, however. We also need some deeper understanding if we want to follow the methods through in behavior. Ours is not meant to be a "sudden" technique. We emphasize following a method, a notion which is foreign to Zen.

How to induce the recognition and giving up of our identifications, arriving closer to that flexible state of full degree of conditionality where we are no longer at the mercy of our past associations, or of the propagandist, and have the inner freedom to size up and respond wholeheartedly to each situation on its own terms? It requires self-knowledge and persistence.

The various formulations which are a part of the structure of the system serve as guides. In becoming aware of the principles in action, we also recognize how we prevent ourselves from following them. We are able to make comparisons, to judge.[4]

We try to get the "feel" of the methodology, to get beyond the verbal principles. We discover the hidden assumptions we really live by. For example, when did we ascribe to some one a stereotype that we learned, rather than look at the individual himself, experiencing that person (i.e. identify a higher order abstraction with a lower)? When did we act as if our inference in a situation had the same value as "the facts," and when did we differentiate? What were the consequences? When did we assume that we knew "all" about something, rather than realizing that more could have been said? With what expectations did we approach a person, and how did they influence our actions? How did these behaviors feel? What did we notice?

Through experimenting, practising, devising exercises and examples, discussing, reading, learning from group experiences, we develop an inner guide.

To give just two or three examples: One way to explore the processes of abstracting would be to classify different types of activities the term refers to. Dr. J. S. Bois, in The Art of Awareness, describes five kinds: EVALUATIVE (referring to a hierarchy of values), CLASSIFYING (according to characteristics we find in common), OBJECTIVE (differentiating between classes and subclasses of actual units), SELF-REFLEXIVE (reacting on a higher level to our reacting on a lower level, e.g. fear of fear), and RELATIONAL (expressing elements of a situation relationally).[5]

How could the feeling of changing our way of using symbolism be conveyed? I recall an occasion in a seminar given by Dr. Bois when we had multiplication problems to solve using only Roman numerals. We experienced something of the struggle of working out a problem with an entirely different set of symbols, of "changing our gears."

On another occasion, Dr. O. R. Bontrager asked the seminar participants to make a detailed drawing of the room they were in. After each person's drawing was put up on the wall and each was seen to be different, they were asked to make a "map of the map," a drawing of one particular drawing left up, a "map of some one else's map." He usually ended his simple, "baby-like" activities with the question, "Is there anything in it for you?" If these "exercises" sound extremely simple and hardly to be taken seriously, it is suggested that they be tried out, in the light of the basic principles of the work. These are efforts to get to experiential levels.

There are degrees of learning and applying our learning. It depends on how far we want to go, in what directions, and where it will lead us professionally or personally. Not everyone wants to go into it deeply; there are benefits along the way.

Sometimes the language aspect is stressed, sometimes the non-verbal. In a previous article I have related our organismic non-verbal behavior to general semantics principles and indicated that this may be a powerful way of learning by experiencing. To "sense" the way we move and let this bring our unconscious assumptions to awareness can be revealing.[6]

I would like to elaborate here on the wholeness or interrelatedness of our reactions. Considering the subtle, close relationship between our inner "thinking" and physiological conditions, it is not surprising that the beneficial physiological effects of the method have been emphasized. It functions through the process of delaying our actions, getting in touch with a situation before reacting to it. Such a delay, which could be for as long as a fraction of a second or longer, involves the ability to be "silent on non-verbal levels," an aspect of training in becoming conscious of abstracting (we have to "shut up" inside to be able to receive messages, to get out of our habitual verbal ruts, our usual modes of reacting). I have for some years wished that an experiment could be set up to test this with currently available instruments, such as an electro-encephalograph, psycho-galvanometer, apparatus for measuring muscular contractions, etc. The difficulties of setting up such experiments and controlling the variables have been too baffling so far to carry out in a "natural" situation.

Perhaps some of the research being done in psychology will present ways of working it out. Much of the current work, for example, experimenting with people on the "control" of (or ability to influence) their brain wave patterns appears to concentrate on their reactions when they are less aware of their environment, "leaving" it temporarily in states of meditation, or achieving a state of inner calm in other ways under laboratory conditions.[7] There do not seem to be specific data dealing with the effects of coming more in contact, more vividly aware of, our immediate surroundings while retaining a condition of inner quiet. I should like to see research done in this direction, which I see as involving being "silent on the objective level," more conscious of abstracting, etc. Thus, the claim in General Semantics of changed physiological or neurological functioning might in time be subject to scientific investigation.

General semantics essentially represents a methodology for bringing us more in contact with our everyday living world, for helping us to get our "feet on the ground" and to feel the ground under our feet. Through the use of the extensional devices – indexing, dating, the "etc.", hyphens and quotes – the static forms of language can more adequately be made to represent the dynamic processes going on. It also recognizes the importance of inferential knowledge, and of our unlimited capacity to build higher order generalizations, to imagine, to visualize, to create. It asks that we learn to differentiate these.

Classifying "knowledge" into two different types, Northrop related them to general semantics in his 1954 Korzybski Memorial Lecture, "Mathematical Physics and Korzybski's Semantics." Referring to what he called "concept by intuition" he said:

> At this point his thinking touched the silence of the Oriental Buddhist's intuition and the Hindu's ineffable immediacy when he becomes one with Brahman . . . It is this silent level of knowledge, the immediate portion of our being and of all things, that we in our Western culture have tended to lose or to dismiss as of secondary importance.[8]

The axiomatically constructed relational thinking – "concept by postulation" – is also very important, he said, for without it we are "deficient with respect to the way of knowing in which Western civilization is unique and most proficient." This is also a part of the underlying outlook of general semantics.

It was more the latter type of knowledge that Korzybski referred to when he wrote about the "circularity of knowledge":

> Our submicroscopic knowledge is hypothetical in character. The world behaves AS IF its mechanisms were such as our highest abstractions lead us to believe, and we will continue to invent theories WITH THEIR APPROPRIATE TERMINOLOGIES to account for the intrinsic mechanisms of the world we live in, ourselves included. We read into nature our own latest highest abstractions, thus completing the inherent circularity of human knowledge, without which our understanding of nature is impossible . . . .
>
> Thus, all our fundamental deeper knowledge must be, and can never be anything but, hypothetical, as what we see, hear, feel, speak about, or infer, is never IT, but only our human abstractions ABOUT "it." What kind of linguistic form our inferential knowledge is cast in thus becomes of utmost importance.[9]

The next non-aristotelian system, or perhaps the next development of the present one must, I believe, more explicitly include and develop our sense of "oneness" with the rest of nature, of communion with it rather than observation of it as the underlying attitude pervading all our actions. This while retaining the clarity and evaluational discipline of consciousness of abstracting. Perhaps this can be worked out in the future.

Now as the barriers between different disciplines fall away, and between different ways of experiencing, and they are seen to be aspects of a larger whole, I am reminded of an old Chinese story of the Spirit of the Ocean speaking to the Spirit of the River:

> You cannot speak of ocean to a well-frog – the creature of a narrower sphere. You cannot speak of ice to a summer insect – the creature of a season. You cannot speak of Tao to a pedagogue; his scope is too limited. But now that you have emerged from your narrow sphere and have seen the great ocean, you know your own insignificance and I can speak to you of great principles.[10]

REFERENCES

1. Ludwig von Bertalanffy, General System Theory: Foundations, Development, Application. George Braziller, New York, 1968, p. 32.
2. Alfred Korzybski, Science and Sanity: An Introduction to Non-aristotclian Systems and General Semantics. International Non-aristotelian Library Publishing Co., distributed by Institute of General Semantics, Lakeville, Conn. 1st ed. 1933, 4th ed. 1958, pp. 379, 380.
3. M. Feldenkrais, Body and Mature Behaviour: A Study of Anxiety, Sex, Gravitation and Learning. International Universities Press, Inc. N.Y., New York, 1949, p. 26.
4. For a comparison of characteristics of aristotelian and non-aristotelian orientations, see Alfred Korzybski, op. cit., (pp. xl-xlii in 4th ed.)
5. J. Samuel Bois, The Art of Awareness: A Textbook on General Semantics. Wm. C. Brown Co. Publishers, Dubuque, Iowa, 1966, pp. 87-90.
6. Charlotte Schuchardt Read, "Exploring Relations Between Organismic Patterns and Korzybskian Formulations." General Semantics Bulletin, Nos. 32 & 33, 1965, pp. 47-52.
7. Robert Keith Wallace, "Physiological Effects of Transcendental Meditation." Science, 27 March 1970, pp. 1751-1754. See also 'Time,' July 18, 1969, p. 67.
8. F. S. C. Northrop, "Mathematical Physics and Korzybski's Semantics." General Semantics Bulletin, Nos. 16 & 17, 1955, p. 14.
9. Alfred Korzybski, "The Role of Language in the Perceptual Processes," Ch. 7 in Perception: An Approach to Personality, by Robert R. Blake and Glenn V. Ramsey (ed.), The Ronald Press, New York, New York, 1951, pp. 200, 201.
10. R. G. H. Siu, The Tao of Science: An Essay on Western Knowledge and Eastern Wisdom. The Technology Press, M.I.T. and John Wiley & Sons, New York, 1957, p. 82.

The General Semantics of Set Theory

C. A. HILGARTNER

I have taken as my topic "The General Semantics of Set Theory". However, I propose to approach the topic crabwise, or rather squid-wise – full speed to the rear – by initially addressing myself to the topic of "A Set Theory Analysis of General Semantics".

Polanyi has pointed out that no man goes to the trouble to put a theory in the form of an axiomatic system unless he already believes the theory "true".

> When certain undefined terms, axioms, and symbolic operations are established as the logical antecedents of mathematics, these are based on the prior assumption that mathematics is true.
>
> Furthermore, I derive an unholy glee from his description of the paradox of a mathematics based on a system of axioms which are not regarded as self-evident and indeed cannot be known to be mutually consistent. To apply the utmost ingenuity and the most rigorous care to prove the theorems of logic or mathematics, while the premises of these inferences are cheerfully accepted, without any grounds being given for doing so, as 'unproved asserted formulae', might seem altogether absurd. It reminds one of the clown who solemnly sets up in the middle of the arena two gateposts with a securely locked gate between them, pulls out a large bunch of keys, and laboriously selects one which opens the lock, then passes thru the gate and carefully locks it after himself – when all the while the whole arena lies open on either side of the gateposts where he could go round them unhindered. A fully axiomatized deductive system is like a carefully locked gate in the midst of an infinite empty area.[1]

I assert that I see myself as that kind of clown – and now this paper serves as my arena. So: I become solemn.

Ten to twenty years ago, N. Bourbaki demonstrated that "all mathematics, and therefore all products of accurate thinking, can be based on set theory".[2] In this sense, modern set theory (so-called 'naive set theory') constitutes THE GENERAL logical language; its standards constitute our paradigm of "the standards of mathematical logic".

In the fall of 1965, I committed myself to perform a logical analysis of a theory of human behavior I had proposed. I had based this theory on the non-aristotelian premises of Korzybski (but had carefully avoided using the usual terminology of general semantics). Since then, in a collaborative setting, I have succeeded in translating that theory of human behavior into the form of an axiomatic system stated in a set-theory notation, and have progressively extended the theory so as to take into account more and more classes of phenomena. And by now a portion of this work has appeared in print.[3]

Recently, I spent a couple of months trying to figure out what (if anything) these theories say – and in the process, by cross-referencing between my own works and a few general semantics textbooks, "rediscovered" the terminology of general semantics. So now I propose to raise two questions:

1. Can we base general semantics on set theory?

2. Having based it on set theory, do we find general semantics free of self-contradiction?

In this paper, (a) I shall discuss the topic of the 'logic of opposites', and the famous "Laws of Thought" of Aristotle; (b) I shall state the non-aristotelian premises of Korzybski, and show the relations between them and Aristotle's "Laws of Thought"; (c) I shall specify the 'universal set' assumed in general semantics; (d) I shall derive, state and prove a theorem of general semantics; and (e) then I shall draw a few conclusions.

I shall not start with extensive explanations of set theory. Instead, for the sake of brevity (and intelligibility), I shall illustrate the set theory notions I need in the process of using them.[4] For example, in order to discuss the 'logic of opposites', I point out that every statement in set theory assumes and implies some defined or clearly understood "universe of discourse" or CONTEXT, which I shall designate as the 'universal set' U. Furthermore, the terms 'set', 'element', and 'belongs' comprise the 'undefined terms' of set theory, where the criterion of correct usage requires that for any 'element' we can determine without ambiguity whether or not it 'belongs' to a given 'set'. If the element x 'belongs' to the set Z, I write $x \in Z$; if not, I write $x \notin Z$.

THE LOGIC OF OPPOSITES

I can display the 'logic of opposites' by reference to some 'universal set' U which contains one proper subset Z. (In set theory, we can regard a set as a subset of itself, an 'improper' subset of itself. A 'proper' subset of a superset, then, does not contain all of the elements contained in the superset.) For example, by analogy to point-sets on a plane, I can show these relations using the venerable device of Venn diagrams:

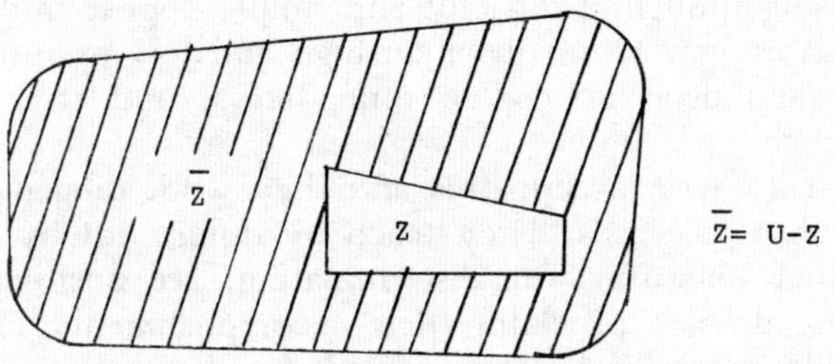

Figure 1

Here every element of the point-set U either belongs to the set Z or else it does not belong to Z, in which case it belongs to its complement $\bar{Z}$ (not-Z).

Thus, according to the 'logic of opposites', any classification (term) symbolically splits the "universe" into two mutually exclusive parts, which I could represent as $(x \mid x \in Z)$ and $(y \mid y \in \bar{Z})$.

But in order to classify we must define or clearly assume SOME PARTICULAR context or 'universe of discourse'. For example, if Z stands for the set designated by the English noun "dog", before we can classify elements as members or non-members of this set we need to know whether the context involves "the edible", or "domesticated mammals", or "humans whose behavior I disapprove of", or "constellations of stars", or "mechanical devices", or "meteorological haloes", or whatever. And given that our 'universal set' involves "the edible", we still need to know whose dietary customs we refer to: e.g., the tribal customs of the United States, where "dog" might refer to a class of small skinless sausages, or those of China, where (I understand) canine meat qualifies as a delicacy.

The complements of many of our common terms themselves have common names, e.g., 'night' vs. 'day'. In the context of nomad hunters roaming about on an allegedly non-spheroidal Earth, this pair of terms might appear "perfectly natural", with no specific assumed context. But for the crew of Apollo 12 en route to the moon, this pair of opposites becomes utterly irrelevant, and so its specific assumed context becomes fully apparent.

For me one of the most beautiful expressions of the 'logic of opposites' comes from the Tao Teh Ching:

> Whenever the most beautiful is perceived, ugliness arises, the least beautiful. Whenever good is perceived, evil exists, its natural opposite.
>
> Thus perception involves opposites: reality and fantasy are opposing thoughts; difficult and simple oppose in degree; long and short oppose in distance; high and low oppose in height; shrill and deep oppose in tone; before and after oppose in sequence.
>
> The truly wise accept this and they work diligently without allegiance to words. They teach by doing, not by saying; are genuinely helpful, not discriminating; are positive, not possessive; do not proclaim their accomplishments, and because they do not proclaim them, credit for them can never be taken away.[5]

I take both this excerpt from the Tao and the things I said above to indicate that the 'logic of opposites' constitutes the basis for the process of naming objects; now I shall deal with some special features of the 'rules for naming' in Indo-European languages, such as Greek or English.

THE "LAWS OF THOUGHT" OF ARISTOTLE

Korzybski wrote:

> Let me recall the 'philosophical grammar' of our language solemnly called the 'laws of thought', as given by Jevons:
>
> 1. The law of identity. Whatever is, is.
>
> 2. The law of contradiction. Nothing can both be, and not be.
>
> 3. The law of excluded third. Everything must either be, or not be.
>
> These 'laws' have different 'philosophical' interpretations which help very little and for my purpose it is enough to emphasize

that: (1) The second 'law' represents a negative statement of the first, and the third represents a corollary of the former two; namely, no third possible between two contradictories. (2) The verb 'to be', or 'is', and 'identity' play a most fundamental role in these formulations.[6]

I can state the "Laws of Thought" in set theory notation by reference to a 'universal set' U which contains one proper subset Z; then they become

IDENTITY	$Z \equiv Z$ or	
	$Z + Z = \emptyset$	(where the symbol '+' indicates 'symmetric difference', viz., $A + B = [(A - B) \cup (B - A)]$
CONTRADICTION	$\forall: x \in Z \cup \bar{Z}$	
EXCLUDED MIDDLE	$Z \cap \bar{Z} = \emptyset$	

Translated into words, these sentences say:

IDENTITY: The set Z proves identically equal to the set Z (where 'identically equal' signifies "Not only consists of the same elements, but also has the same name"); or, the symmetric difference between the set Z and the set Z proves equivalent to the empty set.[7]

EXCLUDED MIDDLE The set Z and the set $\bar{Z}$ have no elements in common (the intersection of Z with $\bar{Z}$ proves equivalent to the empty set).

CONTRADICTION All elements x (which belong to our universal set U) exist such that x belongs to the union of Z with Z.

Thus, demonstrably, the "Laws of Thought" express the 'logic of opposites', and so appear to make explicit the 'rules for naming' in Indo-European languages. But please note two crucial omissions: Unlike the Tao, the "Laws of Thought" ignore the distinction between 'NAME' and 'THING NAMED' (or between 'map' and 'territory'); and likewise ignore the topic of CONTEXT. As we shall see, these special 'omissions' of the "Laws of Thought" turn out to have very curious implications.

However, using the "Laws of Thought" as premises, we can (if we so desire) build up various more or less self-consistent logical systems, perhaps including naive set theory; but we will find that these logical languages cannot handle at least one intrinsically interesting class of relations.

THE NON-ARISTOTELIAN POSTULATES OF KORZYBSKI

Korzybski proposed an alternative set of postulates, which (as he points out) constitute NEGATIONS of the "Laws of Thought".[8] In our original set theory statements of these premises, we made use of several different sets of conventions, which may have served to conceal or at least to obscure the direct contrast between the non-aristotelian and aristotelian premises. Now in order to make this contrast explicit, I shall give several alternative statements of the non-aristotelian postulates.

The undefined terms of a korzybskian system, which we can define in terms of each other but cannot further specify in words, comprise STRUCTURE, ORDER, and RELATIONS. The postulates go by the names of Non-Identity, Non-Allness, and Self-Reflexiveness.[9] In one lexicon, we let Y stand for the set of characteristics of the 'territory', viz., the environmental object $y^i \in E$, which in principle remains unknown to our organism; and we let $\beta a \rho f(y^i) = Z$ stand for the set of characteristics of our organism's 'picture' of $y^i \in E$. (The composition-relation $\beta a \rho f$ summarizes the processes by which an 'organism' forms a 'picture'.[10]) Then the postulate of Non-Identity becomes

NON-IDENTITY No "thing" exists 'identical' with any other "thing"; a 'map' does not consist of the same elements as does the 'territory' which it represents.

$$[y^1, y^2 \in E, t_1 \neq t_2] \Rightarrow y^1 + y^2 \neq \emptyset$$

(Premise 1)

Or, alternatively stated,

$$Y + Z \neq \emptyset$$

(In words, the element y at time t_1 and the element y at time t_2 both comprise elements of the set A, and time t_1 differs from time t_2; if these conditions hold, then the symmetric difference between y^1 and y^2 does not consist of the same elements as does the empty set; or, alternatively stated, the symmetric difference between the set of characteristics of the 'territory' (y) and the set of characteristics of our organism's 'picture' of the 'territory' (z) does not prove equivalent to the empty set. Here we might interpret "not equivalent to the empty set" as "not negligible".)

In a slightly different lexicon, we defined a mapping from Y onto Z,

$$Q \subset Y, \rho(Q) = \{z \in Z \mid \exists\, y \in Q, z = \rho(y)\},$$

in terms of which we defined two difference relations, $Y - \rho^{-1}(Z)$ and $Z - \rho(Q)$. Then the postulate of Non-Allness becomes

NON-ALLNESS No 'map' can show ALL the characteristics of the 'territory' which it represents; no 'map' remains free of elements extrinsic to the 'territory' which it represents.

$$Y - \rho^{-1}(Z) \neq \emptyset$$

(Premise 3)

$$Z - \rho(Q) \neq \emptyset$$

(Premise 4)

(In words, the difference between Y and the inverse-ρ of Z (that subset of Y which our organism can in principle detect by means of his unaided sensory receptors, namely, Q) does not prove equivalent to the empty set; the difference between Z and $\rho(Q)$ (where $\rho(Q)$ designates that subset of Z not extrinsic to the 'territory') does not prove equivalent to the empty set.)

In both Premises 3 and 4, these (sub)sets subtracted from the (super)sets Y or Z (namely, $\rho^{-1}(Z) = Q$ or $\rho(Q)$) at the very least comprise subsets of Y or Z respectively:

$$\rho^{-1}(Z) \subset Y,$$
$$\rho(Q) \subset Z;$$

or at most they prove equal to Y or Z respectively:

$$[\rho^{-1}(Z) \subset Y \cap Y \subset \rho^{-1}(Z)] \Rightarrow Y = \rho^{-1}(Z),$$
$$[\rho(Q) \subset Z \cap Z \subset \rho(Q)] \Rightarrow Z = \rho(Q).$$

Then the postulate of Non-Allness asserts that these subsets do not prove equal to their respective supersets. This fact we can express also in the notation of symmetric differences:

NON-ALLNESS: $Y + \rho^{-1}(Z) \neq \emptyset$

$Z + \rho(Q) \neq \emptyset$

We used a third lexicon in stating the postulate of Self-Reflexiveness, which will require explication.

SELF-REFLEXIVENESS No action or utterance of any organism exists free of self-reference.

$$\text{“}x_i/(E-x)\text{”} \Leftrightarrow \text{”}Sf_i\text{“}$$

(Premise 5)

(In words, the postulate of Self-Reflexiveness asserts, "Gestalt if and only if subsidiary 'awareness' of the intra-organismic processes by which our organism puts together his Gestalt". Here the term 'Gestalt' refers to a structure composed of a FIGURE of FOCAL INTEREST to our organism (via sensory modality i) bounded by a GROUND or CONTEXT more or less empty of interest.)

This statement of the postulate makes use of the conventions by which we formalized the notion of 'awareness', Aw, namely

$Aw = Sf \times Ot$,

$Cp_1(Aw^i) = $ "Sf"i$\subset$Sf, $Cp_2(Aw^i) = $ "x/(E–x)"i = "Ot"i$\subset$Ot,

V("Sf"i) = "Ot"i

Thus I can give a more general statement of the postulate as

SELF-REFLEXIVENESS: $Ot \Leftrightarrow Sf$

or, on the level of 'consciousness' (Cs) rather than 'awareness',

SELF-REFLEXIVENESS: $Oe \Leftrightarrow Se$,

where

$$Cs = Se \times Oe,$$
$$Cq_1(Cs) = Se,\ Cq_2(Cs) = Oe,$$
$$U(Se) = Oe.$$

Paraphrasing in terms of our mapping ρ, we get

SELF-REFLEXIVENESS: $[\rho(Q)] \Leftrightarrow [Z - \rho(Q)]$,

where

$$Z \subset Cs,\ Cq_1(Z) = Se = [Z - \rho(Q)],$$
$$Cq_2(Z) = Oe = [\rho(Q)].$$

As originally stated, ". . . the postulate of Self-Reflexiveness refers to the structure of intra-organismic relations". But to mention 'intra-organismic relations' presumes first that we distinguish between 'name' and 'thing named' (or between 'map' and 'territory'). Thus, considered in an even more general sense, this postulate states the dual requirement that we distinguish between 'name' and 'thing named', and that we distinguish between our organism's Gestalt and the intra-organismic processes ('processes of abstracting') by which he puts together his Gestalt. Putting these requirements into notational form, I paraphrase the postulate as

SELF-REFLEXIVENESS: $Y + \beta a\rho f(Y) \neq \emptyset$

$Se + U(Se) \neq \emptyset$

(In words, the symmetric difference between the set of characteristics of the 'territory' (Y) and the set generated by the processes by which our organism forms a 'picture' of the 'territory' (viz., $\beta a\rho f(Y) = Z$) does not prove equivalent to the empty set; the symmetric difference between the set known as "the subsidiary 'awareness' of the processes by which our organism puts together his Gestalt", otherwise known as "the 'self'-component of 'consciousness' " (viz., Se) and that set generated by the intra-organismic processes by which our organism puts together his Gestalt, otherwise known as "the 'other'-component of 'consciousness' " or "our organism's 'consciousness'-Gestalt" (viz., U(Se) = Oe), does not prove equivalent to the empty set.)

This form of the postulate obviously implies-and-assumes Non-Identity.

So far so good: by stating the non-aristotelian postulates in set theory notation, we have negotiated the first hurdle. Now I propose to show the exact relation between the aristotelian and non-aristotelian premises: Korzybski explicitly distinguished between 'name' and 'thing named' (or between 'map' or 'picture' and 'territory'); whereas Aristotle did not so distinguish. In a set theory formalism which expresses the distinction between 'name' and 'thing named', we cannot in any prima facie sense render the notion of "did not distinguish"; instead, we have to work out some kind of special logical "trick" which will produce the effect of "did not distinguish" without breaking the rules of naive set theory, and then assert that Aristotle, who "did not distinguish", acted AS IF he pulled this logical "trick". Here I can express this "trick" as the 'inverse' of the 'claim to generality' made in behalf of non-Aristotelian systems: Aristotle made special logical assumptions, not tenable for the general case, which if substituted into a non-aristotelian system will "reduce" it to isomorphism with the aristotelian system.

The non-aristotelian postulates of Korzybski reduce to the aristotelian "Laws of Thought" if and only if the symmetric difference between the set of characteristics of the 'territory' (Y) and the set of characteristics of our organism's 'picture' of Y (viz., $\beta\alpha\rho f(Y) = Z$) proves NEGLIGIBLE. In notation, we can indicate 'negligible' by two different conventions, $\overline{L_M}$ (subliminal, "so small I can't detect it") or $\emptyset$ non-existent (the empty set).

Thus, by analogy with our statement of the korzybskian postulates, we can state the "Laws of Thought" in our notation as the postulates of Identity, Allness, and Linearity: [11]

IDENTITY: (Aristotle did not explicitly distinguish 'name' (or 'map') from 'thing named' (or 'territory').)

$$Y + Z \subset \overline{L_M}.$$

(In words, the symmetric difference between the set of characteristics of the 'territory' (Y) and that of 'map' (Z) remains negligible (subliminal, so small as to escape detection): the direct contrary of the postulate of Non-Identity.)

ALLNESS: (Since Aristotle did not explicitly distinguish 'name' from 'thing named', he also failed to take into account the concurrent and related problems of the 'incompleteness' of 'maps' and of the inclusion in 'maps' of elements extrinsic to the 'territory'.)

$$Y - \rho^{-1}(Z) \subset \overline{L_M},$$
$$Z - \rho(Q) \subset \overline{L_M}.$$

(In words, the difference between Y and inverse-ρ of Z proves negligible (subliminal); the difference between Z and $\rho(Q)$ proves negligible (subliminal): the direct contrary of the postulate of Non-Allness.)

By reasoning similar to that employed above, we can also express Allness by the notation of symmetric differences:

ALLNESS: $$Y + \rho^{-1}(Z) \subset \overline{L_M},$$
$$Z + \rho(Q) \subset \overline{L_M}.$$

Some implications of these formulations illuminate the contrast between aristotelian and non-aristotelian premises. For example, in terms of our mapping ρ,

$$\rho^{-1}(Z) = Q \subset Y.$$

According to the postulate of Allness, $Y - \rho^{-1}(Z) \subset \overline{L_M}$, from which it follows that

$$Y + Q \subset \overline{L_M},$$

or, stated in words, the symmetric difference between the (sub)set of characteristics of the 'territory' which our organism can in principle detect by means of his unaided sensory receptors (Q) and the entire set of characteristics of the 'territory' (Y) remains negligible (subliminal).

Furthermore, we already saw that $Cq_2(Z) = Oe = [\rho(Q)]$. But according to the postulate of Allness, $Z - [\rho(Q)] \subset \overline{L_M}$, from which it follows that

$$Z + Oe \subset \overline{L_M}.$$

Then the postulate of Allness becomes

ALLNESS: $$Y + Q \subset \overline{L_M},$$
$$Z + Oe \subset \overline{L_M}$$

So if the postulate of Allness holds, viz., if the 'territory' remains practically devoid of elements undetectable by the unaided sensory receptors of our organism and if our organism's 'picture' of the 'territory' remains practically devoid of elements extrinsic to the 'territory', then in effect our organism's 'picture' of the 'territory' approximates a "point-for-point perfect replica of" the 'territory'. In set theory, we can render the notion of a "point-for-point perfect replica of" (or tautology, or y = y relation) by means of the notion of 'identity-mapping' (which implies a '1-to-1' function: see below):

$$I_E = \{(y,y)y \in Y\} : I_E(Y^i) = Y^i$$

Then, given that $\rho^{-1}(Z) = Q$ and $\rho(Q) = Oe \subset Z$, the relation $Z + Oe \subset \overline{L_M}$ becomes

$$Z + I_E(Y) \subset \overline{L_M}.$$

In other words, the postulate of Allness implies-and-assumes the postulate of Identity. Otherwise stated, according to the postulate of Allness, our organism's 'picture' of the 'territory' approximates a "point-for-point perfect replica of" the 'territory' – a highly questionable conclusion (or premise).

We already saw that the postulate of Self-Reflexiveness states the dual requirement that we distinguish 'name' from 'thing named' and that we distinguish between our organism's Gestalt and the intra-organismic processes by which he puts together his Gestalt.

If, however, we admit the postulate of Identity, then we transform any account we may generate concerning the structure of the transactions between an organism and his environment in such a way that a priori we regard

(a) our organism's 'name' or 'map' or 'gross perception' as a "point-for-point perfect replica of" the 'thing named' or 'territory' or 'thing perceived', and

(b) the activities by which our organism puts together his Gestalt as a "point-for-point perfect replica of" his Gestalt.

Then the postulate of Linearity becomes

$$\text{LINEARITY:}\quad Y + Z \subset \overline{L_M} \Leftrightarrow \beta\alpha\rho f(Y) + I_E(Y) \subset \overline{L_M}$$
$$\Leftrightarrow U(Se) + I_E(Y) \subset \overline{L_M}$$

So the postulate of Linearity comprises the direct contrary of the postulate of Self-Reflexiveness.

Thus in order to ignore the distinction between 'name' and 'thing named', and to ignore the topic of context, we must PRETEND (tacitly assume) that our organism's 'name' or 'map' constitutes a point-for-point perfect replica of the 'thing named' or the 'territory', and that the activities by which our organism puts together his Gestalt constitute a point-for-point perfect replica of his Gestalt. In other words, these two crucial 'omissions' built into the "Laws of Thought" imply that both the relations of the 'organism' with his 'environment' and the relations of the 'organism' with 'himself' prove equivalent to tautologies (prove indistinguishable from an identity-mapping on the environment).

For myself, I could say that on a prima facie basis, the

non-aristotelian premises appear more attractive to me than the aristotelian "Laws of Thought". But this judgment on my part assumes some kind of context from which I judge. So I raise the question: What (if anything) does 'general semantics' refer to? What context does it imply, what 'universal set' does it assume?

To summarize, the 'general semantics' of Korzybski refers to a 'universal set' composed of a human-organism-as-a-whole-in-his-environment-at-a-date.

But what can we do with that?

Recently, in the process of elaborating a fully rigorous definition of the term 'living system', I expressed this central issue as follows: [14]

THE BOUNDARY

> 'Living systems' display an intrinsic 'relatedness': The notion of 'organism' (or 'living system') implies an 'environment' for this 'organism', and vice versa. And both terms imply a dynamic 'boundary' between 'organism' and 'environment' which in a curious way "belongs" to both, in that it serves as the 'organ of interaction' between them. [12] In this context, a holistic viewpoint appears indispensable: If for any reason I try to deal with a 'system' less comprehensive than 'an-organism-as-a-whole-in-its-environment-at-a-date', by that act I annihilate the very phenomena of interest, viz. this 'interacting at the boundary', the 'relatedness' of 'living systems'. For an 'isolated LIVING system' comprises a contradiction in terms: an 'organism' 'isolated' from its 'environment' (e.g., a mammal cut off from air) cannot survive. [13]
>
> In order to mathematize this fundamental 'relatedness', I have set up a Cartesian product space, (O x E), "in which to represent the INTERACTIONS AT THE BOUNDARY in the (ORGANISM x ENVIRONMENT) FIELD". [14] Every element of this space, then, comprises an (o,e) couple, an interaction of organism with environment.

Thus by a very simple mathematical trick, viz., by defining our 'universal set' as composed only of ordered pairs, "interactions of organism with environment", I set up a logical language in which it remains impossible to elaborate even one theoretical construct which does not refer to this fundamental 'relatedness'.

Now I have the wherewithal to define the key constructs of general semantics in set theory terms. So I shall define a small number of its key constructs, then state and prove a theorem, and then draw some conclusions. (Table I gives a listing of these key constructs.)

ABSTRACTING

In order to define this term, on our 'universal set' I shall specify one relation, discussing it both as an 'operator' (verb-form) and as a 'set' (noun-form).

DEFINITION: The verb-form 'ABSTRACTING' designates that set of 'fundamental life-processes' by which an organism SELECTS ("responds to") certain elements of his 'external' and/or 'internal' environment, and IGNORES other elements.[15]

The term 'abstracting' emphasizes the "ignoring" at least as much as the "selecting".

In set theory notation, the operator ρ (Premise 2) stands equivalent to the notion of 'abstracting'.

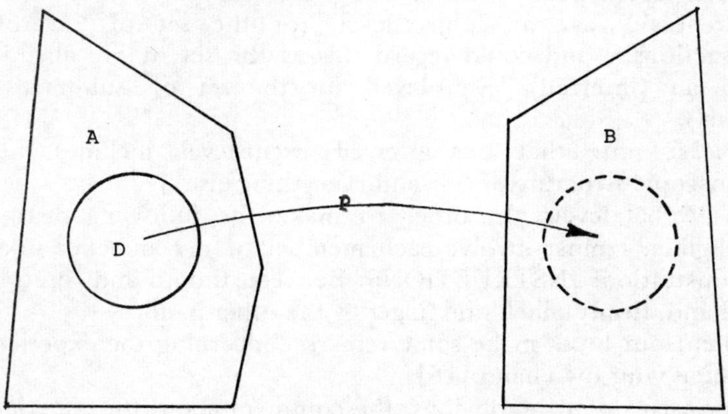

Figure 2

DEFINITION: The noun-form 'ABSTRACTION' designates any set ('map', 'picture', 'construct', 'symbol', 'Gestalt', etc.) produced by the processes of 'abstracting'. (In Premise 2, the set Z stands equivalent to the notion of 'abstraction'.)

ORDER OF ABSTRACTION

Given two related sets, one of which comprises an abstraction from the other: then these two related sets belong to different 'ORDERS OF ABSTRACTION'.

SPECIAL CASES

1. Silent levels: I could describe my non-verbal waking experience as a set of "moments of 'awareness' ", elements of which, from moment to moment, shift in emphasis from one 'sensory modality' to another and shift in focus from aspects of my 'environment' to aspects of 'my own functioning' and back again.

After Korzybski, I could designate this set of "moments of 'awareness' " as the 'object-level' (or the set of 'macroscopic transactions'), and could regard this as the set of my abstractions from an (inferred) 'event-level' (or the set of 'sub-microscopic events').

(Please note that this inferred 'event-level' includes the sub-microscopic structure of 'me-and-everything-else'.)

2. Verbal levels: In order to make the following distinctions intelligible, I must involve each member of my audience in a little demonstration: INSTRUCTIONS: Between thumb and forefinger of one hand, firmly pinch one finger of the other hand.

Then, out loud, make some remark concerning the experience of pinching your own hand.[16]

To insist upon the obvious, the comment about the experience IS NOT the experience itself. Instead, I would say that the comment constitutes an abstraction from the non-verbal ('object-level') experience. Thus the experience and the comment differ in 'order of abstraction', the former belonging to the 'silent levels' and the latter to the 'verbal levels'.

3. Higher verbal orders: Let us say that the demonstration of pinching his own hand elicits the following soliloquy from one person:

(a) "Ouch!

(b) "That kind of hurts.

(c) "That's an odd thing to do, to pinch yourself.

(d) "You must be a little odd yourself, to make us do a screwy thing like that.

(e) "What are you, some kind of sadist or something?"

In general semantics terminology, (a) represents an immediate response or 'label' (L) for the non-verbal experience (O); (b) represents a 'description' (D) of the experience; (c) represents an 'inference' (I_1) from the 'description'; (d) represents an 'inference' (I_2) from the first 'inference'; and (e) represents a still higher-order 'inference' (I_3) from the second 'inference'.

Each one of these "remarks" (sets) differs from the previous occurrences in 'order of abstraction'. And as you can see, when I indicate this sequence of operations by means of Venn diagrams, the resulting figure turns out as one form of Korzybski's 'structural differential'.

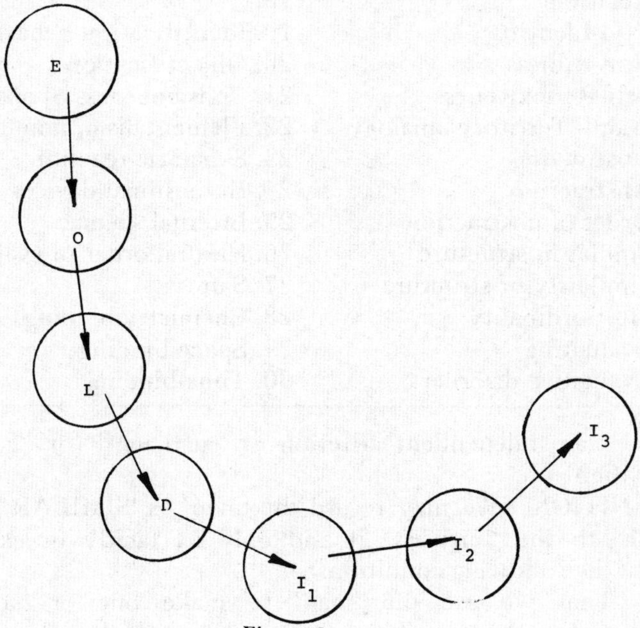

Figure 3

Please note also that this sequence of operations shows a space-time ORDER.

SIMILARITY OF STRUCTURE

In order to define this term, I need the relation of 'abstracting', and then must define a JUDGMENT ('similar in structure'), and finally must show that the term 'similarity of structure' constitutes a CRITERION.

Given: two related sets, one of which constitutes an abstraction from the other, such that we might designate one as a 'territory' and the other as a 'map' of that 'territory'; and

TABLE I Key constructs of General Semantics[35]

1. Structure	16. Identifying
2. Order	17. Identification
3. Relations	18.
4. Non-Identity	19. Identification behavior
5. Non-Allness	20. Objectification
6. Self-Reflexiveness	21. Consciousness of abstracting
7. 'Map'-'Territory' analogy	22. Elementalism, non-elementalism
8. Abstracting	23. Semantic reaction
9. Abstraction	24. Extensional devices
10. Order of abstraction	25. Internal silence
11. Similar in structure	26. Natural order of evaluating
12. Similarity of structure	27. Sanity
13. Multiordinality	28. Chemistry-binding
14. Evaluating	29. Space-binding
15. Evaluative disorders	30. Time-binding

Given: an independent criterion of 'sufficient' (or "good enough fit"); then

DEFINITION We may regard the 'map' as 'SIMILAR IN STRUCTURE' to the 'territory' if and only if, tacitly or explicitly, we satisfy the following conditions:

(a) That we use the 'map' to make one or more definite (disconfirmable) prediction(s) concerning the 'territory';

(b) That we perform one or more tests which in principle, as judged by our criterion, could disconfirm the prediction(s); and
(c) That upon examining the results of our test(s), we find that, as judged by our criterion, the results do NOT disconfirm the prediction(s).

Thus the term 'similar in structure' functions as a JUDGMENT, which I might paraphrase as, "The 'map' appears GOOD ENOUGH for our present purposes". The complement of this term comprises the judgment 'dissimilar in structure', or "A map NOT good enough for our present purposes".

DEFINITION The notion of 'similarity of structure' designates a possible relation between 'map' and 'territory', viz., the logical possibility that predictions based on a 'map' might survive testing without disconfirmation, as judged by specified criteria. Thus the term 'similarity of structure' itself functions as a CRITERION, which I might paraphrase as, "The STANDARD for a 'map' good enough for our present purposes".

The complement of 'similarity of structure' would give "The standard for a 'map' not good enough for our present purposes", which provides no surprises. A more revealing "opposite" to the term 'similarity of structure', which functions as a criterion, comprises the behavior of "Refusing to admit a criterion for a 'map' good enough for our present purposes". This immediately leads into the topic of 'evaluative disorders', and provides some unexpected interconnections and insights.

EVALUATING AND EVALUATIVE DISORDERS

The notion that "predictions based on a 'map' might survive testing without disconfirmation" codifies in ten words an entire 'theory of the structure of human psycho-dynamics'. To state it explicitly, an 'organism' comprises a self-correcting system, generating hypotheses concerning how in its here-now situation to obtain what it 'needs' and how to avoid catastrophe. 'Gross perceptions', in these terms, constitute one kind of 'hypothesis' (or 'map', 'picture', 'guess', etc.). If in any sense the organism makes use of his 'perceptions' to guide his activities or choices, in effect he subjects his hypotheses to test; and (provided the outcome does not prove fatal) he confronts the results of his tests. And even non-human organisms 'learn', rectifying their 'maps' in the light of painful or unsatisfying experiences.

Here, the term 'EVALUATING' refers to the process of 'generating and testing hypotheses' in a survival-related context. And since "The 'map' does not exist 'identical' to the 'territory' which it represents", the process of 'generating hypotheses' proves equivalent to 'making predictions based on a "map" '. So in a very direct sense, our survival depends on EVENTUALLY generating 'maps' which qualify as 'similar in structure' to the 'territory' they refer to.

The notion of 'EVALUATIVE DISORDERS', then, refers to the set of strategies by which a human can MINIMIZE the likelihood that his predictions will survive testing without disconfirmation. (And in order to strive to minimize the reliability of his predictions, an organism must (perhaps for obscure reasons) first commit himself to remain in a state of frustration.) Finally, 'evaluative disorders' imply and assume 'DELUSIONS' ('beliefs' held REGARDLESS of the evidence, in order to 'solve' (stabilize) some important and otherwise intractable 'problem').

Although other strategies for remaining in error may exist, none seems to me more reliable than a systematic refusal to credit the logical possibility that predictions based on a 'map' might survive testing without disconfirmation: for then, I can claim, no efforts of my own can suffice to forestall my inevitable frustration; and by this trick I can conceal from myself-and-others the self-defeating character of my activities.

In order to 'admit no criterion for "a 'map' good enough for our present purposes", one could (1) fail to distinguish between the two sets ('map' and 'territory'); or, (2) deny the possibility of any specifiable or usable relation between the two sets.

1. 'Failure to distinguish between two related sets which differ in order of abstraction': This phrase gives one form of the definition of the term 'identification'.

(a) 'Absolute' failure to distinguish: Taken in an 'absolute' sense ("Inability to distinguish between the members of EVEN ONE pair of related sets"), this condition defines a situation incompatible with life. As the whole thrust of my work over at least the last five years shows, I as an organism can without self-contradiction represent an 'organism' as a mechanism for classifying elements of his interactions with his 'environment' as members or non-members of various sets, in order to achieve the outcome of his own survival for at least a few moments longer. For example, an organism composed of one or more cells maintains intra-cellular (intra-organismic) concentrations of salts, sugars, etc., markedly different from the concentrations of these substances in the medium he lives in. But this implies that the organism "has" some kind of 'picture' of the amounts of these various substances 'inside' and 'outside' his boundary ("has" the ability to MEASURE concentrations), and also "has" a 'picture' of "how much" of these various substances he 'needs' ("has" a set of 'tolerable concentration-limits' with respect to both 'internal' and 'external' environment), and finally, "has" the ability to take in, excrete, differentially metabolize, etc., in order to maintain the "measured" concentration values within the "tolerable" limits. But that means that we must represent even a one-celled organism as operating on at least two different orders of abstraction, e.g., "standards" and "measurements" ('map' and 'territory'); or on at least three different orders of abstraction, if we take into account the difference between "measurement" and "thing measured"[17]. Thus, if an organism succeeds in operating on certain elements of the set of 'substances in the "internal" and/or "external" environment' so as to keep his intra-organismic concentrations of these substances within these limits, then in this terminology I must say that this organism has "distinguished between related sets which differ in order of abstraction" – as evidenced by the fact that he may remain alive at least a few moments longer.

Thus an organism who engages in this 'absolute failure to distinguish' will fail to account for even the most elementary characteristics of 'living systems', including his own continuing survival.

(b) 'Relative' failure to distinguish: To oversimplify, the condition of 'relative failure to distinguish between the members of certain pairs of sets (or certain CLASSES of pairs of sets) which differ in order of abstraction' might lead to two main classes of evaluative errors: (i) that class of errors which comes from regarding 'event-level phenomena' as 'symbols' (existing only on 'verbal levels') which we could "argue away" (in other words, 'denial'); or (ii) that class of errors which comes from regarding a 'symbol' as existing as a 'silent-level' "thing" (in other words, 'objectification').

But to distinguish these two classes of evaluative errors does comprise an 'oversimplification'. On a higher order of abstraction, the mechanism of 'denial' proves EQUIVALENT to the mechanism of 'objectification': both expressions point to the process of "relying on a 'map' dissimilar in structure from the 'territory' it allegedly 'represents' ". Put another way, we cannot show ONE of these mechanisms without in the same act showing THE OTHER also: a kind of "underestimation" of the 'territory' ('denial') comprises a kind of "overestimation" of the 'map' ('objectification'). And finally, the act of "underestimating" the 'territory' and "overestimating" the 'map', taken as a whole, has the effect of reducing the 'organism's' account of his relations with his own 'environment' to a kind of tautology (and of utterly ignoring the 'organism's' relations with 'himself'). In other words, in this kind of act the 'organism' relies on the "laws of Thought" of Aristotle as his fundamental premises.

2. 'Denial of the possibility that two related sets which differ in order of abstraction can bear any specifiable or usable relation to one another': This phrase gives another form of the definition of the term 'identification'.

(a) 'Absolute' denial: Taken in an 'absolute' sense ("Denial that EVEN ONE pair of related sets which differ in order of abstraction may bear any specifiable or usable relation to one another"), this phrase (condition) states an obvious self-contradiction: If no two sets may bear a relation to one another, in what sense then may we call them 'a pair of related sets?' Even if we paraphrase so as to avoid the obvious self-contradiction, by dropping the requirement that the two sets show the relation of 'differing in order of abstraction', and thus reduce our "given" to "any two sets", this condition forbids even the act of classifying elements as members or non-members of sets: Given the 'universal set' U and a set A which constitutes a proper subset of U, then every element of U comprises either an element of A or of $U - A = \bar{A}$. But that defines a relation between A and $\bar{A}$, namely, the relation of complementarity, as forbidden by the condition.[18]

Interpreted at biochemical levels, this condition, like the 'absolute failure to distinguish', proves incompatible with life: An 'organism' unable to classify elements of the set of 'substances in the "external" and/or "internal" environment' as members or non-members of sets cannot maintain the concentration-differences between his 'internal' and 'external' environment which make up a fundamental aspect of his own structure. So an organism who bases his behavior on this 'absolute' denial commits himself to a self-contradiction.

But a human organism intent on denying something will not stop himself from denying on the grounds that the denial leads to a self-contradiction. I could speculate, then, that this 'absolute' denial (like the 'absolute failure to distinguish') underlies the 'philosophy' or 'metaphysics' of a person implacably determined to commit suicide or to murder "everybody else", or of a person in the kind of profound depression which results within a day or so in failure of blood-pressure, irreversible shock and death. (Certain South Sea Islands tribes, I have heard, have a ritual of "boning" an offending member (magically declaring him dead); individuals so "boned" have died even under the care of modern physicians in Western-type hospitals.)

(b) 'Relative' denial: The 'relative' form of this condition becomes "Denial of the possibility that certain selected pairs of sets (or certain CLASSES of pairs of sets) which differ in order of abstraction can bear any specifiable or usable relation to one another".

This condition defines "selective blindness'[19]; or the tendency to "keep things in logic-tight compartments"[20]; or the tendency, while "allowing the parts of oneself in awareness on an equal footing, (to) insulate them from one another, by maintaining that they are "irrelevant" to one another"[21]; or it defines the "world-view" of an aphasic.[22]

But in logical terms the act of denying a relation turns out equivalent to the act of demonstrating that a set remains 'isolated' or 'not related': it requires that I elaborate some special logical construct which serves to "assume out of existence" the 'offending' relation.

Here, in order to deny that my 'map' of some particular 'territory' bears a specifiable or usable relation to the 'territory', I must somehow manage to "lose sight of" the fundamental relations codified in the 'map'-'territory' analogy (and made explicit in the non-aristotelian premises of Korzybski). Whatever tricks I use for this must in a concerted fashion conceal the relations between the 'territory', my 'map', and my own 'inner workings' (processes of 'abstracting').

Perhaps I can best accomplish this feat of concealment by means of some process of self-distracting, e.g. by "overestimating" one set and "underestimating" the other two, and thereby utterly neglecting the relations between them. I could, for example, regard my 'inner workings' as endlessly absorbing, while regarding my 'map' as non-existent, unimportant, or irrelevant and the 'territory' as non-existent, unimportant, or a hopelessly baffling 'chaos'. (Or I could engage in some form of the "failure to distinguish" already described, which would involve a 'denial' of the 'territory', an 'objectification' of the 'map', and an 'ignoring' of the 'inner workings'; etc.)

But on a higher order of abstraction, these various mechanisms or strategies prove equivalent: each one serves to "conceal" the relations of our 'organism' with 'himself' and with his 'environment'. And as I showed, I can represent this "concealing" as the process of 'reducing to a kind of tautology'.

Thus every element of the set of 'evaluative disorders' depends on the process of relying on the "Laws of Thought" of Aristotle as fundamental premises. This conclusion seems so important that I shall paraphrase it as a theorem.

Having defined these few notions, I can now state and prove an important theorem of general semantics, which forms the central tenet of Korzybski's 'theory of sanity'.

Theorem Two related sets which differ in order of abstraction (such that we might designate one as a 'territory' and the other as a 'map' of that 'territory') remain INCOMMENSURATE (have no common measure or basis for comparison: in other words, the sets do not "consist of the same elements" nor exist in any other kind of 'equivalence relation', nor do they exist related by any kind of '1-to-1 function'); therefore, to interchange them or otherwise to regard them as 'equivalent' constitutes a 'fundamental error' (leads to 'evaluative disorders').

Key Relation We have previously symbolized the binary relation which exists between two related sets which differ in order of abstraction as $Z = \beta a \rho f(Y)$. I might state this in words as, "Y comprises the 'territory' represented by the 'map' Z". For the sake of brevity, I shall symbolize this binary relation by $Z = \mu(Y)$, $\mu \subset ExO$, which I can also write as $Z\mu Y$. Then, in words, the inverse relation $\mu^{-1} \subset OxE$ designates the relation "... comprises a 'map' which represents the 'territory' ...".

Criterion EQUIVALENCE RELATIONS: There exists a class of binary relations (relations between ordered pairs of elements, e.g. (x,y) taken in that order), of which the notion of "equals" (or "... consists of the same elements as ...") seems by far the most commonly used. (Examples of other 'equivalence relations': (a) proportional fractions; (b) congruent plane figures; (c) similar geometric figures; (d) figures with equal areas; (e) equivalent velocities, e.g. "runs at the same rate as" for a herd of gazelles; etc.)[23]

Every binary relation which belongs to the set of 'equivalence relations' satisfies three constraints: namely, the relation qualifies as 'reflexive', as 'symmetric', and as 'transitive'. If I use the symbolism y R x to indicate "x stands in the equivalence relation R to y", I can state these constraints as

(i) Reflexive: a R a (e.g., a = a);
(ii) Symmetric: If b R a then a R b (e.g., [a = b] ⇒ [b = a]);
(iii) Transitive: If b R a and c R b then c R a (e.g., [a = b and b = c] ⇒ [a = c]).

Criterion 1-to-1 FUNCTIONS: Up to now I have consistently used lower case letters to indicate elements and capital letters to indicate sets. Here I make an exception to this practice: There exists a class of binary relations $f \subset X \times Y$ such that, given any element x, there corresponds one and only one element y for which the couple (x,y) qualifies as an element of f:

$$[(x,y_1) \in f \text{ and } (x,y_2) \in f] \Rightarrow y_1 = y_2.$$

I refer to f as a FUNCTION, and designate this class of binary relations as SINGLE-VALUED functions. (In elementary mathematics, we customarily indicate a single-valued function by the notation y = f(x).)[24]

In general, the INVERSE of a function comprises a binary relation but does not necessarily comprise a single-valued function: for example, for the function $y = f(x) = x^2$, the inverse of this becomes

$$f^{-1}(y) = \pm\sqrt{x}.$$

However, given a function $f \subset X \times Y$ such that its inverse

$$f^{-1} \subset Y \times X$$

also qualifies as a function, then

$$[(y,x_1) \in f^{-1} \text{ and } (y,x_2) \in f^{-1}] \Rightarrow x_1 = x_2.$$

I designate binary relations of this type as 'ONE-TO-ONE' (briefly, '1-to-1') functions. Then a function $f \subset X \times Y$ which maps from $A \subset X \times Y$ onto $B \subset X \times Y$ qualifies as 1-to-1 on A onto B if

$$[a_1 \in A, a_2 \in A, a_1 \neq a_2] \Rightarrow f(a_1) \neq f(a_2).$$

Moreover, I make use of the notion of FUNCTIONS OF FUNCTIONS: If y = f(x) (where $f \subset X \times Y$) and z = g(y) (where $g \subset Y \times Z$), then the substitution of the first relation into the second gives z = g(f(x)). This operation gives a COMPOSITION FUNCTION, which I designate by $g \circ f \subset X \times Z$. Thus

$$g \circ f = g(f).$$

From this it follows that the composition functions of a 1-to-1 function with its inverse become

$$f^{-1}(f(a)) = a \text{ for each } a \in A,$$
$$f(f^{-1}(b)) = b \text{ for each } b \in B.$$

Or, regarding f and f^{-1} as sets, these composition-functions of a 1-to-1 function and its inverse specify IDENTITY-MAPPINGS:

$$f^{-1} \circ f = [(a,a)\ a \in A] = I_X \subset X \times X,$$

$$f \circ f^{-1} = [(b,b)\ b \in B] = I_Y \subset Y \times Y.$$

Summary of Argument To begin with, I shall restrict discussion to the topic of the 'silent levels'.[25]

I shall begin by specifying a set of operations which, if carried out, would result in the elaboration of actual examples of a 'silent-level' 'territory' and of a 'silent-level' 'map' of that 'territory' (which would exist as a part of a human 'observer' of this 'silent-level' 'territory').

Next, in order to prove the "incommensurate" clause of the theorem, I shall specify a 'verbal-level' set which qualifies as a 'map' of the ('silent-level') 'territory', and another 'verbal-level' set which qualifies as a 'map' of our organism's ('silent-level') 'map' of this 'territory'. Then I shall examine the relations which exist between the elements which make up these two related 'verbal-level' sets, as judged by the criteria specified in detail above. Finally, I shall examine these two 'verbal-level' sets to see whether, as judged on the basis of the best evidence available, either one qualifies as 'similar in structure' to the 'silent-level' set it allegedly represents. If each one does qualify as 'similar in structure' to its respective 'territory', then I have grounds for concluding (predicting) that the elements which make up these two 'silent-level' sets do (will) behave AS IF the (inferred) relations between their respective elements meet, or fail to meet, the same sets of criteria as do the relations between the elements of the two 'verbal-level' sets.

Then, in order to prove the "fundamental error" clause, I shall specify two different 'silent-level' 'maps' of the 'silent-level' 'territory' previously discussed, such that these two 'maps' prove entirely equivalent except for differences introduced as a consequence of the "slightly different circumstances" under which they get elaborated. Again, in order to permit discussion, I shall specify two 'verbal-level' sets, each of which qualifies as a 'map' of one of these 'silent-level' 'maps'. Then I shall examine the 'first and second projections' of each of these two 'verbal-level' sets. I expect that the 'second projections' of these two sets, the so-called 'other'-components or Gestalten, will prove equivalent to each other; but I expect that the 'first projections', the so-called 'self'-components, will differ markedly in structure. An examination of these differences in structure should allow me to prove the "fundamental error" clause.

Finally, on the basis of these two clauses, I shall offer a definition of the term 'sanity'.

Proof As mentioned above, to begin with I shall restrict discussion to the topic of the 'silent levels'.

Please imagine that you observe the following non-verbal HAPPENINGS:

> DEMONSTRATION: I hold a THING as high above the floor as I can conveniently reach, and then drop it. It falls to the floor and hits with a loud "Thunk!"

Anything I may say about these HAPPENINGS will exist on 'verbal levels', whereas these happenings remain non-verbal ('silent-level'). In order to avoid confusing 'name' with 'thing named', I shall use un-primed letters to designate non-verbal HAPPENINGS, and shall use primed letters to indicate 'verbal-level' representations of these non-verbal HAPPENINGS.

As our two related but 'silent-level' sets which differ in order of abstraction, I shall take as 'territory' the set A, which I describe as "the non-verbal series of changing-positions-with-changing-times of our THING ($stone_1$) undergoing a free fall at times t_j"; and as a 'map' of this 'territory' I shall take the set B, which I describe as "my visual 'awareness' of $stone_1$ undergoing a free fall at times t_j".

A. "INCOMMENSURATE"

1. The elements which make up the non-verbal set A comprise a series of non-verbal HAPPENINGS, which I have described as "the non-verbal series of changing-positions-with-changing-times of $stone_1$ undergoing a free fall at times t_j. Using the conventions of classical physics I could represent (symbolize) these HAPPENINGS by the set A', made up of the same elements as the Newtonian equation $s = \frac{1}{2}gt^2$.

2. The elements which make up the non-verbal set B comprise a series of non-verbal HAPPENINGS which I have described as "my visual 'awareness' of $stone_1$ undergoing a free fall at times t_j". Using the conventions of the logical calculus of behavior of Hilgartner & Randolph,[26] I could represent these HAPPENINGS by the set B'. Here I have no "simple" formula from the differential-integral calculus by which to define the set B'; rather, the set B' consists of the same elements as does an array of set-theory sentences which

describe the ('verbal-level') "tracking operation" of a ('verbal-level') 'organism'; and this "tracking operation" in turn consists of the same elements as does a ('verbal-level') series of changing patterns of stimulation of retinal rods and cones (etc.), of changing direction of gaze, of changing binocular projection differences, of changing ocular accommodation, of changing convergence of ocular axes, of whatever scanning of the visual fields occurs in the course of the "tracking operation", of the sequence of impulses from the bony labyrinth of the inner ear, etc.

3a. I now have the wherewithal to make explicit the relations which exist between the elements which make up these two ('verbal-level') sets, A′ and B′. I have previously formalized this relation as

$$[a \in A', b \in B'] \Rightarrow b\mu a, \text{ where } A' \subset E', B' \subset O', \mu \subset E' \times O'.$$

(i) 'Equivalence relations': The inverse relation $\mu^{-1} \subset O' \times E'$ does not qualify as a member of the class of 'equivalence relations':

$$\text{Not reflexive: } [b \in B'] \Rightarrow b\mu^{-1}b.$$

("The element $b \in B'$ as an element of a ('verbal-level') 'map' B′ does NOT "represent" the element $b \in B'$ of the ('verbal-level') 'territory' A′.")

$$\text{Not symmetric: } a\mu^{-1}b \Rightarrow b\mu^{-1}a$$

("If $b \in B'$ comprises an element of a ('verbal-level') 'map' which represents the element $a \in A'$ of the ('verbal-level') 'territory' A′, then the element $a \in A'$ does NOT qualify as an element of a ('verbal-level') 'map' B′ which "represents" the element $b \in B'$ of the ('verbal-level') 'territory' A′.")

Furthermore, the relation $\mu \subset E' \times O'$ itself does not qualify as a member of the class of 'equivalence relations':

$$\text{Not reflexive: } a\mu a.$$

$$\text{Not symmetric: } a\mu b \Rightarrow b\mu a.$$

Therefore, I conclude that we may not regard these two ('verbal-level') sets as "composed of the same elements": $A' \neq B'$. Nor may we in any other sense regard them as 'equivalent'.

(ii) '1-to-1 functions': The inverse relation $\mu^{-1} \subset O' \times E'$ does not qualify as a member of the class of 'single-valued functions': If the element $b_r{}^i \in B'$ (where the superscript i indicates "at any instant ti", and the subscript r indicates 'retinal') designates the ('verbal-level') retinal image corresponding to a single element $a'^i \in A'$ (where a'^i designates the ('verbal-level') position of stone_1 at one instant of its free fall), then $\mu^{-1}(b_r{}^i)$ designates not a single unique 'environmental object' $a'^i \in A'$, but rather designates that set composed of any or all configurations, likely or unlikely, which could cast a retinal image equivalent to $b_r{}^i \in B'$:

$$\mu^{-1}(b_r^i) = \{a'^i \cup a''^i \cup a'''^i \cup \ldots \cup a^{ni} \mid a^{ji} \in (A' \cup A'' \cup A''' \cup \ldots \cup A^n),$$
$$\mu(a^{ji}) = b_r^i\}, \text{ (where } j = 1,2,3, \ldots, n).$$

Finally, the relation $\mu \subset E' \times O'$ itself does not qualify as a member of the class of 'single-valued functions': If the element $a'^i \in A'$ designates the ('verbal-level') position of ('verbal-level') $stone_1$ at one instant of its free fall, then $\mu(a'^i)$ designates not one unique ('verbal-level') 'image' $b_r{}^i \in B'$, but rather designates the set composed of at least one element from each of the classes of ('verbal-level') 'intra-organismic activities' ('processes of abstracting') which make up the ('verbal-level') "tracking operation" specified above: viz., patterns of stimulation of retinal rods and cones, direction of gaze, etc.:

$$\mu(a'^i) = \{b_1^i \cap b_2^i \cap b_3^i \cap \ldots \cap b_m^i \mid b_k^i \in B'\}, \text{ (where } k = 1,2,3, \ldots, m).$$

Or, paraphrasing these relations in terms of the composition relations of μ with μ^{-1},

$$\mu^{-1}(\mu(a'^i)) = \{a'^i \cup a''^i \cup a'''^i \cup \ldots \cup a^{ni} \mid a^{ji} \in (\bigcup_{j=1}^{n} A)\} \neq I_E,$$
$$\mu(\mu^{-1}(b_r)) = \{b_1^i \cap b_2^i \cap b_3^i \cap \ldots \cap b_m^i \mid b_k^i \in B''\} \neq I_O.$$

In other words, neither A′ nor B′ proves indistinguishable from the appropriate identity-mapping:

$$A' + I_E \neq \emptyset,$$
$$B' + I_O \neq \emptyset.$$

Therefore, I conclude that we may not regard these two ('verbal-level') sets as related by a 1-to-1 function (nor even by a single-valued function).

3b. The elements which make up the set B′ comprise elements of an (organism x environment) Cartesian product space designated as (O x E), every element of which stands for one logically possible 'interaction at the boundary', and consists of an ordered pair, an (o,e) couple (an interaction of 'organism' with 'environment'). As judged on the basis of the best evidence available at present,[27] the elements which make up the set B′ qualify as 'similar in structure' to the fundamental non-verbal 'relatedness' displayed by 'living systems'. Thus I conclude that the set B′ qualifies as 'similar in structure' to the ('silent-level') set B.

3c. On the contrary, the elements which make up the set A′ (defined by the Newtonian equation quoted above) comprise elements of the 'phase space' of Newtonian mechanics, and thus refers to an 'isolated' physical 'system': the theoretical constructs ("Laws") of Newtonian mechanics defined on this 'phase space' (like the theoretical constructs of classical thermodynamic theory [28])

> specify only the relations between the various 'states' of our 'isolated' 'system', and say nothing whatever about the relations of our 'isolated' 'system' with the rest of the 'universe'. [29]

Thus these elements do NOT qualify as 'similar in structure' to the fundamental non-verbal 'relatedness' intrinsic to 'living systems'. However, almost three centuries of experiences have shown that (as a first approximation, at least) the Newtonian equation quoted above does qualify as 'similar in structure' to the INFERRED non-verbal, non-biological, non-social 'territory' which (we say) consists of actual free-falling objects (which we postulate on the basis of our non-verbal 'abstracting').

Thus I conclude that the set A′ qualifies as 'similar in structure' to the ('silent-level') set A.

3d. Since the relations between the elements which make up these two 'verbal-level' sets A′ and B′ fail to meet the two sets of criteria specified in detail above, and since these two 'verbal-level' sets qualify as 'similar in structure' to their respective 'silent-level' sets A and B, therefore I conclude (predict) that the two 'silent-level' sets A and B do (will) behave AS IF the (inferred) relations between their respective elements also fail to meet these two sets of criteria.

In other words, I conclude that we may not regard the 'silent-level' sets A and B as related by 'equivalence relations' nor by '1-to-1 functions' (nor even by 'single-valued functions').
Therefore "incommensurate".

B. "FUNDAMENTAL ERROR"

The "fundamental error" clause of this theorem implies (at the very least) a comparison between two 'maps' of a given 'territory', one of which qualifies as 'similar in structure' to the 'territory' and the other of which does not.

In order to prove this clause, I again take as ('silent-level') 'territory' the set A, which (as before) I describe as "the non-verbal series of changing-positions-with-changing-times of our THING ($stone_1$) undergoing a free fall at times t_j"; and I hypostatize two different ('silent-level') 'maps' of this 'territory', B_1 and B_2. I describe both of these sets as "our organism's visual 'awareness' of $stone_1$ undergoing a free fall at times t_j".

Specifically, a single organism under one set of conditions 'observes' the HAPPENINGS designated as the set A, and elaborates the ('silent-level') set B_1 as a 'map' of A; or, under slightly different circumstances, this organism 'observes' the HAPPENINGS designated as the set A, and elaborates the ('silent-level') set B_2 as a 'map' of A. These assumed conditions make the 'silent-level' sets B_1 and B_2 exactly equivalent except for the differences introduced as a consequence of the "slightly different circumstances" under which our organism 'observes' the single set of 'silent-level' HAPPENINGS A. The proof of the "fundamental error" clause, then, will center about a comparison of the structure of the set B_2 as compared with B_1.

1. As before, the elements which make up the non-verbal set B_1 comprise a series of non-verbal HAPPENINGS which I could describe as "our organism's visual 'awareness' of $stone_1$ undergoing a free fall at times t_j". Here the "circumstances" of our organism's 'observing' operate so that he does not in any way 'interfere' with his own functioning, viz., with the process of 'generating and testing hypotheses in a survival-related context'. In other words, in the ('non-verbal') set B_1, our organism's activities approximate what we have elsewhere called " 'unimpaired' human behavior".

Making use of the conventions of the logical calculus of behavior of Hilgartner & Randolph,[30] I could (as before) represent these happenings by the ('verbal-level') set B', which (as before) consists of the same elements as does a ('verbal-level') "tracking operation" which in turn consists of the same elements as does a ('verbal-level') series of changing patterns of stimulation of retinal rods and cones, etc., etc.

2. The elements which make up the non-verbal set B_2 comprise a series of non-verbal HAPPENINGS which I could also describe as "our organism's visual 'awareness' of $stone_1$ undergoing a free fall at times t_j". I shall represent this set of non-verbal happenings by the ('verbal-level') set B''.

Here I take the "slightly different circumstances" of our organism's 'observing' to operate so that he does 'interfere' with his own functioning, viz., with the process of 'generating and testing hypotheses in a survival-related context'. Specifically, in the non-verbal set B_2 our organism engages in (non-verbal) activities which I might describe as "acts of selective inattention", "acts of sustained tensing of certain pairs of antagonistic muscles", "acts of 'staring' (viz., restricting the free play of eye-motions enough to interfere with the process of 'seeing')", etc., performed in such a way that he manages to 'FIXATE' on what he 'SEES' and to 'NEGLECT' what he DOES IN ORDER TO 'OBSERVE'.

(i) I reemphasize that regardless of whatever special set of "circumstances" may surround any set of 'observations' ('silent-level' 'abstractions'), the fundamental structure of the non-verbal 'processes of abstracting' remain unaltered: they still consist of non-verbal HAPPENINGS, activities of our organism-as-a-whole-in-his-environment-at-specific-instants.

(ii) Moreover, as displayed in the statements of our premises (and our counter-premises) above, we can resolve any 'awareness' or 'consciousness' into its 'first and second projections', the 'self'-component and the 'other'-component or Gestalt.

(iii) And as Gestalt psychologists have so convincingly shown, a given organism can organize quite different Gestalten from a single 'environmental object', by very subtly altering the "circumstances" which surround his 'observing'.

3a. Each of the ('verbal-level') sets, B' and B'', constitutes a symbolic representation of one or the other of our two sets of non-verbal HAPPENINGS, B_1 or B_2 (both of which I have described as "our organism's visual 'awareness' of stone$_1$ undergoing a free fall at times t_j"). Thus the 'other'-component of each of these ('verbal-level') sets comprises a ('verbal-level') Gestalt which I can describe as "a visual Gestalt of stone$_1$ undergoing a free fall at times t_j", and which I can formalize as consisting of the same elements as does the Newtonian equation $s = \frac{1}{2}gt^2$:

$$Cq_2(B') = Cq_2(B'') = Oe' = Oe'' = [s = \tfrac{1}{2}gt^2].$$

That makes the 'other'-component or Gestalt of each ('verbal-level') set equivalent to the previously-defined ('verbal-level') set A'.

3b. In my view, no organism can have any kind of 'awareness' or 'consciousness' of any 'existence' at all, his own or any other, except as a consequence of 'interactions between 'organism' and 'environment' occurring at the boundary'. In other words, the 'territory' which the non-verbal set B_1 somehow 'represents' must consist of interactions between two kinds of non-verbal THINGS: 'the falling of stone$_1$ at times t_j', and 'an organism which 'observes' (the falling of stone$_1$ at times t_j)'.

As a consequence of the special set of "circumstances" surrounding the set of non-verbal 'observations' B_1, our organism remains alert and oriented, able to recount to himself-or-others not only what he 'sees' but also what he DOES in order to 'observe', and how it comes about that he bothers to notice the falling of a stone here-now. (And as I pointed out above, as judged on the basis of the best available evidence, my representation of the set B_1, namely, the 'verbal-level' set B', qualifies as 'similar in structure' to the set B_1.

That means, then, that the 'self'-component of B' comprises a specifiable array of ('verbal-level') 'processes of abstracting' by which our ('verbal-level') 'organism' puts together his ('verbal-level') Gestalt:

$$C_{q1}(B') = Se', Se' + U(Se') \neq \emptyset.$$

And this formulation obviously implies-and-assumes the non-aristotelian premises of Korzybski.

3c. As already noted, as a consequence of the "slightly different circumstances" surrounding the set of non-verbal 'observations' B_2, our organism engages in "special activities" by means of which he manages to 'fixate' on what he 'sees' and to 'neglect' what he DOES in order to 'observe'. Thus our organism manages to keep himself more or less 'disoriented', and consequently manages to find difficulty in recounting to himself-or-others what he DOES in order to 'observe', or how it comes about that he bothers to notice the falling of a stone here-now. (And as judged on the basis of the best available evidence, my representation of the set B_2, namely, the ('verbal-level') set B'', qualifies as 'similar in structure' to the set B_2.)

That means, then, that these "special activities" of our ('silent-level') organism serve to conceal, to eliminate from consideration, the structure of his relations with himself and with his own environment. In other words, I can symbolize these non-verbal 'special activities' which form part of the 'silent-level' set B_2 by showing the 'self'-component of the 'verbal-level' set B'' as 'negligible':

$$[C_{q1}(B'') = Se'' = [Z'' - \rho(Q'')]] \subset \overline{L_M}.$$

As we have already seen, this formulation proves equivalent to the verbal statement that "Our organism's 'picture' of the 'territory' remains practically devoid of elements extrinsic to the 'territory' "; or, in notation,

$$Z'' - \rho(Q'') \subset \overline{L_M} \Leftrightarrow Z'' + Oe'' \subset \overline{L_M}.$$

But that condition can hold if and only if our mapping ρ qualifies as a 1-to-1 function:

$$\begin{aligned} Z'' - \rho(Q'') \subset \overline{L_M} \quad & \Leftrightarrow \rho^{-1} \circ \rho = I_E, \\ \rho^{-1} \circ \rho = I_E \quad & \Leftrightarrow Y'' + Z'' \subset \overline{L_M} \\ & \Leftrightarrow Z'' + I_E(Y'') \subset \overline{L_M} \end{aligned}$$

And further, that condition can hold if and only if the operator U qualifies as a 1-to-1 function:

$$\begin{aligned} Z'' - \rho(Q'') \subset \overline{L_M} \quad & \Leftrightarrow Se'' \subset \overline{L_M} \\ & \Leftrightarrow U \circ U^{-1} = I_{Oe} \\ & \Leftrightarrow Oe'' + I_{Oe} \subset \overline{L_M} \\ & \Leftrightarrow U^{-1} \circ U = I_{Se} \\ & \Leftrightarrow Se'' + I_{Se} \subset \overline{L_M} \end{aligned}$$

3d. Thus beyond question this formulation implies and assumes the "Laws of Thought" of Aristotle as fundamental premises.

But if we grant those premises, then we no longer have any way to distinguish between 'map' and 'territory', and so $B'' \mu B_2$ becomes

$$B'' = B_2,$$

and $B_2 \ \mu \ A$ becomes

$$B_2 = A.$$

Therefore these ('silent-level') "special activities" of our organism, viz., the "acts of selective inattention", the "acts of sustained tensing of certain pairs of antagonistic muscles", the "acts of 'staring' ", etc., by means of which our organism manages to 'fixate' on what he 'sees' and to 'neglect' what he DOES in order to 'observe', (and which thus serve to eliminate from consideration the topics of the organism's relations with himself and with his environment) constitute the 'silent-level' equivalent of the 'verbal-level' notion of "the act of taking the "Laws of Thought" of Aristotle as fundamental premises".

But I have already proved that 'map' and 'territory' remain 'incommensurate', viz., that the relations between 'map' and 'territory' do not qualify as any kind of 'equivalence relations' nor as any kind of '1-to-1 functions'.

Therefore the set B'' qualifies as 'similar in structure' to its 'territory', namely, to the set B_2; but the set B_2 does NOT qualify as 'similar in structure' to ITS 'territory', since it contains special logical (or psycho-logical) constructs which on an a priori basis serve to eliminate from consideration one or more key relations which make up the 'territory'.

4. To treat related but incommensurate sets AS IF the relations

between them qualified as 'equivalence relations' and/or '1-to-1 functions' (symbolically to 'interchange' 'map' and 'territory') violates the rules of naive set theory and will immediately or eventually lead to a paradox, self-contradiction, or other solecism.

5a. Any 'map' which includes even one paradox, self-contradiction, or other solecism will not survive logical scrutiny (and thus comprises a member of the class of 'inadequate' formulations).

5b. An 'inadequate' formulation qualifies as 'similar in structure' to no 'territory' whatsoever.

6. The act of relying on a 'map' which does not qualify as 'similar in structure' to the 'territory' it allegedly 'represents' comprises an element of the set of 'evaluative disorders' (or 'fundamental errors'). Therefore "fundamental error".

C. RELATION TO "THEORY OF SANITY"

'Sanity' then comprises that class of behavior not based on the fundamental error of 'interchanging or otherwise regarding as equivalent the members of pairs of related sets which differ in order of abstraction' (the class of behavior free of 'evaluative disorders').

SUMMARY

Using a small sampling of the key constructs of general semantics, I have demonstrated that I can indeed state them in set theory terms (with an impressive economy of means); I further claim that I can do this with ANY construct of general semantics and, moreover, claim that in other contexts I have already done a rather complete job of it, although not yet entirely explicitly.

Furthermore, having translated it into set theory terms, I find 'general semantics' free of self-contradiction, as judged by the standards of modern set theory.

But when we consider what these efforts have accomplished, we suddenly find ourselves in the position of the man who says to his dog, "Speak, Fido!" Whereupon the dog replies, "To whom, about what, and for what purpose?"

For as I have taken pains to point out, the unregenerate 'general semantics' of Korzybski (the first non-aristotelian system) constitutes an account of the structure of the transactions occurring between a 'human-organism-as-a-whole-and-his-environment-at-a-date'. Moreover, in the non-aristotelian set theory calculus of behavior of Hilgartner & Randolph,[31] in which I took pains to avoid using the traditional terminology of 'unregenerate' general semantics, we explicitly set up a model of an 'organism' interacting with his 'environment', who manifests his 'fundamental life-processes' by classifying the elements of his own interactions with his environment as "belonging" or "not belonging" to various sets, and who does this 'classifying' in order to keep himself intact-and-growing for at least a few moments longer. Recently I extended this set theory calculus of behavior so as to account for the experimental findings concerning the mechanisms by which a red blood cell 'regulates' its own carbohydrate metabolism. In the process, I made of the term 'living system' an acceptable set theory operator – which constitutes a degree of rigor in biology not previously attained, to the best of my knowledge.[32] And finally, the present paper shows some of the results of my efforts to base the traditional terminology of 'unregenerate' general semantics on set theory.

By way of contrast, naive set theory (outside the context of non-aristotelian systems) conceals (TAKES FOR GRANTED) the process whereby elements get classified as members or non-members of sets. It codifies the rules for operating on these sets so as to draw logically correct inferences but has nothing whatever to say concerning "who" or "what" DOES the 'classifying', nor "how", nor "for what purposes". In other words, outside the context of non-aristotelian systems, set theory cannot deal with – it ASSUMES OUT OF EXISTENCE – the intrinsically interesting class of relations between an 'organism-as-a-whole' and his 'environment', the so-called 'interacting of the boundary' at specific moments.

Thus, perhaps unexpectedly, set theory ignores the topic of CONTEXT (e.g., the context of set theory). Furthermore, although it came as a surprise to me and I realized it only in the process of writing this paper, the fact that set theory distinguishes between 'element' and 'set' does NOT mean that set theory distinguishes between 'name' and 'thing named': We can with equal facility use the term 'element' to refer to a non-verbal, non-symbolic 'external' THING or to a symbolic representation of it, e.g., my 'perception' of it. Set theory makes no provision for indicating which sense we intend.

I explain these findings by concluding that naive set theory implies and assumes the "Laws of Thought" of Aristotle as fundamental premises, and thus remains committed to the pretense that the relations of our 'organism' ('classifier') with his 'environment' and the relations of our 'organism' with 'himself' prove equivalent to tautologies (identity-mappings on the environment).

Since mathematicians have based every branch of mathematics on set theory, these findings permit of further conclusions. As Korzybski points out, "the majority of mathematicians take mathematics as a clear-cut entity, 'by itself'."[33] Within the limited circle of mathematicians I know, this attitude still seems almost universal.

However, from a non-aristotelian viewpoint, mathematics does not exist as an 'isolated' domain sui generis (a manifestation of 'the mind'), but rather constitutes a most characteristic and important form of human behavior. To hold otherwise requires the "pretenses" discussed above.

In other words, by addressing ourselves to the one topic which lies outside the customary domain of 'isolated' mathematics, viz., the logical structure ('psycho-logics', to use Korzybski's term) of human behavior, and then by using the tools of mathematical set theory to perform a logical analysis of the resulting theories, we exponents of general semantics have disclosed and eliminated an untenable hidden assumption (an unavoidable self-contradiction) in the logical foundations of the mathematics existing prior to 1970. Thirty-seven years ago, Korzybski predicted that "the problem of mathematical foundations does not belong to mathematics but to psycho-logics."[34] By elaborating a psycho-logics in a set theory notation, I appear to have tested that prediction; and as I have just shown, I have not disconfirmed it.

To make what I've said if possible even more offensive, I assert that any mathematician who regards 'mathematics' as an 'isolated' domain sui generis comprehends neither his own field of study nor the relations between his field of study and his own humanity. And worse yet, I claim that general semantics in general, and my own work in particular, provides an adequate and well worked out basis for a completely general theory of human evaluating, which would include mathematics, physics, chemistry, biology, and the human psycho-social sciences as special cases.

Behold: A better gateway! saith the clown.

But those whose stomachs prove strong enough to allow them to examine my claims will (I predict) find them definite disconfirmable predictions; and (at least in my better-oriented moments) I welcome the efforts of others to test and perhaps disconfirm these claims.

REFERENCES

1. Polanyi, M. Personal Knowledge: Toward a Post-Critical Philosophy. Chicago: University of Chicago Press, 1958, 2nd edition 1962. New York: Harper Torchbook, 1964, p. 191.

2. Ashby, W. R. "The Set Theory of Mechanism and Homeostasis." Technical Report No. 7, Electrical Engineering Research Laboratory, University of Illinois, Urbana, Illinois, 1962.

3. Hilgartner, C. A. & J. F. Randolph. "Psycho-Logics: Human Behavior in the (Organism x Environment) Field. 1. A Logical Calculus of Behavior." Journal of Theoretical Biology 23: 285-338 (1969a); "2. The Structure of 'Unimpaired' Human Behavior." Journal of Theoretical Biology 23: 347-374 (1969b); "3. The Structure of Empathy." Journal of Theoretical Biology 24: 1-29 (1969c).

4. I regard this as a paper ABOUT mathematics rather than a mathematical paper. But in order to back up the conclusions stated in the SUMMARY, I include, in addition to the simple set theory analysis of terms from general semantics (which I predict will tax no one's mathematical ability), four sections stated in an algebraic set theory notation: A) a statement on one 'universal set' of the "Laws of Thought" of Aristotle; B) a statement on a different 'universal set' of the non-Aristotelian premises of Korzybski; C) a restatement of the "Laws of Thought" on this latter 'universal set' which also specifies their relations to the non-Aristotelian premises; and later, D) a formal proof of a 'theorem of general semantics'. These formal sections make up more than half the bulk of this paper (partly because, in hopes of making this material intelligible to non-mathematicians, I have given an English translation of each set theory sentence).

For those already familiar with naive set theory, I include a brief Appendix which states the main constructs I make use of in three different forms: in words, in terms of Venn diagrams, and in our Bourbaki algebraic notation. A fuller account of our notation appears in Hilgartner & Randolph, 1969a, Appendix I.

For those with no previous exposure to set theory, I offer the following encouragement: The constructs which make up set theory correspond closely to intuitive notions about the process of 'classifying things'. Readers will find, I believe, that they can follow the reasoning of at least the non-formal parts of this paper without previous exposure to set theory, and will need no constructs other than the 'undefined terms' of set theory ('element', 'set', and 'belongs') and the notions of 'universal set', 'subset', 'complement', 'equivalent' ("composed of the same elements as"), and perhaps 'mapping', all of which I develop in the text. Finally, in the formal sections, I hope the verbal translations will help in following the argument.

For those who wish to explore set theory on their own initiative, almost any text on "Finite Mathematics" will contain a chapter on symbolic logic and one on set theory. I recommend the Schaum's Outline Series text: Seymour Lipschutz. Set Theory and Related Topics. New York: McGraw-Hill, 1964.
5. MacHovec, F. J., translator. The Book of Tao. Mt. Vernon, New York: The Peter Pauper Press, 1962.
6. Korzybski, A. Science and Sanity: An Introduction to Non-Aristotelian Systems and General Semantics. International Non-Aristotelian Publishing Co., 1933. 4th edition, 1958, Institute of General Semantics, Lakeville, Conn., p. 749; Jevons, W. S. The Elements of Logic. New York: Sheldon & Co., 1883.
7. To anyone who experiences some kind of outrage or distress over the fact that I use the symbol '+' in a sense different from "add" (as in 2 + 2 = 4), I point out to you that you have the opportunity to make this into a very valuable experience. I deem that Humpty-Dumpty had it aright: "When I use a word," Humpty-Dumpty said in a rather scornful tone, "it means just what I choose it to mean, neither more nor less." To act as if a symbol 'has' some 'one right meaning' clearly qualifies as a disordered evaluation. I contend that the process of going through years and years of formal psychoanalysis might not turn up even one equally convincing example of a clearly DISORDERED evaluation. And here, the "cure" appears very simple (if somewhat painful): AT LEAST ONCE, continue to explore this 'outrageous' symbol and its context, while continuing to feel uncomfortable and to tolerate your own discomfort.
8. Korzybski, A., op. cit., p. 10; Korzybski, A. "General Semantics, Psychiatry, Psychotherapy, and Prevention." M. Kendig, editor: Papers from the Second American Congress on General Semantics. Chicago: Institute of General Semantics, 1943.
9. Hilgartner, C. A. & J. F. Randolph. A Logical Calculus of Behavior, pp. 295-7.
10. Hilgartner, C. A. & J. F. Randolph. A Logical Calculus of Behavior, p. 313.
11. Hilgartner, C. A. & J. F. Randolph. The Structure of 'Unimpaired' Human Behavior, p. 355.
12. Perls, F. S., Ralph Hefferline & Paul Goodman. Gestalt Therapy: Excitement and Growth in the Human Personality. New York: Julian Press, 1951, p. 303.
13. Hilgartner, C. A. "Metabolic 'Control' in Mature Erythrocytes." (Submitted for publication.)
14. Hilgartner, C. A. & J. F. Randolph. A Logical Calculus of Behavior, p. 295.
15. In order to complete the picture for humans, I point out that we remain able to detect still other elements of our 'external' and/or 'internal' environments only with the assistance of extra-neural instruments (e.g., Geiger counters, microscopes, etc.), and we remain utterly unable to detect still other elements, which therefore have purely inferential 'existence' (e.g., 'molecules', 'atoms', 'electrons', etc.).
16. Korzybski, A. "The Role of Language in the Perceptual Processes." R. R. Blake and G. V. Ramsey, editors: Perception: An Approach to Personality. New York: Ronald Press, 1951.
17. Hilgartner, C. A. "Chain-Indexing in Experimental Design." General Semantics Bulletin. No. 35, pp. 44-53, 1968.
18. In logical terms, we cannot demonstrate that a set defined on some

'universal set' remains 'isolated from' or 'not related to' other sets defined on that 'universal set', at least not in this prima facie sense. But we can find some special logical construct which renders the sense of 'isolated' or 'not related' in a fashion compatible with the relational language of set theory, as you will see.

19. Perls, F. S. Gestalt Therapy Verbatim. Lafayette, California: Real People Press, 1969.

20. Cartier, F. A. "Three Persistent Misconceptions of Communication." ETC.: A Review of General Semantics 20: 135-145 (1963).

21. Hilgartner, C. A. "Feelings, Orientation, and Survival: The Psychological Dimension of the Current Human Crisis." Presented at the Ninth International Conference on General Semantics, San Francisco State College, August 1965.

22. Lee, L. L. "Some Semantic Goals for Aphasia Therapy." ETC.: A Review of General Semantics. 18: 261-274 (1961).

23. Nahikian, Howard M. A Modern Algebra for Biologists. Chicago: University of Chicago Press, 1964, pp. 16-17.

24. Hilgartner, C. A. & J. F. Randolph. A Logical Calculus of Behavior, pp. 334-5.

25. An earlier version of this proof appears in Hilgartner, C. A. & J. F. Randolph, "The Structure of 'Unimpaired' Human Behavior," pp. 357-60.

26. Hilgartner, C. A. & J. F. Randolph (same as note 3).

27. Hilgartner, C. A. & W. R. Johnson. "Encounters in a Corridor: Human Inter-Personal Transactions." Presented at the Tenth International Conference on General Semantics, Denver, August 1968. Also Hilgartner, C. A. "Metabolic 'Control' in Mature Erythrocytes."

28. Landsberg, P. T. Thermodynamics. New York: Interscience Publishers, 1961.

29. Hilgartner, C. A. "The 'World-View' of Classical Thermodynamics." (Submitted for publication.)

30. Hilgartner, C. A. & J. F. Randolph (same as note 3).

31. Hilgartner, C. A. & J. F. Randolph (same as note 3).

32. Hilgartner, C. A. Metabolic 'Control' in Mature Erythrocytes.

33. Korzybski, A. Science and Sanity, p. 133.

34. Korzybski, A. Science and Sanity, p. 752.

35. Murray, Elwood. The listing of general semantics constructs came from the preliminary Agenda for the Conference on Research Designs in General Semantics held at Pennsylvania State University, March 1969.

General Semantics as a General System which Explicitly Includes the System Maker

ROBERT P. PULA

Among the Korzybskian formulations which many students seem hard put to 'get' and INTERNALIZE are those which get at the heart of system building in particular and the language-'reality' relationship in general; namely, multiordinality, self-reflexiveness, consciousness of abstracting and non-identity. These formulations make such a demand on the learner's ability to see himself as a focus in the organism-environment continuum (consciousness of abstracting) that we must often resort to techniques of silence to break through the enthralling web of language. We will not be able to do that in print. Nevertheless, I want to talk about the untalkable because these formulations seem to me to be ones that will help us to begin to build a bridge between general semantics and general systems theory. And they can do this precisely because they apply to the process by which systems are built.

By MULTIORDINALITY Korzybski meant that characteristic of many words that they can be used at different levels of abstraction WITH NO CHANGE IN THEIR DICTIONARY MEANINGS. Another way to say this is that while the nominal ('symbolic' in its simplest sense) 'meaning' remains constant, the functional or operational 'meaning' changes with the context – the critical context here being the level of abstraction at which the word is being used. It must also be kept in 'mind' that 'level of abstraction' cannot be understood in a spatial sense; that is, it is not necessarily a detectable PLACE, although future research may allow us to locate specific levels of abstraction as neurological events.[1] For now we must simply understand levels of abstraction as referring to degrees of complexity and/or simplicity ('distance' or 'nearness') relative to what a general systems theoretician would call the 'boundary' of a given system and what a restricted semanticist would call a referent.

Unconsciousness of the multiordinality of terms readily leads to confusion and therefore conflict in daily life. More importantly for us, lack of awareness of this mechanism can lead to confusion in system building. Perhaps a general systems theoretician would say that the 'same' word is ISOMORPHIC at different levels of abstraction, but that people generally react as though isomorphism (SIMILAR form, a kind of reverberation among levels of abstraction) 'means' SAME (identical) form.

Another characteristic of multiordinal terms is that (apparently reflecting neurological structures) they are SELF-REFLEXIVE. By self-reflexiveness as used here I 'mean' two things: 1. the multiordinal term can be used at indefinitely large 'numbers' of levels of abstraction; 2. the multiordinal term can usually be applied to 'itself.'

This first aspect of self-reflexiveness seems characteristic of all human abstracting, including our number system. It has often been observed that, no matter how large a number you may suggest, I can always say "plus one." So, too, it is theoretically impossible to exhaust the number of possible statements that can be made in any series; no matter what you may say, I can always make a statement ABOUT (i.e., at a 'higher' level of abstraction) your previous statement. There is no escape from this aspect of self-reflexiveness except by fiat: I simply DECIDE to stop the process noting that my instruction to myself to stop the process is yet another step in the process. But there comes, finally, blessed silence.

That language structure reflects neural structures and, by feedback mechanisms, may ALTER neural structures, is one of the eminently plausible speculations of Korzybski in support of which we have, as yet, insufficient data. (See my paper, "Neglected Formulations," in Communications: General Semantics Perspectives. New York: Spartan Press, 1970.) We see a possible reflection of this in the second self-reflexive aspect of multiordinal terms, their ability to be applied to 'themselves.'"Never say never" is one example. The first "never" is at a 'higher' level of abstraction, is s ABOUT the second "never," yet, because it is at a 'higher' level of abstraction, cancels out the second "never."

While all this is going on, the first ('higher' level) "never" may be said to be commenting on itself as a cancelling-out kind of word. (Does this reflect neural structures?) If we move to yet a third level of abstraction, we get "$never_3$ say $never_2$ say $never_1$" which, by cancelling out the previous cancellation yields a new statement which might read, "Always say never" – all of which suggests that we might want to change the injunction to read "Sometimes say never," or, more in the direction of sanity, "Never play the 'never' game."

What we see here is the magician practising before a mirror so that he won't get fooled by his own tricks.

Any statement is self-reflexive in that it tells us something about the statement maker. Royce[2] first pointed out that an ideal map would have to include a map of itself; Korzybski[3] extended this to include the map-maker. And, in fact, by self-reflexiveness, any map is at least by implication a map of the map-maker: his assumptions, world-view, even (witness the illustrations on early maps) his cosmology.

Awareness of the self-reflexiveness of one's maps is an aspect of that sine qua non of general semantics, CONSCIOUSNESS OF ABSTRACTING. The general semanticist strives for on-going awareness of himself as a map-maker or system-builder; as an active focus of the system under construction; as, in a very Heisenbergian sense, imbedded in any measurement he may make.

The formulation of NON-IDENTITY seems the easiest to accept at presumably 'cortical' levels and most difficult at presumably 'thalamic' levels. The formulation simply states that different levels of abstraction are not identical and therefore not to be confused. Most fundamentally, this applies to the gross levels of abstraction labeled by general semanticists 'verbal' and 'non-verbal.' In one of his few metaphors, Korzybski[4] asserted that "The map is NOT the territory"; that is, no map is any territory (except in the sense that a map may function as a 'territory' for another 'higher' level map). For our purposes, the emphasis must be a vigorous, "The system is not what it systematizes." That includes, of course, both constructs being discussed in this book.

A secure internalization of the non-identity formulation seems our best guarantee to assure that any system-system-builder continuum will remain open.

It has been asserted by at least one writer on general systems theory (F. Kenneth Berrien[5]) that all systems are open; that a closed system is, by observation, impossible, since it must and does receive inputs from 'outside'. He further observes that systems which are denied needed inputs stagnate and then disintegrate, that is, become non-systems or indistinguishable dispersions within some suprasystem. An example might be the Bulgars who gave their name to Bulgaria, but have long since 'disappeared' into the Slavic culture-system of present-day Bulgaria.

I think it appropriate that we avoid an either-or commitment to totally open or totally closed systems. I agree with Berrien that it makes more sense to regard all systems as open, including those to which we rightly apply the second law of thermodynamics. However, I DO urge retention of the open-closed continuum, a formulation of relative open-ness and closed-ness.

General semanticists are very interested in the behavior of that system of ultimate concern for man: namely, man. We find it very helpful, therefore, to formulate (for example) the so-called catatonic schizophrenic as a (relatively) closed system; a scholastic philosopher, a strict constructionist of the U.S. Constitution, or a non-revisionist Marxist, as a closed-but-not-necessarily-pathological system; and, hopefully being demonstrated in this book, general semanticists and general systems theoreticians as unabashedly open systems.

Perhaps the greatest compliment that can be paid when two systems (x and y) meet, is for the representatives of system x to speak in the vocabulary of system y. Korzybski, a mathematically trained organism and therefore committed to what is usually called 'rigor,' warned repeatedly against tampering with the terminology of his system. He knew very well about the effect of terminology on results obtained. "If you change its vocabulary," he is quoted as saying, "you destroy it."[6] (He had before him the famous example of the substantive "heat" in blocking the emergence of thermodynamics. He also had, much closer to home, the example of the dilutive effect produced by some popularizers of his own work.) Nevertheless, the terminologies of general semantics and general systems theory seem so isomorphic that I think that a matching of terminologies can be attempted without prostituting either. And if it doesn't turn out well, we can simply say, "The experiment didn't work," and scrap it. After all, is this not one of the key benefits to be derived from an open commitment to an open system?

Actually, what I am going to do goes in the other direction: general systems theory → general semantics. But, once done, the flow can be reversed to produce a hint as to how general semantics formulations might be expressed in the general systems vocabulary.

For my general systems model I will use the summary of Chapter II, "Basic Definitions and Assumptions" from F. Kenneth Berrien's General and Social Systems.[7] I will quote Berrien's definitions, assumptions and propositions in the order in which he presents them and follow each with a general semantics 'translation,' noting disparities, etc., where that seems needed. The definitions are numbered; each translation follows the definition to which it refers. Many of my observations will reflect the general semantics formulations presented in the first part of this paper. Special effort will be made to avoid what general semanticists call the 'is of identity': forms of the verb 'to be' which imply identification of levels of abstraction, particularly identification of verbal and non-verbal levels.

As used below, SYSTEM and STRUCTURE are equivalent, interchangeable, multiordinal terms, the 'meanings' of which are functions of their levels of abstraction.

DEFINITIONS AND 'TRANSLATIONS'

2.1 A system is a set of components, interacting with each other, and a boundary which selects both the kind and rate of flow of inputs and outputs to and from the system.

We formulate a system as a structure, apparent aspects of which we see as functions of each other, while guarding against reacting to any of these aspects-of-convenience as elementalistically isolated, even in the sense of being 'parts'; we further formulate the system-structure as abstractive-projective in accordance with the functioning of as yet insufficiently understood electro-colloidal processes.

2.2 A component is a unit of a system that in combination with other units functions to combine, separate, or compare the inputs to produce outputs.

Depending on the level of abstraction at which we choose to make an examination-formulation, an aspect or focus of the system-structure functions AS IF 'it' were singly or in combination with other foci, participating in the abstracting, reacting and projecting 'manifested' by the system.

2.3 The boundary of a system is that region separating one system from another whose function is to filter or select inputs and outputs. It can be distinguished by some difference in the relationships existing among the components WITHIN the boundary compared with relationships which occur ACROSS the boundary.

In considering the system-not-system (or system_1-system_2) continuum, we can formulate a transitional 'area' where the abstracting-reacting-projecting performed by the system can most easily be 'detected.'

2.4 Exogenous conditions are those external to the system which bring components to a proximity where interaction is possible.

With regard to any ALREADY FORMULATED STRUCTURE, we can further formulate an area 'outside' the structure (environment) 'with which' the subject structure interacts, i.e., abstracts-reacts-projects.

2.5 Some boundaries are separated by an interface region between systems that merely transports inputs or outputs from one boundary to another.

'Between' intense foci or related conglomerates of foci (structures, systems) we may formulate a region where no STRUCTURES OF PRESENT INTEREST are said to exist, EXCEPT those structures which permit communication ('contact') 'between' the structures of interest.

2.6 A system may exist in various states. A state of the system is a particular pattern of relationships existing among the components and the particular filtering condition of the boundary.

It seems most appropriate to me to assume that any system can best be understood as a process, i.e., a dynamic structure. The abstracted 'state' of the system (structure, pattern of relationships) will be a function of the TIME at which the 'reading' (abstraction) is made, the 'state' of the abstractOR, etc. This reminds us that the abstractOR must be regarded as a 'part' of any system about which he makes statements. In general systems terminology, we might say that the system-describer (observer) is simultaneously exogenous, interfacial, within the system and at 'its' boundary and, from the point of view of DESCRIPTION the most meaningful locus of inputs and outputs.

2.7 Inputs are energies absorbed by the system.

a. Maintenance inputs are those which prepare or maintain the system to function.

b. Signal inputs are those which are processed by the system.

All systems abstract. Depending on the system's position in a hierarchy of complexity, the abstracting process may be placed on a gradient from simply reactive (e.g., 'rust,' 'combustion,' 'reflexes,' etc.) to SEMANTIC reactive, where EVALUATIONS expressing varying degrees of CONSCIOUSNESS are made. The distinction between 'maintenance' and 'signal' inputs seems misleadingly elementalistic (verbally splitting what cannot be found split at non-verbal levels).

2.8 Outputs are those energies which are expelled by the system as a consequence of its operations and are different in some significant way from inputs.

Projection may be seen as the resultant phase following abstracting-reacting, wherein the system 'returns the compliment' to its field (environment). They are distinctly different from the abstracting phase in that they represent TRANSFORMATIONS reflecting the character of the abstracting system, and in that they are LATER IN TIME than the abstracting (input or 'inpull') phase. This introduces the vital formulation of ORDER in its dynamic, 'serial' sense. The natural order of abstracting-projecting requires that, with relation to a selected, limited field of observation CONSCIOUSLY ARRIVED AT, abstracting should precede projecting in such a way as to allow for a 'natural flow' of neural impulses and, at a verbal-evaluational level of abstracting, an APPROPRIATENESS of the projecting with relation to the abstracting-environment continuum. Expressed merely verbally, this seems forbiddingly complex. Expressed VISUALLY on Korzybski's structural differential, it seems simple:

ASSUMPTIONS AND 'TRANSLATIONS'

2.9 All systems are open.

a. Open systems are those that exchange energies with their surroundings.

See comment above re the closed-open continuum.

2.10 Systems exist within systems.

The extent of a system is a joint-function of the system's 'place' in some presumed hierarchy and the level of abstraction (submicroscopic, microscopic, macroscopic, etc.) at which the system-maker DECIDES to enter into transaction with the presumed system. The hierarchy will likely represent a movement from relative simplicity to relative complexity and from relative closedness to relative openess.

2.11 Systems which defy current description, or are inconvenient to describe, may be treated as "Black Boxes" without invalidating the analysis of suprasystems in which they are embedded.

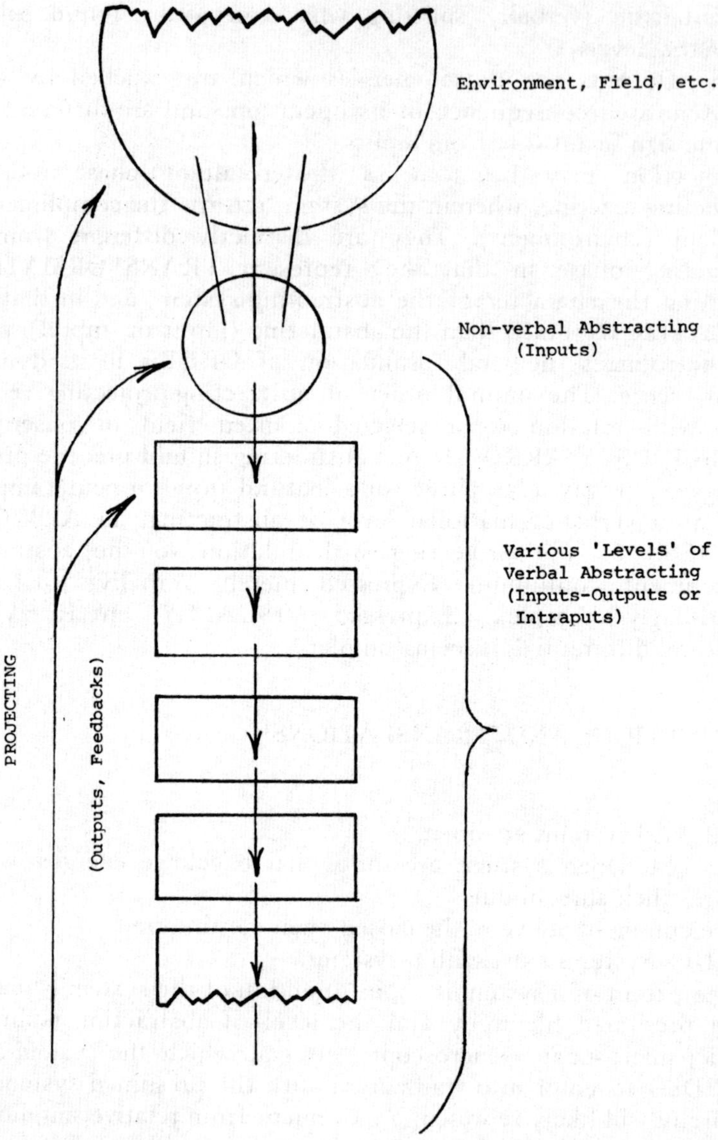
Environment, Field, etc.
Non-verbal Abstracting
(Inputs)
Various 'Levels' of
Verbal Abstracting
(Inputs-Outputs or
Intraputs)
PROJECTING
(Outputs, Feedbacks)

All systems are attempts (formulations) at feeling around in so-called "Black Boxes." The results of this groping are never 'it' and need not be expected to be 'it': they need only give us some degree of predictability with regard to further statements we may wish to make about the system or some as yet unexplored (unformulated) system. The formulation of non-identity frees us from the futile search for 'it' – the completely illuminated 'box,' sans shadows, sans unexpected structures, sans unplanned for contradictions among aspects of the system, etc. To paraphrase Korzybski, WHATEVER you might SAY the system 'is', well it IS NOT.

Note: If systems defy description (are of the Black Box type WITHOUT any input switches, terminals, etc.) and are concurrently parts of a suprasystem, they may VERY WELL invalidate "The analysis of suprasystems in which they are embedded," unless that analysis is tentative and rigorously non-absolute. It seems a regression to a prescientific elementalism to posit black box raisins submerged in the leavened dough of suprasystems such that they have no effect on the taste of the baked result. I am reminded of the old (and very futile) dispute about the 'discreteness of matter' versus non-mathematical 'infinity'. (Does matter's discreteness, which makes it separate from non-matter, set limits to the 'infinity' in which it is placed?)

2.12 The functions of a system are dependent upon its structure.

Recognition of the multiordinal character of the key terms here and of the different dictionary uses of FUNCTION (multi-meaning) allow ready agreement here. However, I think it might be valuable to observe that the structure of a system and its functions are variables of a function. The functions of a system may be said to be its structure under the aspect of TIME. Korzybski's profoundly inexhaustible statement applies here (and throughout this discussion): ". . . structure is THE only possible CONTENT of science and of all human 'knowledge'."[8]

2.13 A critical proximity is necessary for components to interact.

Again, the level of abstraction (observation) would be critical here. If, as Professor Berrien observes and we have already agreed, all systems are open, they must ALL be in 'contact' and therefore interacting, no matter how subtly. Failure to allow for this may account for the many phenomena presently considered to be inexplicable. Nevertheless, it seems clear that for most macroscopically observable phenomena, the 'critical proximity' assumption would hold.

PROPOSITIONS AND 'TRANSLATIONS'

2.14 Initially, systems exhibit greater resistance to destruction than the subsystems or suprasystems which evolve subsequently. Later, the reverse may be true; . . .

a. Components are both mutually attractive and repellent.

I find it hard to separate this 'initially' from the 'initially' which would represent the 'time' at which the observer (inference-maker) came upon the scene. The phraseology seems to grant primacy to the 'system' over the subsystem or suprasystem to which it is related; yet, obviously, all three terms (and whatever their non-verbal correlates might be) DEFINE EACH OTHER. I know of no time at which we were able, even with regard to evolutionary theory, to make observational formulations about 'initial' states of systems, nor to distinguish what constituted a system PURE and therefore its relative sub and supra systems. That the reverse seems plausible as applied to social sub and supra systems, Professor Berrien makes clear later on.[9] But even at this later stage, we have come upon the hierarchy of perceived-projected systems in medias res. We must keep that in mind if we are to avoid a scholastic-platonic regression into 'certitude'.

For Proposition 2.14a I merely suggest that, if we note that 'components' are both mutually attractive and repellent, that we are not dealing with 'components' in any aristotelian sense – all to the good.

2.15 The state of the system is one determinant of the output.

No 'translation' seems necessary here; a good Heisenbergian point.

2.16 Both signal and maintenance inputs are necessary for long-term survival.

See comments under 2.7.

2.17 Systems may produce useful or useless outputs as determined by the suprasystem.

Indeed, suprasystems seem often to be the determiners of what may be said to constitute useful or useless outputs. I suggest only the cross-cultural notion that a system may produce useful or useless outputs with relation to ITSELF. This seems a particularly important amendment considering the effect of feedback mechanisms which may operate WITHOUT REGARD TO THE SUPRASYSTEM.

Let's take an example:

A person (system) talks to himself in a way which is perfectly consistent with the culture (suprasystem) to which he 'belongs'. But the (neuro-linguistic) feedback effect of what he tells himself, for example, "I must compete," induces such stress within the person (system) that some of his subsystems begin to break down: at the level of popular medical abstraction called ulcers, hypertension, etc.

It seems important to me to introduce this point about INTRASYSTEMIC uselessnesses: outputs which through neuro-linguistic feedback become damaging inputs without reference to the suprasystem – what I have previously called intraputs. Granted, this applies so far as we know only on human levels, and may not apply to systems in general. It may be seen as implied in Korzybski's definition of man as time-binder.

> 2.18 Suprasystems select the outputs of their subsystems which are useful; contrariwise, the suprasystem rejects useless outputs from its own subsystems.

Here again I see difficulties, precisely on the human level. The danger centers, I suggest, on that particularly potent form of projection called anthropomorphism. What seemed to me an excellent example of this sort of error was manifest throughout a profoundly gloomy paper written by Anatol Rapoport for the International Conference on General Semantics in 1968. In that paper Rapoport IDENTIFIED a suprasystem ("super-organism") which he called "STATUS BELLIGERANS, the war-waging state." Rapoport first de-humanized the "super-organism" by pointing out that it was NOT merely a collection of individuals and therefore NOT sensitive to inputs by individuals; that it (I will paraphrase Berrien here) "selects the outputs of its subsystems which are useful and rejects those which are useless." Rapoport pictures this 'selecting' happening quite automatically, in accordance with the laws of suprasystems. Having created this monster, Rapoport proceeds to utterly mythologize it, finally referring to 'it' as "the beast." He concludes, like some existentialist dragonslayer afflicted with impotence: "The beast must be subdued. How to do it is the problem relevant to 1968."[10]

I would revise 2.18 to read, "Certain subsystems select outputs which are to be acceptable to the suprasystems of which they are a part, BUT MAKE IT LOOK LIKE THE SUPRASYSTEM DID THE SELECTING."

2.19 Surviving systems are those in which useful outputs exceed useless outputs.

No translation seems necessary here.

2.20 A critical proximity of components brought about by exogenous forces is a necessary but insufficient condition for system operations.

Nor here, except to refer again to points made under 2.13, and the always necessary consciousness of abstracting as particularized in the formulations: multiordinality, self-reflexiveness and non-identity.

There, then, is one general semanticist's attempt at building a support for a bridge across the interface that presently separates general semantics and general systems theory. I see the attempt as exploratory and tentative in the extreme. I have, after all, limited myself to a discussion of the formulations of only one general systems theoretician. Nevertheless, I feel that the support offered is a critical one, one that no responsible system-building enterprise should be without – and that is the epistemological support.

I doubt that any system can be stronger than its epistemology. I see this epistemological awareness as an input that general semanticists can (modestly) make on an on-going basis, rather like a gadfly-in-residence to the general systems theoretician, a 'voice from the back of the room' constantly reminding him to be prepared for the time when, too many ACTIVE unknowns having accumulated in the 'black box,' the whole system collapses.

We may imagine that a lot of confident systems-planning preceded the recent collapse of the Lockheed-Rolls Royce complex – a collapse so fundamental as to require the exposure of the active presence of two necessary but bashful components of the system who very likely would have preferred to remain in their 'black boxes': the American and British governments. Perhaps some of the difficulties might have been anticipated if there had been present a greater degree of CONSCIOUSNESS OF SYSTEM BUILDING – I don't pretend to know. But an attitude of EXPECTATION OF THE INCREDIBLE might have made the (inevitable?) 'adjustments' (an earthquake may be called an adjustment) more palatable, if only because the 'unthinkable' had been built into the system from the beginning.

We need in all planning and system building sessions an Israeli parliamentarian of the type featured in this post-World War II story:

The Knesset, agonized by the difficulties that go with founding a new state among hostile neighbors, is debating policies, possibilities, etc. The delegates have reached a condition of general despair when a member brightly suggests that they declare war on the United States. He reasons, amid the growing enthusiasm of his colleagues, that the United States would crush Israel promptly and then pour in billions of dollars worth of aid, create a modern industrial state, secure its boundaries, etc. Just when the general mood is becoming euphoric, a voice from the back of the hall is heard to say: "Yes. But suppose we win?"

REFERENCES

1. See A.R. Luria, "The Functional Organization of the Brain," Scientific American, Mar., 1970, Vol. 222, Nr. 3, pp. 66-78, and C. R. Legendy, "On the Scheme by Which the Human Brain Stores Information," Mathematical Biosciences, 1967, 1, pp. 555-597.
2. Royce, Josiah, The Principles of Logic, cited in Korzybski, Science and Sanity, p.58.
3. Korzybski, Alfred, Science and Sanity: An Introduction to Non-Aristotelian Systems and General Semantics. Lakeville, Conn.: International Non-Aristotelian Library Publishing Co., 1933; 4th ed., 1958, p.58.
4. Ibid., pp. 58, 498 and 750.
5. Berrien, F. Kenneth, General and Social Systems. New Brunswick, N.J.: Rutgers University Press, 1968, pp. 15-17.
6. Hayakawa, S.I., "Introduction," Margaret Gorman, General Semantics and Contemporary Thomism. Lincoln, Neb.: University of Nebraska Press, 1962, p.v.
7. Berrien, op. cit., pp.32-33.
8. Korzybski, op. cit., p.348.
9. Berrien, op. cit., p.87.
10. Rapoport, Anatol, "The Question of Relevance," in Lee Thayer (ed.), Communication: General Semantics Perspectives. New York: Spartan-Macmillan, 1970, pp. 163-176.

Homo Sapiens as a Model Maker

J. SAMUEL BOIS

The main purpose of the present paper is to bring into the clear a difference of opinion between my fellow general semanticist and personal friend, Elwood Murray, and myself. In a recent letter he had written:

> On page three of your letter of February 10, you say: "I present general semantics as the science-art of system building.' Would you narrow this to 'system building through our perceptions?' It is not our business to build systems; it is only our business to perceive them wherever they may be. Where these semantic systems are not congruent with the existences (systematic or otherwise), we should be correcting them. (?)

The difference, as I see it, comes to this: do we build systems from what we perceive in the real world outside, or do we invent, produce, create, build, etc., systems out of our imagination, later to check if we can construct a reality set-up that fits with them? Are we simply drawing a sketch of a landscape that stretches in front of us, making, as the saying goes, a map of the territory?

I do not agree fully with the map and territory analogy so often repeated with reverence by the faithful disciples of Korzybski. I find that it fails to cover the whole of the discipline he began to build.

His third premise, 'The map is self-reflexive,' gives the map-territory analogy a semantic twist that changes the whole picture. Homo sapiens does not simply perceive as a camera perceives what it is focused upon; he perceives SELF-REFLEXIVELY. As a semantic transactor, time-bound in a cultural heritage expressed mostly in the language he has learned to speak, he has first-order experiences that emerge in a definite pattern. This pattern has the multi-dimensional structure of his semantic state at the moment. It involves attitudes, feelings, purposes, habits, electro-chemical conditions, anticipations of the future and pressures from the past. It is not merely a process of making a more or less accurate map of an assumed territory, the features of which would determine the characteristics of the map he is tracing.

When he wants to describe the origin of creative scientific work,

Korzybski does not refer to what he calls elsewhere the 'natural' order of abstracting. He says:

> Creative scientists know very well from observation of themselves, that all creative work starts as a 'feeling,' 'inclination,' 'suspicion,' 'intuition,' 'hunch,' or some other unspeakable affective state, which only, at a later date, after a sort of nursing, takes the shape of a verbal expression, worked out later in a rationalized, coherent, linguistic scheme called a theory.'[1]

Adelbert Ames, Jr., the rigorous experimenter in phenomena of perception, speaks in the same vein:

> The permanence of forms of my objective world is due to my selection, from the never exactly repeating environmental world, of these particular aspects whose relative continuance is of significance in carrying out my purposes.[2]

In revising Bachelard's original epistemological profile in my own treatise on general semantics, I labelled the creative experience the UNIFYING stage of human functioning, because I wanted to emphasize the fact that creativity involves a certain degree of cosmic consciousness, of felt participation in the energy process that sustains the existence of the whole universe.

> I call it unifying to set it in sharp contrast to the divisive experience of strict objectivity, which we consider desirable at the layman's level, and to the scientific detachment that we take as the only proper attitude for the sophisticated research worker.
>
> Because we take it for granted that the object of our observations, thoughts, constructs, etc., is 'out there' for us to see and theorize about, we miss the fact that the original singling out of the object is an act of our choice and creation. The choice may not be entirely personal and conscious; it may be determined in part by the culture in which we are reared, by the purposes we have set ourselves, and by the circumstances in which we happen to be. But it is our choice in the sense that we could, if we wanted, abstain from it or make a different choice. For instance, if I say: 'I accept such-and-such a theory because it sounds sensible to me,' I am actually choosing a particular theory because it fits with the 'growing tip' of my experience. It is the 'natural' expansion of what I am at the moment, as the flower in spring is the natural expansion of the bud on the branch, as the fruit is the natural expansion of the flower that

disappears in producing it. [3]

In the spiralling advance from one level of understanding to the next, I, as a semantic reactor, am the throbbing center of activity of the whole transaction that engulfs me and the 'object' I am relating myself to. I am not simply sketching a map of a territory by abstracting one by one its significant characteristics. I internalize an experience construed according to my already acquired knowledge, and I project out of my total self a new tentative model of similar experiences I may encounter later, each one unique and unrehearsable. When, by relational abstracting, I have gathered from a series of such experiences what I consider a set of invariant relations, I devise by abduction a mental model that I express in a diagram, a verbal statement, or a mathematical formula.

There are types of relatedness that I have for a long time accepted as standard. They are the systems, Euclidian, Newtonian, and Aristotelian that were prevalent in the cultural milieu in which I obtained my undergraduate and graduate degrees. Since then, I have accepted systems that contradict flatly these former ones, which remain useful in simple situations. I call the new systems non-aristotelian. They cover more than the strictly logical aspects of my functioning. They correspond to new ways I have adopted for my being-in-the-world. 'Right at the beginning,' writes Heisenberg, 'we stand in the center of the confrontation between nature and man, of which SCIENCE, (caps mine) of course, IS ONLY ONE ASPECT.' [4]

Korzybski reminds us that, back of our theories, models and systems, there is the actual process of semantic reacting, which logical terms fail to bring out in the clear.

> The present book is written entirely from the semantic reaction point of view; and so the treatment of the material, and the language used, imply, in general, a psychological response to a stimulus in connection with meanings, this response being expressed by a number of words as 'implies,' 'follows,' 'becomes,' 'evokes,' 'results,' 'feels,' 'reacts,' 'evaluates,' and many others. [5]

> In fact, science, mathematics, 'logic,' etc., may be considered from the NON-ELEMENTALISTIC point of view as GENERALIZED RESULTS of semantic reactions acceptable to the majority of informed and not heavily pathological individuals. [6]

Michael Polanyi sees the progress of science as the transmission of semantic attitudes and beliefs.

> The discoveries of science have been achieved by the passionately sustained efforts of succeeding generations of great men who overwhelmed the whole of humanity by the power of their convictions. Thus has our scientific outlook been molded, of which. . . logical rules give a highly attenuated summary . . . . Science will appear thus as a vast system of beliefs, deeply rooted in our history by a specially organized part of our society. . . Science is a system of beliefs to which we are committed.[7]

Years ago, Erwin Shroedinger had already highlighted the affective component of our scientific enterprise.

> Our interest in any investigation. . . is due to some further consideration which we intend to attach to the result that we try to get hold of. And here it is immaterial whether the anticipated consideration, or line of thought, be already existent in the shape of a clearly defined and elaborate theory or whether it be still in the embryonic stage of being a mere vague intuition of some genius of experimental research.
>
> The psychological truth of what I have said becomes manifest the moment we are faced with the difficulty of explaining to the layman just WHY one is carrying out this or that investigation. When I speak of layman here I do not mean the term to apply just to those people who do not give their minds to the consideration of impractical things. . . I mean the term to extend much wider. In the circle of a learned society which unites representatives of the various branches of science and literature in order to cooperate in research work, everyday one finds oneself a layman in the sense quoted above. Each of one's fellow members finds himself to be a layman in the same sense. For having attended a lecture given by a colleague he frequently cannot help asking himself. . .: 'What, in the name of Providence, is the fellow making such a fuss about?'. . .Let us say that it was your own lecture. It may happen that a colleague comes up to you and says: 'Look here, do tell me why that particular thing interests you? To me, it seems quite immaterial whether, etc., etc. . . .' Then you will endeavor to explain. You will try to show all the connections your theme has with others. You will try to DEFEND your own interest in the matter. I mean that you will try to defend the reason WHY you are interested. Then you will probably notice that your feelings are

> much more ardently aroused in this discussion than they were in the lecture itself. And you will become aware of the fact that only now, in your discussion with your colleague, have you reached those aspects of the subject that are, so to speak, nearest to your heart. [8]

Nearest to your heart! That's it! Science is born of semantic reactions, of what a situation, question or problem means to you at a particular moment. And your mental models, your scheme of thinking, your planning of an experiment, all this is an invention of yours for the purpose of getting where you want to be. You make models and systems to insure your survival in your chosen profession and to press forward to your own self-actualization.

By accepting this version, engaged as I am in the sciences of man, I find myself in comfortable agreement with our brothers in the sciences of things. Their achievements are outstanding, and ours might also be, once we accept a similar epistemic transformation.

A simple textbook, used in a college where they train electronic engineers, puts it this way.

> Taken together, Planck's and Einstein's discoveries constitute the prelude to a re-evaluation of the epistemological status of physics, a re-evaluation which was the most essential result of the advance of physical science in our century. Where eighteenth and nineteenth century scientists thought they were exploring Nature 'as such,' twentieth century physicists find the reflection of their own prejudices.
>
> Light is not a wave motion, nor is it made of particles. It is something else again, to which there is no macrocosmic analogy. This something else we cannot clearly imagine. But nevertheless we can partially understand and explore its characteristics. We have the right to make pictures and models, but we must be careful not to identify these pictures of reality with reality itself. Today we fully realize this, but at the beginning of the century physics had not yet passed through its catharsis.
>
> Planck's and Einstein's discoveries therefore brought about much agony of the soul. [9]

This agony of soul, according to Leon Brillouin, the French physicist, calls for a drastic treatment.

> This picture of an objective real world around us is what we have inherited from the Greeks, and we must get rid of it through surgery of the mind, however cruel it may be. [10]

My experience has been that, when that surgery has removed the vestigial organs of our earlier epistemic make-up, we feel relieved and free. We then accept without hesitation such statements as the one attributed to Niels Bohr by Aage Petersen:

> There is no quantum world. There is only an abstract quantum physical description. It is wrong to think that the task of physics is to find out what nature is. Physics concerns itself with what we say about nature.[11]

We do invent models and systems. Thus we build science as we go. Modern epistemology, continuing in the tradition of general semantics, becomes the science-art of system building.

REFERENCES

1. Alfred Korzybski, Science and Sanity, Institute of General Semantics, 1948, p. 22.
2. Hadley Cantril, Ed., The Morning Notes of Adelbert Ames, Jr., Rutgers University Press, 1960, p. 163.
3. J. Samuel Bois, The Art of Awareness, Brown, 1966, p. 172.
4. Quoted in Current, June 1960, p. 52.
5. Korzybski, op. cit., p. 25.
6. Korzybski, op. cit., p. 25.
7. Michael Polanyi, Personal Knowledge, University of Chicago Press, 1958, p. 171.
8. Erwin Schroedinger, Science Theory and Man, Dover, 1957, p. 94.
9. Tor Ragnar Gerholm, Physics and Man, The Bedminster Press, Totowa, N. J., 1967, p. 218.
10. Leon Brillouin, Scientific Uncertainty, and Information, Academic Press, 1964, p. 52.
11. Aage Petersen, "The Philosophy of Niels Bohr," in The Bulletin of Atomic Scientists, Sept. 1963, p. 12.

The Language of E-Prime

D. DAVID BOURLAND, JR

PROLOGUE: A PLAYLET

ACT 1

Time: 1916
Place: Vienna (Austria)
ASSEMBLED POPULACE: Professor Einstein, tell us about $E_0 = mc^2$ and relativity!
EINSTEIN: Forget it – it is too difficult.

ACT 2

Time: 1969
Place: Vienna (Virginia)
ASSEMBLED POPULACE: Mr. Bourland, tell us about $E' = E - e$ and E-Prime!
BOURLAND: Forget it – it seems too simple.

INTRODUCTION

Recently we have witnessed an increasing, if belated, concern devoted to long-standing ecological problems. We have seen leaders in the field take the first hesitant steps aimed at alleviating some of the more serious problems associated with river pollution, air pollution, noise pollution, and so on. While improvement in our external environment remains an urgent and serious problem, one of equal gravity faces us in the UNIQUELY human dimension: our linguistic environment.

M. Kendig, a former director of the Institute of General

Semantics, has pointed out that most people act about as conscious of their linguistic environment as fish appear conscious of the water in which they swim. If we expect to react HUMANLY to our linguistic environment, rather than ANIMALISTICALLY, we must become more aware of our linguistic environment and find means for improving conditions within that environment.

Quite apart from deceptive argumentation, sophistry, advertising excesses, etc., our linguistic environment, from the earliest times, has suffered increasingly from one particularly static verb: "to be". A form of this verb appears in written languages as early as Sanskrit, which includes bhavati, meaning in English "(he) is". Our very word for what will come subsequently ("future") has its etymological roots in futurus, from the Latin for "about to be". Upon reflection, it becomes apparent that this verb tends to weave in and out of our nearly every utterance. Its conjugated forms include the following linguistic elements: be, is am, are, was, were, been, being, the scruffy negative ain't, the dialectical amn't plus various rarely encountered archaic forms: war, wast, wert, ben, art, beest, and bist. It even goes underground through the contractions 'm, 's, and 're. If this verb simply provided a dispensable auxiliary (which it does), or some structural necessity of the English language (which it does not), no one would worry about it very much. But, to the contrary, this verb happens to function as an arch villain that debilitates our statements, contaminates our picture of the "outside world", and can help us to talk ourselves into un-sanity! That may sound like a strong charge, but it merely summarizes the observations some noted people have made over the centuries. Thomas Hobbes, Augustus de Morgan, George Santayana, Bertrand Russell, and most particularly Alfred Korzybski have warned us about this verbal hazard, but uniformly to little avail.

Their warnings became (I hate to admit it) vitiated because they found themselves – those titans, those preeminently linguistically and philosophically sensitive men, who made such huge contributions – even they found themselves unable to kick the habit of using this addictive verb, "to be."

WHY NOT USE "TO BE"?

You may very well wish to demand at this point, "What IS wrong with using 'IS'?" Let us address ourselves to the four primary semantic uses of the verb "to be" upon which Korzybski, Irving J. Lee, Wendell Johnson, etc., focussed our attention. In our normal speech we tend to use the various forms of "to be" in the following ways:

*Identity. (John is a success. Ivan is a communist. A rose is a flower. She is a revolutionary.)

*Predication. (John is ugly. Ivan is handsome. A rose is red. She is sexy.)

*Auxiliary. (John is reading. Ivan is plotting. The rose is wilting. She is misleading you.)

*Existence. (Descartes was. I am. You MAY be, but then again . . .)

Alfred Korzybski, in his major work, Science and Sanity: An Introduction to Non-Aristotelian Systems and General Semantics, provided a detailed exposition of how we tend to diminish our sanity by reacting to the WORDS we use to speak about happenings in the "outside world" as if the words consisted of happenings themselves. As Korzybski put it, we have great difficulty realizing that "The word IS NOT the thing". Korzybski believed firmly that two fundamental aspects of our Aristotelian heritage required urgent revision: the "is" of Identity and the subject-predicate structure of language which involves the "is" of Predication. Unfortunately, he also believed it impossible or awkward to speak or write in English without the Auxiliary and Existence uses of "to be".

Korzybski's views concerning the Identity and Predication uses of "to be" proceeded from his background, of course. From his experience as an observer of psychotic patients at St. Elizabeth's Hospital in Washington, D.C., in the late 1920's he learned how often the mechanism of Identity becomes involved in cases of extreme psychosis, as well as in more tractable forms of "mental" illnesses. Time after time Korzybski saw these patients rigorously treating quite different persons, circumstances, etc., as if they had the same identical characteristics as some other person, circumstance, etc., as in the classical "I am Napoleon," or whoever.

From his broad study in the biological, physiological, and physical sciences (in the late 1920's and early 1930's). Korzybski recognized the EVERYDAY SIGNIFICANCE of such matters as the uniqueness of structures on the event level, the ceaseless change of each structure in the "real world", and our consequent inability to say "all" about any happening event, situation, etc. These complexities testify to the fallacy and inadequacy of the "is" of Predication, which leads to such inanities as "Roses are red", "Buildings are high", "Dividends are good", and "Revolution is useful".

Let us pursue this point, which may still seem almost silly, in terms of the following three statements:

1. The earth is flat.
2. The earth is round (spherical).
3. The earth is somewhat pear-shaped.

The verb "to be" carries with it a huge intellectual momentum of completeness, finality, and time independence. From what we know today, it seems clear that each of those statements by itself does not give an adequate basis for conducting the diverse affairs of mankind, although each statement does describe the situation properly for some restricted purposes. This dual condition of adequacy-inadequacy appears characteristic of the Predication use of "to be" and provides the source of both its charm and its danger.

We can agree, I trust, that the Identity and Predication uses of "to be" do not reflect factual circumstances in the world as we experience it. For those die-hards among us who have some doctrinaire bias, or who otherwise did not pay attention, I shall recapitulate:

*Everything in the "real world" changes: sometimes so rapidly that we cannot experience the changes directly (as in the case of a table), sometimes so slowly that we can (as in the case of a river).

*Every person, as well as every "thing" undergoes such changes.

*One particular verb in English – "to be" – carries with it archaic associations and implications of permanence and static existence that we do not find in the "real world".

WRITING IN E-PRIME

We have heard of the allegedly disastrous effects produced by the infamous "Person from Porlock" on English Literature. A similarly unknown "Person from Tolland (Connecticut)" produced a more benefical effect upon me when, in 1949, I held a Fellowship for study under Alfred Korzybski at the Institute of General Semantics in Lakeville, Connecticut. This Person, whose name I can no longer retrieve, wrote a voluntary letter to the Institute suggesting that, in view of the problems Korzybski had discussed which inhere in the use of the "is" of Identity and Predication, perhaps we should just abandon ALL uses of the verb "to be". While no one else at the Institute seemed particularly interested in this suggestion, it struck me as having a significant amount of merit. Having a paper in preparation for the then forthcoming Third Congress on General Semantics, I decided to revise my paper (once again) to see if I could say what I wanted to say without using any form of "to be". It turned out that, in the process of this revision, I acquired an intermittent, but severe, headache which lasted for about a week; and the paper that resulted presented a peculiar clarity (despite the fact that one WAS crept in).[1]

But the question arises, in general, how well can we write without "to be"? Can we function adequately with the sub-set of English which does not contain any form of "to be", which I have called E-Prime?[2]

For a part of the answer, let us turn to Thomas Hobbes. In 1651 he wrote, in introductory comments to his Leviathan,

> As we use the verb is, so the Latins use their verb est, and the Greeks their 'esti through all its declinations. Whether all other nations of the world have in their several languages a word that answers to it or not, I cannot tell; but I am sure they have not need of it. . . . And if it were so that there were a language without any verb answerable to est or is or be, yet the men that used it would be not a jot the less capable of inferring, concluding, and of all kind of reasoning than were the Greeks and Latins.

For another part of the answer, it pleases me to report the following instances: I can manage reasonably well with both written and spoken E-Prime, as do several of my associates. In addition to the twelve more or less technical papers I have written in E-Prime (including this one), others have appeared by Dr. C. Andrew Hilgartner and T. W. Hefferon. Dr. Daniel A. Schwartz has recently earned the PhD in physics from the University of California at San Diego; his dissertation, written in E-Prime, had the title, "The Spatial Distribution of the Diffuse Component of Cosmic X-rays."[3] In the field of technical writing, C. H. Morgan of the Hughes Aircraft Company and K. L. Ruskin of the Control Data Corporation find

E-Prime highly useful in bringing clarity to complex communication and data system problems. In Wellington, New Zealand, R. W. Fox has recently initiated a well-received course in business writing that centers around the application of E-Prime to such prosaic, though important, matters as the preparation of inter-office memoranda.

The usefulness of E-Prime even extends to a form of trouble-shooting. Several years ago some of my associates in an operations research firm had spent quite a bit of time writing and polishing and re-writing and re-polishing a report that they could not seem to bring to a satisfactory completion. Then C. H. Morgan volunteered to re-write the report in E-Prime. After he had done so it became obvious why the report had seemed unsatisfactory: the usual linguistic forms had obscured the shallowness of the analysis and the inadequacies in the discussion of results. This made it comparatively simple for him and others to proceed to produce a sounder report. From a preventive viewpoint, Dr. Schwartz has reported that the amount of ATTENTION required to purge ALL uses of "to be" seemed to help him to write a better manuscript.

Let me assure you at once that I do not hold that E-Prime can cure ALL of the linguistic, semantic, philosophic, etc., problems that exist in the extraordinarily rich and convoluted English language. From the viewpoint of general semantics, we must recognize that such fallacies as ascribing equal values to words and to what the words represent, plus other undesirable linguistic behavior, can still occur, even with the habitual use of E-Prime. However, by removing the supreme irritant, "to be", we CAN alleviate a huge amount of potential difficulty.

Correspondingly, from the viewpoint of theological discourse, if one insists on speaking dogmatically, one could still manage to do this in E-Prime: but probably with much greater difficulty. You have probably seen, as I have, extraordinary exercises in philosophical argumentation concerning "being", "identity", and "essence" by such diverse writers as Jean-Paul Sartre (Being and Nothingness), Ayn Rand (Introduction to Objectivist Epistemology), and Etienne Gilson (Being and Some Philosophers). In my opinion those, and similar, writers could have presented their contributions far more incisively had they not allowed themselves the questionable luxury of using the verb, "to be".

E-Prime, which avoids ALL uses of "to be", may appear slightly extreme in that it removes the comparatively benign Auxiliary and Existence uses, as well as the malignant Identity and Predication uses. (Those who might feel that the term "malignant" overstates things, might wish to reflect upon the fact that "The People In Charge" formerly put men to death for refusing to subscribe to statements of the order of "The earth is flat". Furthermore, Dr. John L. Schimel, a prominent New York psychiatrist, in recent correspondence has also confirmed the appropriateness of this charge. Dr. Schimel cites the work of such students of schizophrenic processes as Ronald D. Laing and Frieda Fromm-Reichmann, as well as Harry Stack Sullivan.) Unfortunately, I have come to believe that one must avoid this verb ENTIRELY in order to evade the two undesirable forms. I feel that we find this requirement demonstrated in the writing of two extremely linguistically sensitive people: Korzybski himself and the noted lexicographer, Professor Allen Walker Read of Columbia University. Korzybski "knew", and Read "knows", that they should avoid the Identity and Predication uses, but both have continued to do so. Evidently, if one allows oneself the extravagance of using "to be" at all, the powerful undertow of the linguistic environment of our culture lures one into using the forms of Identity and Predication – to the detriment of ourselves and others.

Before proceeding, it seems necessary to clarify a point that arises frequently: some few languages do not have a verb form equivalent to the "to be" of English. Firstly, let me emphasize that we speak here about E-Prime and the semantic problems which have become apparent in the ENGLISH language. Other languages undoubtedly have different semantic problems concerning Identity and Predication, but I cannot address myself to them. Russian, Hebrew, and (so I understand) certain Chinese dialects do not have a verb that fits the English "to be" exactly. I submit that this situation supports my contention that we do not require this verb to communicate. The benefits that accrue from using E-Prime may or may not translate well into those (or other) languages. I must leave that problem to others with a more detailed understanding than I possess of the linguistic nuances that arise in the use and mis-use of languages other than English.

SOME CONSEQUENCES OF E-PRIME

Suppose we DO try to write and speak in E-Prime. What happens then? At least the following differences appear suddenly:

A. VANISHING QUESTIONS.

One simply cannot ask a number of questions (actually, pseudo-questions) that have preoccupied many people: What IS man? What IS art? What IS an object? What IS my destiny? Such questions, by virtue of their structure, set the stage for some kind of Identifications: specifically, confusions in orders or levels of abstraction. They tend to lead to the kind of discourse in which the likelihood of useful information exchange becomes severely reduced. Not IMPOSSIBLE, mind you, but significantly reduced. One might much better ask functional questions such as these: What characterizes man uniquely? Does this object belong to a known, highly regarded (but perhaps currently sterile) art form? To what level of abstraction does this structure belong? What can I do NOW to improve my future possibilities?

B. VANISHING INTERNAL INSTRUCTIONS.

Various schools of psychotherapy have recognized the importance of the silent assumptions which we hold about the world and ourselves, and other schools (especially the "rational therapy" practised by Dr. Albert Ellis) recognize the importance of what we tell ourselves, vocally and sub-vocally. "Self-suggested nonsense," Dr. Ellis calls this in its undesirable forms. Most of us have encountered people whose life patterns have decayed as they keep repeating to themselves such comments as these: "I AM a success, therefore . . ." "I AM a failure, consequently . . ." "She IS a Catholic, so . . ." "He IS a Jew, hence . . ." "I AM a teacher, so what I AM doing must BE teaching." "Since I AM head of this household . . ."

C. ABBREVIATIONS.

Forms of "to be" encourage and facilitate abbreviated statements that may turn out to convey little or no information. For example, we often see such empty comments as: "It IS clear that . . ." "Well, business IS business." "The problem IS just a matter of semantics." Let us discuss that last assertion. While of course most human problems involve important (and usually unexplored because unperceived) semantic issues, these issues do not evaporate just because someone has labeled them thusly. Frequently people offer, "Aw, IT'S just semantics" as an analysis stopper. More productively, one might respond to such a comment, "Certainly; at least in part. Now let's try to clarify some of those semantic problems". I also like the example of the abbreviation, seemingly innocuous and circumspect, that goes like this: "We do this thing because it IS right". This statement gets applied to such efforts as: (a) becoming involved in Viet Nam; (b) immediate withdrawal from Viet Nam; (c) reducing welfare payments; (d) increasing welfare appropriations; (e) fighting inflation (by acts that also increase unemployment); (f) increasing employment (by acts that increase inflationary pressures); and so on. When we say "It IS right", we potentially imply whole volumes of political, sociological, and theological discourse, inspiration, disputation, etc. I usually dislike to engage in the parlor trick of recasting statements that bristle with forms of "to be" into E-Prime, because to do this properly one needs to know the complete semantic context of the utterance. Otherwise, one risks falling into the fallacy of assuming that sentences "mean" something apart from the meanings humans ascribe to them when used in a given context or situation. The already significant contributions to transformational grammar made by Professor Noam Chomsky and his followers will probably increase in value when they stop making that assumption. With that caution, I will offer the following as an example of one way to restate "We do this thing because it IS right" in E-Prime: "We do this thing because we sincerely desire to minimize the discrepancies between our actions and our stated ideals". That revision does not have the snap of the original statement, and probably sounds somewhat pedestrian. However, the E-Prime version openly admits that some creed or set of beliefs held by some humans participated in the over-all situation.

This brings us to the spate of current slogans that seem to require "to be": "Tell it like it IS", "That'S where it'S at", "This IS what it'S all about". One gains great clarity at the price of lowering the level of arrogance, by more simply asking people to: "Tell it the way it seems to you, NOW", or stating that: "I believe that the central issue consists of this . . ."

Confusions due to improperly abbreviating with "to be" even occur in primary schools. All too frequently we hear teachers insisting that children drill in arithmetic by saying, "One plus one IS two; one plus two IS three;" etc. The perfectly correct mathematical expression "equals" certainly need have no more inherent mystery for the young than "plus" and the mis-use of "is" in this context probably has some responsibility for the difficulties some children experience with fractions. They can readily see differences between 1/3 and 2/6, say. The first fraction may EQUAL the second, but obviously some trouble could arise for those taught to translate "=" as "is".

The basically misleading character of "to be" shows up clearly in intellectual slogans. Let us consider one of the more prominent intellectual slogans of our day: "The medium IS the message". Much of the misunderstandings and ill-tempered reactions to Professor McLuhan's fascinating contributions, in my opinion, stem from habitual reactions to this "is" of Identity statement. After attracting our attention with his famous slogan on the opening page of 'Understanding Media,' Professor McLuhan spent the balance of his book documenting and discussing the less gripping, but perhaps more challenging, notion that "Media have hierarchical relations with other media", and conjugating the cultural consequences of those relations.

D. RETURN OF THE ROLE PLAYERS.

Undoubtedly related to the abbreviational aspects and sloganeering discussed above, speaking and writing which employ forms of "to be" seem to suppress the presence and influence of the people who play roles, even highly significant roles, in various situations. E-Prime, in contrast, tends to encourage a writer to present the agents involved rather than, for example, retreat into the passive voice. Some who write scientific and technical papers seem to forget that objectivity, for example, must reside in the person conducting the various experiments, etc., rather than in the passive forms used in reporting the results. One would say in the usual way, "The Constitution WAS signed in 1787", but in E-Prime, "Political leaders representing widely diverse commercial and other interests eventually signed the Constitution in 1787". With E-Prime we become practically FORCED to mirror the fact that certain humans, still living or formerly living, perform or performed the various activities under discussion. It becomes explicitly evident that some person(s) wrote the given paper: it did not appear through some magic. This may lead to the beneficial implication that the person who produced the given non-magical paper may have included, wittingly or not, some inaccuracies or misleading statements. Hence I suggested some time ago that the use of E-Prime would help to foster the critical review of written material. And precisely this has occurred when teachers have used E-Prime recently in courses in History and English.

The popular song by Peggy Lee entitled, "IS That All There IS?" illustrates clearly how the suppression of OURSELVES as key role players can become an important factor in our own problems. That song, of course, documents the disappointments of a girl who did not say to herself, "I expected more" out of (i) a fire, (ii) a circus, (iii) love, and (iv) life. By ignoring the involvement of her personal expectations, and hence the transactional role she played, she placed the whole burden on the external situation and generated the despair portrayed in the song.

When we recognize our own heavy involvement as one of a set of human involvements in almost any situation, we in effect lay the groundwork for possibly beneficial revisions of our semantic reactions. E-Primes offers one additional device to help us keep from confusing ourselves and others. It HELPS, but I do not mean to suggest that E-Prime can prevent confusion entirely.

FREQUENCY STUDIES AND SENTENCE ANALYSES

Turning quantitative now, we may ask how extensively people actually use forms of "to be" in standard English. I have conducted frequency studies sampling a variety of books, newspapers, and magazines. This included material currently written and that which goes back to earlier centuries. I found a high degree of stability with forms of "to be" accounting for about 3% to 5% of the words used. Interestingly, I found a slight, but definite and persistent, difference in the two primarily important portions of newspapers: the lead news story (normally carried in the extreme right-hand column) tends to contain fewer forms of "to be" than the editorials. The observed difference probably would not show statistical significance, but to me it seems semantically interesting.

If we study not just the sheer frequency of forms of "to be" in the stream of speech, but rather the frequency of sentences in which the various uses appear, even more interesting results become apparent. Table I gives the definitions of different uses which I employed in the sentence analyses. Table II shows the results of studying various books.[4] The range from 37.4% to 55.6% documents the progression in the books studied that range from Korzybski's attempt to deal constructively with semantic problems to the dogmatic pomposity of Miss Rand's contribution.

Table I Definitions of "To Be" uses

USE	STRUCTURE	EXAMPLE
Identity	Noun Phrase – "to be" – Noun Phrase	John is a sailor.
Predication	Noun Phrase – "to be" – Modifier	John is good.
Auxiliary	Noun Phrase – "to be" – Verb Phrase	John is reading.
Existence*	Noun Phrase – "to be"	John is.

*I have chosen to classify uses which indicate Time and Location within the Existence category. For example, "It was before they fought that they had fun", and "John is at home".

Table 2 Sentence Analysis of Various Books

BOOK	NUMBER OF SENTENCES IN SAMPLE[a]	MEASURE #1[b]	MEASURE #2[c]
A. Korzybski's Science and Sanity	206	58.7	37.4
J.S. Bois' The Art of Awareness	220	66.8	41.8
K. Popper's Conjectures and Refutations	231	69.7	46.8
A. Rand's Intro. to Objectivist Epistemology	189	72.5	55.6

NOTES
a For each book, twelve pages selected randomly yielded the given number of sentences.
b Measure #1 gives the per cent of sentences having one or more forms of "to be".
c Measure #2 gives the per cent of sentences having one or more uses of Identity or Predication.

In the case of newspapers, I studied the uses of "to be" in the editorials and lead stories of the Washington Post and the Washington Evening Star for the period November 22 to December 2, 1969. On the basis of sentence analysis, the differences between the editorials and the lead stories become quite pronounced. For example, on November 23, the Washington Post editorials contained at least one Identity or Predication use in 40% of the sentences, while the lead story for that day had those uses in only 28% of the sentences. On the same day (November 23) the results for the Evening Star showed 43% and 16% for the editorials and lead story, respectively. In only two instances – one for each newspaper – during the period studied did the lead story contain a higher percentage of sentences with Identity and/or Predication uses than the editorials. To me this shows that the structure of the language employed by the writers for those papers having quite different political biases became quite different when they "reported" and when they "interpreted" the most important developments of the day.
The differences described above for books and newspapers become even more important when found in matters of greater moment, as I will present next.

POLITICAL AND SEMANTIC ISSUES

In the years immediately following World War I Alfred Korzybski observed the stark differences between the consequences of engineering and scientific activity and the fruits of political activity. When engineers build a bridge, Korzybski put it simply, the bridge normally functions as designed. But when politicians "build" a treaty or government, it usually collapses, amid great human suffering.

Korzybski's analysis led him to conclude that the fundamental factor responsible for that discrepancy in performance consists of the structure of the languages used by those who design bridges and those who design treaties. The engineers and scientists use a language (mathematics) which has a structure similar to that of the bridge which must get built, and hence the language produces great predictability. The politicians, however, normally employ a language of archaic structure which primarily uses static terminology while trying to come to grips with dynamic human socio-economic issues.

To the extent that a treaty, constitution, etc, becomes subject to this kind of static-dynamic discrepancy, Korzybski pointed out, one may expect undesirable and unstable consequences.

To put this somewhat differently, Korzybski asserted that social institutions based upon static premises must ultimately, and probably rather quickly, collapse.[5] And if we inquire into this matter semantically, we find that the seemingly innocuous verb, "to be", constitutes the fundamental linguistic purveyor of static premises and assertions. With this verb one may make – indeed, becomes forced to make – statements that promulgate a spurious constancy, implying : "It WAS like this; it IS like this; it will always BE so". But of course, no "it" in the real world, ourselves included, exhibits such a simple organization or orientation.

Having recognized the insidious role which "to be" may perform in socio-political contexts, it occurred to me to analyze several important and basic political documents. The purpose consisted of determining to what extent the language in the documents exhibited a static character, as indicated by their comparative reliance upon forms of "to be".

The political documents chosen for study consisted of the Constitution of the United States of America, the Communist Manifesto, Machiavelli's The Prince, Aristotle's Politics, and Robert Welch's The Blue Book. The investigation included both the over-all frequency of forms of "to be" and the undesirable uses of Identity and Predication, plus a similar study by sentences in those documents.

The results for over-all frequency of appearance of forms of "to be" appear in Table III. Table IV shows the results of the sentence analysis. Some distortion in the results undoubtedly exists, due to the fact that Machiavelli and Aristotle did not write in English. However, the original language in both bases belongs to the Indo-European family, and so the distortion probably does not amount to too much. English consisted of one of the original languages in which the 'Communist Manifesto' appeared, of course, and Engels participated in the editing of the version I analyzed.

In my assessment, the results in Tables III and IV, ranked in accordance with the increasing appearance of the uses of Identity and Prediction, also range from the great flexibility and power of our Constitution to the sterility of Mr. Welch's nightmare and the rigid dogmatism of Aristotle. I submit that these results give quantitative substantiation for Korzybski's thesis.

Table III Political document study: frequency of "to be".

DOCUMENT	SAMPLE SIZE (IN WORDS)	MEASURE #3[a]	MEASURE #4[b]
1. Constitution of the U.S.A.			
a Main Body[c]	4,946[d]	2.5	22.2
b Complete	7,586[d]	2.5	24.0
2. 'Communist Manifesto'	1,926	3.8	34.3
3. 'The Prince'	1,962	4.6	39.6
4. 'Politics'	2,289	4.4	48.0
5. 'The Blue Book'	1,836	4.0	48.6

Notes a. Fraction of the incidence of "to be" forms divided by the total number of words in the sample, expressed as a per cent.

b. Fraction of the incidence of Identity and Predication uses divided by the total number of forms of "to be" expressed as a per cent.

c. Exclusive of amendments.

d. Complete documents, not a sample.

CONCLUDING REMARKS

Although the language of E-Prime has its peculiar part to play in what some have called "applied epistemology," it also provides a device for everyday application with beneficial results. I suggest that you produce your next letter or article or office memorandum in E-Prime. You will probably find this an interesting exercise. You will also probably find yourself turning increasingly to what some English teachers call "action verbs" and those copulatives that describe what goes on more modestly than "is": APPEARS, SEEMS, BECOMES, FEELS, GETS, etc. You may bring in your own involvement in what you describe, rather than seemingly imitating one who provides dicta chipped onto stone tablets. You will ascribe more honestly the various reasons for various actions to those who had a part to play in them.

Table IV Political document study: sentence analysis.

DOCUMENT	SENTENCES IN SAMPLE	MEASURE #1[a]	MEASURE #2[b]
1. Constitution of the U.S.A.			
a. Main Body[c]	103[d]	57.2	18.4
b. Complete	159[d]	59.7	23.3
2. 'Communist Manifesto'	444[d]	45.4	26.2
3. 'The Blue Book'	207	74.0	48.8
4. 'The Prince'	175	77.8	53.6
5. 'Politics'	188	82.0	60.1

Notes
a. Per cent of sentences employing some form of "to be."
b. Per cent of sentences having one or more uses of Identity or Prediction.
c. Exclusive of amendments.
d. Complete document, not a sample.

I challenge you: try this technique first on some matter of importance to you – THEN talk about E-Prime. But, please, IN E-Prime!

REFERENCES

1. Subsequently published as "Introduction to a Structural Calculus: A Postulational Statement of Alfred Korzybski's Non-Aristotelian Linguistic System", General Semantics Bulletin, Nos. 8 & 9, 1952.
2. The name comes from the semantic equation: $E' = E - e$, where E represents the some 490,000 words of standard English, and e represents all the forms of "to be".
3. This feat by Dr. Schwartz constitutes, quite in addition to the scientific excellence of his contributions, a rather major politico-linguistic victory. Most universities have dissertation style guides that do not allow the author to use the term "I," which E-Prime forced Dr. Schwartz to use, quite honestly, in describing the results that HE had achieved. These manuals or guides also encourage the passive voice in a vain search for objectivity.

4. I finally standardized upon sample of twelve pages, determining that the percentages became stable for samples that size or larger.
5. Korzybski discussed this matter in some detail in his Manhood of Humanity, E. P. Dutton, 1921; second edition distributed by the Institute of General Semantics, Lakeville, Connecticut.

Coping with Semantic Problems in System Development

K. L. R. BOURLAND

INTRODUCTION

In his paper, "General Semantics As a General System Which Explicitly Includes the System-Maker," Robert Pula stated: "Unconsciousness of the multiordinality of terms readily leads to confusion and therefore conflict in daily life. More importantly for us, lack of awareness of this mechanism can lead to confusion in system building."

I wish to describe to you several instances in which unconsciousness of multiordinality, plus several other key formulations of general semantics hampered the development of a large computer-based information storage and retrieval system that I participated in recently. I ask you to restrain such curiosity as you may possess relative to precisely WHICH computer company performed work on this project, and precisely WHAT system this work concerned. Let it suffice to say that the effort involved over 100 programmers and systems analysts, and that (if this means anything to you) we automated an AUTODIN switch.

More specifically, I wish to describe to you how I found it helpful in coping with system development problems to employ the orientations of consciousness of abstracting, Alfred Korzybski's extensional devices, multi-valued aspects of terms, multiordinal aspects of terms, and the semantic tool of E-Prime.

Whereas most of the contributions to this conference have dealt with the basic issues of general systems theory and general semantics on comparatively high orders of abstraction, this effort more modestly provides some extensional examples of the formulations of general semantics operating in an environment more frequently considered the province of general systems theory.

E-PRIME: A SEMANTIC TOOL

I have written this paper in the language of E-Prime, a subset of the English language and differing from ordinary English only in that it contains no form of the verb "to be." Invented by D. D. Bourland, Jr.,[1] and described by him here in another paper, E-Prime has a plethora of general semantics formulations inherent in it.[2] Its use requires more acute awareness by the user of his own speech patterns, written or vocal, increases his awareness of the speech patterns of others, and seems to facilitate consciousness of abstracting. It tends to force one initially to describe situations on comparatively low orders of abstraction (by specifying role players) before sweeping on to summarizing higher orders of abstraction.

WHY NOT USE E-PRIME?

In the process of developing technical documents, I have found that I frequently need to include lists of items labeled "Tables," and sketches and diagrams of items labeled "Figures." Since I write in E-Prime, when I refer to a Table or a Figure in the text material, I ordinarily write: "Table I contains a description of . . . ," or "Figure 1 shows . . . ."

Readers of text written in this way have stated: "I think you should express it as 'A description of . . . is contained in Table I,' and '. . . is shown in Figure 1'." To such comments I can only respond: "Why NOT do it my way?" So far, no one has come up with a valid reply.

Journalists have long endorsed writing in the active voice "because it makes better copy." Another journalistic rule pertaining to news stories requires the writer to state "who, what, why, when, where, and how" in the leading paragraph of the article. Some personal journalistic experience and a recent visit to the School of Journalism at the University of Missouri at Columbia lead me to note that, however pragmatically useful, these journalistic techniques have no underlying methodology; as "gimmicks" these methods simply "work." The journalists at the University of Missouri seemed fascinated to discover a methodology, E-Prime, backing up the newspaper tradition of writing in the active voice "because it makes 'better' copy."

E-PRIME AND A REAL-TIME SYSTEM

Not only in newspaper stories does E-Prime make better copy. Obfuscation becomes very difficult when using E-Prime in varied writing contexts. In this paper I have chosen to describe some of the benefits that ensued from persistently using E-Prime in the development of a large, complex real-time computer system.

Visualize a scenario of the operational environment of a project staffed by computer professionals: people seated at desks, file cabinets and bookcases placed near the lucky, an occasional semi-wall partition, with the computer room down the hall. I have a centrally-located desk, a file cabinet, and a typewriter (the only member of the professional staff to have one). On a typical day another project member brings a sheaf of papers to my desk, saying: "This is the system initialization description." I quickly scan the front page, finding only the title "SYSTEM INITIALIZATION." In a tactful way I ask something to the effect: "Have you just written this, or do you know who wrote it?" "I just wrote it," responds the donor. "Thank you for bringing this to me," say I, "and I may call on you later with questions." I immediately write on the cover page, "Received from [name of donor] on [that day's date]."

Subsequently I review the material, etc., but most importantly for purposes of this paper, I have indexed and dated the system initialization papers on an impetus which I ascribe to the awareness of abstraction enforced by vigilant use of E-Prime. If a junior programmer has prepared the system initialization description, I would have to view it as potentially imperfect. If the Programming Manager has brought the description I would consider it (potentially even more imperfect, but probably) much more authoritative. At this point I know several of the journalist's requisites for a news story's lead paragraph: WHO – Programming Manager; WHAT – system initialization description; WHY – integral part of the basic makeup of the system, hence important to document; WHEN – that day's date, as written on the cover page; WHERE – in the project environment; HOW – not yet determined. Later I will have to ask the Programming Manager on what sources he had based his writeup. Had he designed these procedures himself? Had he made a few modifications to a standard procedures package? To assure optimal legitimacy of the system initialization information, I would need to learn at least these things.

This example provides just one context in which I found E-Prime an invaluable eliciter of questions, resulting in obtaining answers to at least some of the questions, and knowing important questions which remained unanswered.

CREDENTIALS OF THE MAP MAKER

E-Prime often forces one to express oneself in the first person, or at least to ascribe a source or basis for an assertion. One might call this "naming the map maker." I see in this process an addition to Korzybski's classical three-stage "map-territory" analogy which in E-Prime becomes:

1. A "map" belongs to "a level of abstraction" different from that of the "territory" it represents.
2. A "map" does not contain all the structural characteristics of the "territory" it represents.
3. A "map_1" may self-reflexively contain a "map_2" of higher order which shows the relation between the "$territory_1$" of "map_1" represented in some detail to the surrounding "$territory_2$" of "map_2".
4. One needs to know as much as possible about the "map maker" (and when he made the "map") in order to evaluate properly the credibility of the "map".

It MATTERS to know who made the map for any given territory, as the map maker includes in his map the product of his particular background. Hence, if a marketeer gives me information (a map) supposedly describing some equipment or systems capability (the territory described by the map), I would probably check at least another source to verify the correctness of the marketeer's map. Reasons for this become quickly apparent: the marketeer has selling as his principal goal; he does not necessarily care if he makes his sale by representing his product accurately, since his measure of effectiveness or success lies in making the sale, however he can. Responsible system developers have different measures of effectiveness, as their superiors or clients will probably judge them on the basis of timely, meticulous, professional work performed. A system developer's map and a marketeer's map of a given territory could, and probably would, differ significantly. I cite the case of marketeers, associated with the company responsible for the project discussed in this paper, assuring the customer that adding an optical character reader (OCR) to the computer system in the future would require no additional computer programming efforts. In fact, adding an OCR in the future would entail enormous, expensive new programming efforts.

DESCRIBING COMPUTER PROGRAMS AND THEIR INTERACTION

It pleases me to report that use of E-Prime in writing descriptions of computer programs and systems imposes clarity and specificity. In our multi-programming, interrupt-oriented processing environment, computer programs continuously gained and relinquished control of the core memory units in the central processor on a priority basis.

In this environment, if someone writes as a final description: "When Condition X occurs, Program A is called," we find ourselves missing a piece of crucial information; namely, we have to know WHICH program out of the fifty possible "called" Program A. E-Prime makes such a quandary impossible. A writer must state instead something like the following: "When Condition X occurs, Program K calls Program A to assume control of the processing."

Using the structure described above (". . ., Program A is called.") makes it possible to omit extremely important information while using seemingly perfectly acceptable English. Any use of the "Is-of-Predication" (and its insidious counterpart, the "Is-of-Identity, e.g., "This is the system initialization description") carries with it significant danger of expressing incomplete, misleading, or even unauthoritatively wrong information. However, this example shows how even the comparatively benign Auxiliary use of "to be" can lead to unnecessary linguistic woes. Because of the structure of the English language, we have grown accustomed to the use of such linguistic forms and undoubtedly most people accept them unquestioningly. Whether a person chooses to speak and/or write in E-Prime depends upon individual intellectual decision. Despite how one chooses in that regard, knowledge of the availability of the linguistic tool of E-Prime can make possible a useful, more critical scrutiny of the meanings or lack thereof of oral and written English. Further, E-Prime "builds in" the consciousness of abstracting so strongly emphasized in Korzybskian general semantics.

LEXICOGRAPHIC BREAKDOWN IN COMPUTER TERMINOLOGY

The increasing rate of development of computer technology has provided our language with increasingly many new additions for our dictionaries. However, due to the paucity of lexicographers in the computer field, we have few of these new computer-oriented terms standardized and documented, with no general consensus on the meanings for many terms. Instead, we have a sizable number of multi-valued terms which may mean many different things depending upon their users. Various computer installations carry with them characteristically unique terminologies. It appears that these terminologies differ radically from installation to installation. I found that it becomes extremely important to recognize this fact in order to "understand" what one reads and what one hears people say.

Examples of multi-valued, computer-oriented terms include the following: program, routine, subroutine, calling sequence, programming, module, [computer] word, block of [computer] words, file, table, software, etc. I will describe some meanings with which I have familiarity for one of the more commonly used multi-valued terms, the term "software." When I first encountered the word "software" in a major way in 1966, I worked for a company that dealt with "software" exclusively. The company brochures and promotional materials described it as a "software firm." In this early context, we used the term "software" to mean that we performed studies of various kinds, and that our SOLE product consisted of reports on the findings of the studies. We used the word "software" in the trade language, so to speak, to contrast it with the word "hardware". "Hardware" companies build airplanes or sell or manufacture nuts and bolts. You might have called a software company a "think tank".

Today, the word "software" has assumed varied, complex, and specific implications, depending upon its context. Table I contains a list of some of the different meanings that I have encountered for the term "software".

Table I Definitions of "Software"

$Software_1$	=	non-hardware
$Software_2$	=	computer programs (only).
$Software_3$	=	computer Operating System and "Utility" computer programs; also known as "systems programming".
$Software_4$	=	computer programs and documentation for them, excluding any systems analysis.

In the context of the project described in this paper, I "coped" with wide use of multi-valued terms by not using those terms. Given more time, it might have become practical to make a dictionary and attempt to achieve consensus among project members on meanings of some of these multi-valued terms. However, due to time constraints, I could not attempt such an undertaking.

UNCONSCIOUSNESS OF ABSTRACTING

As described above, developing computer-oriented terminology has resulted in multi-valued terms. In my computer system development project the lack of documentation for words' multiple meanings created problems, and the lack of recognition of the existence of this multi-valued condition created even further problems for members of the project team. One might describe this as multiordinal misunderstanding.

The contractually-required project specifications included several documents, among them the client's Request for Proposal document and the "winning" company's proposal. Correlating the information in these papers with real-world implementation became confused due to lack of differentiation between levels of abstraction.

Certain project members confused information in the project specifications with the quite different actual tasks that the programmers had already completed in developing working, end-product computer programs. The project members afflicted with unconsciousness-of-abstracting assumed that because the project specifications called for supplying Function A in a certain way, the responsible programmers had completed Function A in that specific way. This invalid assumption led to incorrect system descriptions; the persons unconscious of abstracting wrongly assumed that the words written on the paper of the project specifications represented real-world accomplishments of actual computer programs.

APPLICABILITY OF KORZYBSKI'S EXTENSIONAL DEVICES

I found Alfred Korzybski's extensional devices particularly useful as semantic tools, lending great support in ordering the large mass of written information that ensued from the project discussed in this paper.[3] Five of the six extensional devices had significantly greater applicability in this instance than the remaining one: use of the term "etc." as a reminder that one cannot hope to know or say "all" about a given topic. The relevant five devices – indexing, dating, using quotation marks, and hyphenating – proved remarkably effective in ordering and expressing information in the clearest way possible, given constraints of time pressures and great bulk of work requiring attention.

INDEXING

In the strictest Korzybskian sense, in order to use the extensional device called "Indexing," one should apply consecutive numeric subscripts to the same term to express differentiation of each item in the string from its predecessor and its successor. In the context of my project, I could have made a chart identifying Programmer_1 as T. Watson, Programmer_2 as G. Dantzig, programmer_3 as W. Zangwill, etc., which might have proven somehow useful – but I did not. Instead I wrote the name of a given programmer wherever necessary for identifying responsibilities or authorship for work complete, work in progress, work not yet begun, and so on. This relatively crude form of indexing contributed significantly to development of effective management tools for this project, used virtually invariably with an accompanying date as described below.

DATING

Coping with the complexities of the large, technically-difficult project in which I participated and which I describe in this paper included use of the extensional device "dating." In practical application this involved putting the date of THAT day on every paper that I wrote or received. It became my practice to put my initials and the day's date, among other identifiers (such as "working paper") on each page that I wrote, eventually amounting to somewhat more than three hundred pages of technically-specific material, mainly written in E-Prime. I insisted on dating each page because in a developing project such as ours, document cannibalization occurred frequently; in this way it became relatively simple to discover if the material at hand remained current, or if some new version had superseded it.

My system of applying dates (and source of authorship) to ALL papers that I could affect made my project sector a deliberately-enforced "process world," and we could differentiate the clearly-labeled outdated versions of papers from the clearly-labeled current versions of papers. Without my insistence on the use of dates and labels (essentially "indexes" as described above), I feel almost certain that the entire project would have drowned in an ocean of anonymous working papers.

USING QUOTATION MARKS

I found quotation marks helpful in clarifying the relationship of a part of a sentence to its context, but in this milieu, focusing on writing computer programming specifications, use of quotes seemed needed relatively infrequently.

HYPHENATING

Hyphenation often helped clarify otherwise potentially muddled sentences. I include the following example as one of the more complex instances of hyphenation applied usefully:

> The send-interrupt, read-to-disk, check-for-end-of-message-indicator, thread-new-disk-sector-address, display-DD173-continuation-form, issue-FREAD, exit-to-dispatcher cycle continues until the routine "CRT" senses the end-of-message indicator.

Because of the preliminary nature of the particular document that included the above sentence, we had not yet assigned paragraph numbers; deprived of that standard method of referencing, we found that with the use of hyphens we could summarize a long, complex process relatively briefly.

CONCLUSION

In conclusion, I should like to mention that in my opinion the preceding comments provide explicit evidence in support of the contention that general semantics has immediate relevance for the development of complex computer systems. We really do not need to concern ourselves further with whether we should regard general semantics as a special case of general systems theory or conversely.

Indeed, in a manner conceptually similar to Gödel's famous proof that one cannot prove every theorem implicit in a set of postulates, Professor Dodd has shown that – provided we take a sufficiently broad approach – we must of necessity operate within a closed system.[4] Accordingly, we need not devote further energy to the matter of who has the most open system.

Perhaps we can use the time conserved as just described in developing more effective General Systems and more effective General Semantics.

REFERENCES

1. D. D. Bourland, Jr., "A linguistic Note: Writing in E-Prime," General Semantics Bulletin, Nos. 32 and 33, 1965/1966.
2. D. D. Bourland, Jr., "The Semantics of a Non-Aristotelian Language," General Semantics Bulletin, No. 35, 1968.
3. A. Korzybski, Introduction to the Second Edition, Science and Sanity, International Non-Aristotelian Library Publishing Company, March 1941.
4. See "On Producing the Epicosm Model Reiterantly – To Mirror the Cosmos to Men", in this volume.

The Computer System User: An Information Need Model

GLYNN HARMON

INTRODUCTION

The purpose of this paper is twofold: (1) to propose a dynamic, cognitive model of interactive computer user information need; (2) to re-examine the findings of two key experiments on the complex problem of information relevance via the proposed model. The paper is based on the assumption that a fundamental objective of information science is to seek better ways to accurately identify and fulfill user information needs which arise throughout the course of inquiry.

Various conceptions of human information processing can serve to clarify the nature of user information needs. For example, Boulding suggests that information seeking behavior depends on personal imagery. Such images possess different degrees of "coherence, consistency, survival power, stability, and organizing power."[1] Information, or components of messages that reach an individual's image after being mediated through his value system, can modify his image. When a Key structure or nucleus of the image is affected, the image can be altered or even reconstellated. Applying Kelly's theory of behavior, information needs might be viewed as products of change within a system of personal constructs. The system of constructs evolves as information is processed and separate constructs are tightened and loosened. However, such constructs cannot be "stretched" too far beyond relatively convenient applications.[2]

Reitman's work in human information processing appears to be particularly significant: "From an informational viewpoint we may regard the whole problem of cognitive structure as a matter of sets of cognitive elements. . . ."[3]

Concepts or cognitive sets serve as information filters and organizers. Such cognitive sets make it possible to focus on and use selected elements from the mass of available detail in one's own memory or in external natural or artificial memories.

MODEL OF USER INFORMATION NEED

Inquiry may be viewed as a series of efforts to acquire a set of needed information elements and to find appropriate set-defining and set-ordering criteria. The culmination state of successful inquiry can be modeled as a complete, ordered set of information elements which the user forms cognitively within his own domain of inquiry: $\{a,b,c,d,e,f,g\}$. The information elements may themselves be sets or subsets of prior inquiries. Inquiry may be depicted as set formation and transformation, and four fundamental stages are now illustrated.

STAGE I: INSUFFICIENT AND UNORDERED INFORMATION

At the outset of inquiry, a null or empty set, $\{\ \}$, often exists. The domain of inquiry and the problem might be undefined or vaguely defined. Few, if any, informational elements are available, or if available, have not been specifically designated as relevant to the domain of inquiry. Hence, the problem is one of acquiring information rather than one of ordering information. However, once two or more information elements have been acquired, ordering problems appear. The set may be ordered differently. Two elements, for example, may be ordered in the following ways: a,d or d,a. But the number of available information elements may be insufficient to suggest an appropriate ordering or a possible definition of the set.

STAGE II: INSUFFICIENT BUT ORDERED INFORMATION

A number of information elements have been tentatively designated as relevant to the inquiry. The number of elements is sufficient to establish set ordering relations and to imply the bounds of the cognitive set: $\{a, d,e, g\}$. The task becomes primarily one of ordering available information. The magnitude of the ordering problem increases with the number of available information elements.

STAGE III: SUFFICIENT BUT UNORDERED INFORMATION

As information acquisition continues, a sufficient number or even a surplus of information elements becomes available: $\{a,x,f,c,d,e,b,g,k\}$. At this point the bounds of the set become more and more apparent and may be delimited as the information elements are permuted. Informational gaps may be seen. However, a state of disarray still exists.

STAGE IV: SUFFICIENT AND ORDERED INFORMATION

The elements are reconstellated until a satisfactory image or cognitive set is obtained: $\{a,b,c,d,e,f,g\}k,x$. The bounds of the set have been delimited so that the redundant or irrelevant information elements k and x may be excluded from the set or discarded. Once the cognitive set is established, it may be elaborated, refined, or challenged by the inquirer himself or by other inquirers. With the acquisition of new information or the revival of such previously excluded elements as k or x, the cognitive set may be restructured or decomposed. If the set is decomposed, inquiry might then revert to stages II or III or in some cases, to stage I. The cycle of information acquisition and ordering is then wholly or partially repeated in order to form a new cognitive set.

In summary, inquiry may be viewed as a cyclic or modulo process involving both the acquisition and ordering of information. Through this process, the changing information need configuration influences the nature of the questions generated. In turn, answers to questions modify the information need configuration as shown in Figure 1.

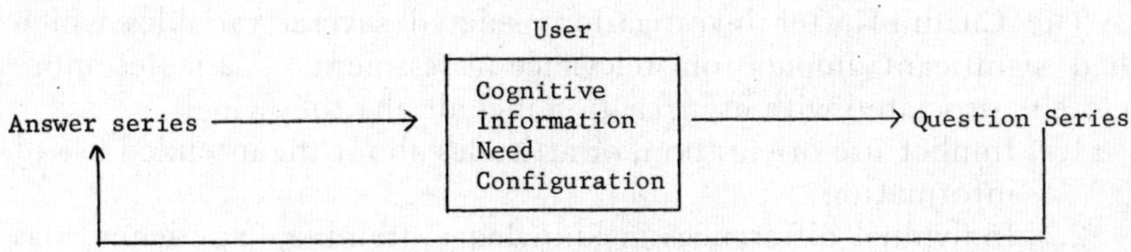

Figure 1 Information need configuration

The model in Figure 2 magnifies the user information need configuration to represent the four basic stages of inquiry. Inquiry may progress from or revert from any one of the four basic stages to any other. For each stage, a corresponding specie of question would probably be generated.

	Unordered Information	Ordered Information
Insufficient Information	I	II
Sufficient Information	III	IV

Figure 2 Information need configuration during various stages of inquiry

The model has been successfully tested in another context on data representing inquiry in symbolic logic.[4]

USER RELEVANCE REINTERPRETED

A useful distinction has been made between USER relevance and information SYSTEM relevance.(5) User relevance may be regarded as a user's evaluation of the strength of the relationship between a system response and user need. Information system relevance may be regarded as the system's assessment of the degree of relationship between system knowledge and the user's query. The following discussion is limited to user relevance and represents the author's attempt to reinterpret a few findings of two leading relevance investigations.

The Cuadra-Katter investigation isolated several variables which had significant impact on relevance assessments. Variables more closely associated with user reaction include the following:

1. Implict use orientation, or attitudes about the intended use of information;
2. Individual differences in knowledge, training, experience, bias or cognitive style;
3. Situational context influences such as socioeconomic conditions or time pressure;
4. Delimitation or specification of information sought; material attributes which evoked such reactions as surprise.[6]

In terms of the user information need model presented earlier, implicit use orientation would involve a directional alignment of the cognitive set. A change in purpose would require a revised data alignment. Individual differences could be similar to predisposing influences on user cognitive sets, while user situational contexts could be comparable to precipitating influences. Through delimitation of information sought, the user would restrict the domain of inquiry and specify the bounds of his cognitive set. If the user encounters stimulus material which has surprise value, his cognitive set would possibly snap together or be broken.

In the Rees-Schultz study, individual user differences – such as cognitive style, personality, education, and work orientation – were also found to be important relevance variables. Moreover, subject experts tended to assign low relevance ratings to document information while non-experts tended to assign relatively high relevance ratings. In research, user relevance ratings tended to drop off from the initial, formulation stage through the executory and interpretative stages. [7] Thus, as research progressed, users tended to become increasingly selective in their acquisition of information.

With respect to individual user differences, the Rees-Schultz and Cuadra-Katter findings are similar. The myriad predisposing and precipitating factors which influence user cognition generally would influence the mode of information acquisition and organization through the course of inquiry. Experts, by virtue of their education and training, would generally possess a repertoire of alternative, partially complete and ordered cognitive sets. In contrast, non-experts would tend to lack the cognitive schema and requisite information for the appropriate selection of cues for problem definition and resolution.

Similarly, as a user progressed from preliminary to advance stages of research, he would tend to become more "expert" about his research problem. Through the experience of successively acquiring and ordering information, his level of sophistication would increase to the stage where he could generate highly pointed or "relevant" questions. His underlying information needs would tend to be increasingly well circumscribed and more highly specified as inquiry progressed.

CONCLUSION

The process by which an individual constructs a mental jigsaw is analogous to the proposed user need model. The user acquires and orders separate pieces or information elements until he conceives or adopts an adequate gestalt, or holistic cognitive mapping. Gaps in the mapping are then filled and adjustments made according to the user's individualized specifications and purposes. In the advanced stages of inquiry the user becomes expert enough to be increasingly selective because his information needs are better circumscribed, and his problem tends to be one of gap-filling rather than one of synthetic ordering.

One might speculate that the user would tend to generate different species of questions reflecting his need for either additional information (content) or organization (form), or for both. An on-line information system might be designed, then, to help the user to cognitively add and organize information according to his purpose. Conversely, the system could help the user to break his cognitive set and to subtract information not suited to his purpose.

REFERENCES

1. Kenneth E. Boulding, The Image, Ann Arbor: University of Michigan Press, 1956, p. 174. 2. George A. Kelly, The Psychology of Personal Contructs: A Theory of Personality, Vol. 1, New York: Norton and Company, Inc., 1955, pp. 77-94. 3. Walter R. Reitman, Cognition and Thought: An Information Processing Approach, New York: John Wiley and Sons, Inc., 1965, p.91.
4. William Goffman and Glynn Harmon, "Predicting Scientific Discovery: A Mathematical Approach," Proceedings of the Sixth International Congress on Cybernetics, Namur, Belgium, 1970.
5. Donald W. King, "Design and Evaluation of Information Systems", in Carlos A. Cuadra, (ed.) Annual Review of Information Science and Technology, Vol. 3, Chicago, Encyclopedia Britannica, Inc., 1968, pp. 70-74.
6. System Development Corporation, Experimental Studies of Relevance Judgments: Final Report, Vol. 1, by Carlos A. Cuadra, Robert V. Katter and others. Santa Monica, Calif., 1967, pp. 75-122.
7. Alan M. Rees, Douglas G. Schultz and others, A Field Experimental Approach to the Study of Relevance Assessments in Relation to Document Searching: Final Report, Vol. 1, Cleveland: Center for Documentation and Communication Research, Case Western Reserve University, 1967, pp. 271-284.

Complexity in the Classroom: Notes on Teaching General Semantics

JAMES BROADUS

> To begin with, lose all fear. You know, in Dante's 'Inferno' there was an inscription. 'Lose all hope; you who enter here'. I would have on the door of a class in General Semantics, 'Get all hope; you who ęnter'. We have to get acquainted to begin with. Lose your tension about these seminars. We will be at home; we will be very chummy. If you don't understand something, tell me and I will repeat it endlessly. Don't have the feeling either that you are students and that I am a teacher. No. You are human beings, $Smith_1$, $Smith_2$, $Smith_3$, etc. – and so I am, say, $Smith_n$. Human beings talking to human beings. Relax to begin with. Don't take it too dramatically.[1]

Thus did Alfred Korzybski begin Seminar Lecture One at Olivet College in 1937.

I suppose anything that might be said about the teaching of anything, might be said about the teaching of general semantics; about studying and learning Korzybskian methodology. Some of these observations about teaching could include: readiness of the learner, readiness of the teacher, recognition that teacher and learner both function more effectively when they realize that they both function as both teacher and learner, and rejoice in the realization that they can teach-learn with each other.

I accept the assignment: teach about the teaching of General Semantics. How can I experience the mutual teaching-learning suggested above? How can I elicit from you, the reader, some of the thinking-feeling which I would have us share? As I write and as you read, might we both think-feel that we are sharing this experience, that we are expanding our mutual 'reality', that we are influencing each other? I suppose the best way to secure your cooperation is to promise mine and to ask yours, which I herewith do.

I offer the above paragraph, with its implied values, its invitation to share and to explore together as one of the explicit examples of approach to the teaching-learning of anything, especially, I think, general semantics. Of course, how much better if we could be together in the same room so we could see each other, watch changing expressions, listen to new inflections. I will make assumptions about your values, reactions, expressions, and inflections. I choose to believe (think-feel) that you interact with me – sometimes nodding agreement, sometimes not, but always seeking to speak to and hear each other with greater clarity.

If I had to settle for only one admonition for the teacher of general semantics, I guess it would be: Do your best to exemplify that which you are teaching, make sure you practise what you preach, believe so intensely in and work so hard at what you are doing, that the mutual teaching-learning is recognizable as such. Make the theory operational.

He who learns and learns and yet DOES not what he knows, is one who plows and plows yet never sows. [2]

I suppose we have all seen zealous advocates of Korzybskian methodology who in their own personal lives, in their relations with others, behaved as if they thought general semantics was something ONLY to be written about, talked about, thought about. If the teacher is to interact convincingly with the learner, he must demonstrate in rather marked fashion, that which he is advocating.

Not only does the teacher of general semantics rejoice in the inseparability of teacher-learner, teaching-learning; but he can not put apart the process: teach-teacher.

The teacher accepts the function of verb.

It seems to me that in the work of Korzybski, we have a methodology which facilitates survival values which serve as reference points for both generalized and specific behavior. The theme and the teacher become one.

I believe the spiral approach is the best one for me. I do not attempt to impose it on others; I am quite sure it is not best for everyone. I am quite sure the spiral approach is not best for me under all conditions. I believe that linearity facilitates, for example, alphabet learning.

But for me, when teaching-learning Korzybskian methodology, I like the spiral. I have a feeling, akin to integrity, which I lose if I separate that which is not separable, if I fragment that which exists only together, if I impose sequentiality on that which may better be described as randomness.

I recently talked with a colleague from another college about the present trend toward 'modularization', or the reducing of

whole bodies of knowledge to smaller and smaller modules of units. We remained skeptical of the benefits from such fragmentation, and we agreed that probably only those areas we might label as 'cognitive' lend themselves to such treatment. It seemed to us that the main function of any subject being taught was that of a vehicle for the kind of activity that was carried on while the teacher-learner and the learner-teacher were teaching-learning.

I just don't feel quite honest when I say: "Now, here are three things", when I believe there is only one thing.

I suppose in my talking or writing I would like to speak or write of 'all' at the same time. I cannot. But I can say, think, feel, so as to reduce to a minimum the false-to-fact structure which says: these barriers exist outside the nervous system of man; these compartmentalizations were discovered, not invented, by man. Maybe we can even come closer if we avoid a fight about the degree of relationship (or lack of it) between DISCOVER and INVENT.

An assignment I suggest to you is to seek some instances where you believe man is superior to the law made by man.

I feel that the spiral approach, as opposed to the linear one, offers me a great deal more freedom. I feel more comfortable when discussions lead where they lead. I become excited when, having only loosely structured expectations, there develops the unexpected. Not only do I see our mutual experience as having more avenues for exploration, but also a greater willingness on both our parts for accepting and encouraging the other's contributions, creativity, relaxation, involvement.

'Science and Sanity' is a good example of the use of the spiral approach.

This approach gives me a sense of greater freedom in the selection of examples. I accept the responsibility for the selection of cases from many sources. A breadth of "for instances" gives us a statistically better chance for mutual understanding. And, it seems to me, exploration of the spectrum of human experience leads to fresh awareness of relevance and to the desire for the removal of categories and barriers which inhibit the process of integration.

Every rule has an exception. Is this a rule? (Reference to the structural differential helps in this instance, AS IT SO OFTEN DOES.)

The behavioral component of man, which we for convenience label 'freedom-responsibility', suggests, I think, that the teacher must constantly search within himself for new evidences, new examples, of relevance and integration in himself and in his teaching. The cyclic phenomenon goes on: I experience integration which furthers my integration which facilitates my discernment leading to the further experiencing and further integration.

I choose to teach-learn via spiral methods. I choose to select my examples freely. I choose to feel responsible for my own further growth. I choose to feel responsible for selecting and providing for both of us those experiences which are primarily reactive, 'emotional', personal, as well as cognitive.

In how many ways can you define yourself?

Somehow, in any way, I seek for us those experiences which help us together to discern order, relationship, and structure, I want us to question how things (and/or people) fit together (or do not fit together); how things (and/or people) influence each other (do you know of any exceptions?). I want us to ask how our mutual and independent existence and independent and mutual behaviors become one, which becomes increasingly recognized as part of the larger and larger structure. We, individually and ultimately collectively, structure our own integration.

With increasing awareness of integration comes increasing ease in the establishment of relevance. As we discern the structures, the patterns of interweaving forces, of the relatedness of processes, we recognize the impact of each upon the other. We are part of each other's environments; we do change each other's lives.

If I truly believe I change you, your life, I can begin to see my responsibility to you and yours to me. The recognition of the breadth and depth of my influence carried with it the recognition the acknowledgment of great strength – strength enough to change lives and strength enough to accept, and gladly accept, the related responsibility. I am suggesting a positive correlation between the awareness of mutual influence and the acceptance of a proud responsibility.

In this city there is a man whom you have neither seen nor heard. In how many ways have you influenced each other today?

Further, the incorporation of responsibility into the structure one holds of and for himself justifies a confident dignity, self-assurance which permits and encourages the kind of involvement which reinforces the self-image that you, I, everyone can change things, ourselves, others – that we do have a large voice in our own destinies.

This involvement with a purpose leads to extra dimensions of excitement, joy, and ecstasy, which in turn facilitates the further interaction which expands awareness of mutual realities by which integration is achieved, etc.

Thus, it seems to me that one teaches the methodology of Korzybski pretty much as he teaches anything else. That is, by learning it (coming to know-feel more about himself), hopefully by living it with others.

Can you think of any reason for wanting to know better the dullest person imaginable?

REFERENCES

1. Korzybski, Alfred, General Semantics Seminar 1937, Second Edition, 1964, Institute of General Semantics, Lakeville, Connecticut, p. 1.
2. Korzybski, Alfred, Science and Sanity, Third Edition, 1948, Institute of General Semantics, Lakeville, Connecticut. p. xxxi.

Teaching General Semantics: Training Methods for Non-Aristotelian Evaluating and Behaving

ALLEN FLAGG

General Semantics considers a human being an organism-as-a-whole-in-an-environment. Men live in many environments, both physical and symbolic, each with its own positive and negative influences. In responding to these environments, the human organism functions on a number of levels. Samuel Bois lists the levels of thinking, feeling, moving, and electro-chemical, with subconscious, conscious and supraconscious components, with memories of the past and expectations of the future.[1] Harry Weinberg called them "levels of knowing and existence", the hierarchies of symbolic and physical structures.[2]

Using general semantics as a general system of evaluation, one may study these levels

1. By analyzing the symbolic and behavioral elements composing each level, and their relationships;

2. By separating into components an element usually considered homogeneous. For example, a word, instead of having one meaning, may have many components of meaning, including grammatical, linguistic, semantic, semiotic, referential, multiordinal, personal, temporal, visual, aural, etc. One may clarify some of these meanings by using such general semantics extensional devices as indexing, dating, the etc., quotes, hyphens, "to me", etc.;

3. By combining two or more levels usually considered separate; for example, talking-feeling-touching, in an experiencing group;

4. By testing the interrelationships between levels. For example, one might ask

How does what one says ("I am a failure") affect one's emotions and behavior?

How does excitement and expectation affect the moving and electro-chemical levels?

How do cultural assumptions in the past or present environment affect behavior?

How do past perceptions affect present perceptions?
How do subconscious images affect conscious behavior?;

5. By investigating differences and similarities among environments; for example, physics, chemistry, history, literature, politics, propaganda, religion, myth, etc.;

6. By keeping up to date with the findings of the physical and behavioral sciences, and modifying teaching methods.

If we limit teaching to a single method, we impose the structure of that method on the staff and participants. General semantics training calls, instead, for an intermixture of lectures, seminars, workshops, "experiences", exercises, demonstrations, experiments, etc., keeping the training open-ended for adding new teaching methods.

Earl Kelley discussed the importance of the assumptions and behavior in teaching.[3] A course or seminar-workshop given by the Institute of General Semantics or the New York Society for General Semantics starts with assumptions different from those circumscribing the normal classroom situation. One may express some of the general semantics assumptions in the following ways:

1. Instead of "teacher" and "students", the staff and participants consider themselves students on a multidimensional spectrum of experience and abilities;

2. Instead of straight rows and assigned places, seating may range from chairs to mats, on the floor, in a circle, etc.;

3. Instead of separation by categories into groups, all participants form one group. Ages, occupations, "intelligence", and other variables may differ widely;

4. Instead of time spent exclusively on cognitive learning, training also includes levels of feeling, perception, moving, awareness, etc.;

5. Instead of one-way and verbal interaction between staff and participants, the training emphasizes informal participation, questioning and inquiring, with exercises, experiments and demonstrations. Neil Postman and Charles Weingartner stress the importance of this orientation.[4]

Aristotelian training and conditioning, including concomitant learning on all levels, has gone on for most of the participants' lives, with little in their environment to bring awareness of a non-aristotelian approach. At an early part of a course a number of de-conditioning exercises and demonstrations explore the inadequacies of exclusively aristotelian assumptions. Participants often express their reactions as surprise or chagrin at "being fooled" by

1. Ames perception experiments: the trapezoidal window, the

distorted room;

2. Perception paradoxes: the young - old woman diagram, the hidden man, the hidden cow, the double-triple tuning fork diagrams;

3. Inference-description confusions: the (rubber) pencil, the (changing) banana;

4. Behavioral maladjustments: news clippings of "unusual" situations, such as a heart attack and death caused by a reaction to a word.

Finding similarities between these classroom reactions and their daily behavior, the participants usually become more open and receptive to "new" non-aristotelian ideas that offer explanations for their reactions and propose more adequate methods for evaluating and behaving.

Participants learn and react as totalities. One learns non-elementalistically, that is, thinking, feeling, muscles and internal organs function together in an environment. A general semantics course stresses relationships between levels:

1. Organism-environment: perception demonstrations, awareness exercises, etc. Charlotte Read has done extensive work and writing on organismic experiencing;[5]

2. Thinking-feeling; encounter groups, D-groups. William Schutz' work;[6]

3. Thinking-moving: three weights demonstration, Weber weights, etc.;

4. Thinking-behaving: the semantitherapy of Harry Weinberg;[7]

5. Interpersonal relations and interdependence: here-now experiencing groups. Rachel Lauer has written of her work and concern for including this area in education;[8]

6. Thinking-feeling-electro-chemical: Neal Miller's experiments in controlling involuntary visceral responses, with therapeutic possibilities.[9] Buryl Payne has developed equipment and techniques for alpha brain wave research;[10]

7. Moving-electro-chemical: Laura Huxley's proposal of using muscular contractions to counteract glandular secretions produced by symbolic fears;[11]

8. Subconscious dream mind-thinking-feeling: the Dream Re-education work of Kilton Stewart.[12] As a part of the training on all levels, participants should learn to control and utilize the symbolic capabilities of their dream minds;

9. Supraconscious mind-thinking-feeling: Bois[13] and Maslow[14] indicate that a peak experience may occur when a

person reacts on several integrated levels; for example, having a sense of "awareness-feeling-understanding of the symbolic-physical interrelationships between the organism and the universe."

Alfred Korzybski, in his chapter "On Non-Aristotelian Training,"[15] stresses the importance of involving students on neurophysical levels. He suggests that in "learning" to use the Structural Differential a student should point at the various levels with his hand as he talks about it. In a seminar a student might say, "This descriptive level (pointing) differs from this inferential level (pointing). Check this level of generalizations (pointing) by experimentation at this process level (pointing)".

One may diagram the importance of training on several integrated levels as follows:

WAYS OF GETTING YOUR IDEAS ACROSS

A guide to Teaching and Learning in an approximate order from the least efficient to the most efficient.

Least
Efficient Generally Verbal

1. Write – Communicate one-way, with limited feedback.
2. Generalize – Rely upon abstract words, definitions stereotypes, categorized ideas, maxims, etc.
3. Simulate – Use a case history; study a concrete situation in a story; discuss, analyze, relate.
4. Illustrate – Recall experiences in common, use examples, illustrations, references news items, etc.
5. Relate – Use operational language, descriptive language, actional language, functional language, relational language, etc., isomorphically related to the subject matter.

Generally Nonverbal

6. Gesture – Use face, hands, body; make sounds; etc.
7. Diagram – Draw diagrams and graphs; isolate key points, show relationships, etc.
8. Picture – Show photos, slides, motion pictures, paintings, posters, scrap books, etc.
9. Demonstrate – Prepare models, act out skits and sociodramas, use demonstrations, exercises, etc.

Most
Efficient 10. Participate – Work together, guide, use participative experiments, feedback analogues, etc.

No one method above gives "best" results. For maximum effectiveness, a variety of methods enhances any one method.

We get the most understanding when we work together in some creative activity in which we can sense and talk to others about our ideas and feelings.

Some specific training exercises and demonstrations include:

1. Facts – inferences confusion: Haney's "Uncritical Inference" test, in which most students jump to conclusions based on assumptions beyond the facts;

2. Cooperation – competition continuum: five squares demonstration, an exercise in which students find that a temporary loss results in an eventual gain;

3. Plurals, the search for multiple answers: nine dots puzzle (more than 10 answers); push dime through hole puzzle (more than 15 answers); determine height of Empire State Building with a barometer (more than 5 answers); etc.;

4. Words as variables: ask students to respond to a poem and the proverbs "Rose is a rose. . . " and "A rolling stone. . . "; a variety of responses result;

5. Map – territory relations: Weiss' "Is of Identity" test, showing how words sometimes represent reality poorly;

6. Assumptions and perceptions: T puzzle, pyramid puzzle, double cone on an inclined plane; assumptions behind perceptions lead students away from answers.

One aspect of teaching situations not usually given much consideration concerns self-fulfilling prophecies: the teacher's expectation of student change or accomplishment. Recently, however, a study by Robert Rosenthal and Lenore Jacobsen has received a considerable amount of publicity.[16] Students whom the teacher expected to spurt ahead had test results that showed they indeed did.

Wendell Johnson studied the reverse of this, negative expectations. He found that a parent's idealized and exaggerated hope for the child's speech fluency often manifests itself in a negative reaction to normal repetitions, hesitancies and pauses. Such a re-action often brought about an increase in repetitions, which the parent called stuttering. Johnson considered stuttering a diagnosogenic disorder; the diagnosis of stuttering itself helps bring on the disorder. The child accepts to some degree the parent's negative evaluations of his speech behavior and increased speech disorders follow.

In general semantics training, supportive expectations and evaluations tend to develop open-ended attitudes, leading to participant's behavior that can be characterized as expanding, enhancing and seeking higher potentials.

The teaching methods, ideally, train the whole person toward his greatest capabilities. The participant can then more competently apply his aptitudes to his specific areas of interest.

Alfred Korzybski wrote: "One of the benefits of the present method of training in sanity consists in the fact that we do NOT dwell upon the personal affairs of the individual, but that we give, instead, a general structural semantic METHOD, by the aid of which every one can solve his problems by HIMSELF."[18]

REFERENCES

1. J. Samuel Bois, The Art of Awareness, Wm. C. Brown Co., 1966, pp. 20, 202.
2. Harry L. Weinberg, Levels of Knowing and Existence, Harper & Brothers, 1959.
3. Earl C. Kelley, Education for What is Real, Harper & Brothers, 1947.
4. Neil Postman and Charles Weingartner, Teaching as a Subversive Activity, Delacorte Press, 1969.
5. Charlotte Read, "Exploring Relations Between Organismic Patterns and Korzybskian Formulations", General Semantics Bulletin, Nos. 32 & 33, 1965/1966, p. 47.
6. William C. Schutz, Joy, Grove Press, 1967.
7. Harry L. Weinberg, op. cit., p. 177.
8. Rachel Lauer, "General Semantics and the Future of Education", ETC., Vol. XXIV, No. 4, December 1967, p. 391.
9. 'Science News', Vol. 97, 3/14/70, p. 274.
10. Psychophysics Lab, 31 Townsend Terrace, Framingham, Mass. 01701.
11. Laura Archera Huxley, You Are Not the Target, Farrar Straus & Giroux, 1963; Fawcett Publications ed., p. 47.
12. Kilton Stewart, "Dream Theory in Malaya", Altered States of Consciousness, ed. by Charles T. Tart, John Wiley & Sons, 1969, p. 159.
13. J. Samuel Bois, op. cit., p. 164.
14. Abraham H. Maslow, 'Toward a Psychology of Being', D. Van Nostrand Co., 1962, p. 67.
15. Alfred Korzybski, 'Science and Sanity', International Non-Aristotelian Library Publishing Co., 1933; 4th ed. 1958, p. 469.
16. Robert Rosenthal and Lenore F. Jacobsen, Pygmalion in the Classroom, Holt, Rinehart and Winston, 1968.
17. Wendell Johnson, People in Quandaries, Harper & Brothers, 1946, p. 439.
18. Alfred Korzybski, op. cit., p. 528.

Strategies for Assessing the Effects of "Training" on Semantic Reactions

FRANKLIN H. SILVERMAN

I

The semantic reaction is fundamental to the general semantics model as formulated by the late Alfred Korzybski. As a matter of fact, general semantics, as a discipline, can be viewed as the study of the semantic reaction. Korzybski[1] stated that he wrote Science and Sanity, his major work on general semantics, entirely from the semantic reaction point of view. Further evidence of the central role of the semantic reaction in the general semantics model is provided by the fact that one of the primary objectives of general semantics training, if not the primary objective, is to ameliorate disturbances of the semantic reaction.[2]

What phenomenon was Korzybski referring to by the term, semantic reaction? In Science and Sanity, Korzybski describes the semantic reaction as ". . . . the psycho-logical reaction of a given individual to words and language and other symbols and events IN CONNECTION WITH THEIR MEANINGS . . .".[3] In the first phrase, ". . . the psycho-logical reaction . . .", Korzybski inserted a hyphen between "psycho" and "logical". He used this device to indicate that the semantic reaction is NON-ELEMENTALISTIC because it conjointly involves "emotional" and "intellectual" factors. That is, the semantic reaction is neither "emotional" nor "intellectual" but an INTERACTION of the two. As such, it would be mediated by BOTH the autonomic nervous system and the central nervous system; hence, the semantic reaction would be preceded by a change in the activity level of both systems. The relative contribution of the two systems to the mediation of specific semantic reactions would be expected to vary.

By referring to the ". . . psycho-logical reaction of a GIVEN individual . . .", Korzybski pointed to another "attribute" of the semantic reaction: The semantic reaction to a given stimulus is apt to be different for different people. This is in part because the "meaning" of a given stimulus to different people is probably different. Obviously, an individual's "experiential history" is an important determinant of the meaning he attaches to a given stimulus. Since no two persons have exactly the same experiential history, it follows that no two persons would attach exactly the same meaning to a stimulus. And, hence, no two persons would be expected to have exactly the same semantic reaction to a stimulus, whether symbol or event.

II

As the stimuli to which we attach meanings – "words and language and other symbols and events. . . " – change as a function of time, our semantic reactions should also. The greater the rate of change of the meanings of such stimuli, the greater the need for developing what Korzybski referred to as "full conditionality" of the semantic reaction. By this he meant that semantic reactions to words and other stimuli should be contingent upon the specific meaning(s) of such stimuli at given points in space-time. For example, the "meaning" of the word and event "war" at $point_1$ in space-time (e.g., $war_{Europe\text{-}1940}$) may be quite different from that at $point_2$ in space-time (e.g., $war_{Vietnam\text{-}1970}$). Since the word and event "war" have no general meaning(s), they should not elicit a stereotyped semantic reaction.

One of the primary objectives of general semantics training has been to assist the individual to increase the degree of conditionality of his semantic reactions. Assuming that this is a desirable goal and that more than one approach to general semantics training is possible, questions such as the following can be posed:

1. What is the effect of $training_1$ on the degree of conditionality of the semantic reaction?
2. Do $training_1$ and $training_2$ differ in their effect on the degree of conditionality of the semantic reaction?
3. Is $training_1$ more effective in increasing the degree of conditionality of the semantic reactions of some persons while $training_2$ is more effective for others?
4. If the answer to question three is "yes", what are the relevant organismic variables?[4]
5. What influence, if any, does the "personality" of the trainer exert upon the effect of specific types of training on the degree of conditionality of the semantic reaction?
6. How long-lasting are the effects of specific types of training on the degree of conditionality of the semantic reaction?

To the best of my knowledge, there are next to no studies which have attempted to answer these questions systematically. This is unfortunate since without at least tentative answers, general semantics training will remain an art-science rather than evolving into a science-art as some members of the discipline desire.

III

In order to answer the questions I have posed about increasing the degree of conditionality of the semantic reaction, it would be necessary to devise strategies for making the required observations. There are a number of considerations in devising such strategies. The remainder of this paper will be devoted to a discussion of some of these considerations.

The problem of devising strategies for assessing the effects of training on the degree of conditionality of the semantic reaction is essentially similar to that of devising strategies for evaluating the effectiveness of therapies – particularly, psychotherapies – and of educational methods. The questions which we would like to be able to answer in all three cases are quite similar. This is a fortunate state of affairs for investigators of the semantic reaction for there exists quite a large literature in clinical psychology, psychiatry, and education devoted to considerations in devising strategies for answering questions such as I have posed.[5]

IV

One of the first considerations in devising strategies for evaluating the effect(s) of "training" on the degree of conditionality of the semantic reaction is the specification of the behavioral events which will serve as the dependent variable or variables. Obviously, such behaviors would only be "aspects" or "components" of the semantic reaction of interest: From the "allness" principle of general semantics[6], it seems reasonable to conclude that it would be impossible to specify the full set of behaviors which constitute a given semantic reaction. The types of behavioral events included in this set could be partially delineated by means of the structural differential:[7] that is, some may be "verbal" (words) and some may be "nonverbal" (not words); some may be "macroscopic" (observable through one or more sense modalities directly – without instrumentation) and some may be "microscopic" (observable through one or more sense modalities with instrumentation – a microscope, an amplifier, etc.); of the verbal events, some may be "descriptions" ("low" level verbal abstractions) and some may be "inferences" ("high" level verbal abstractions); etc.

Since it would never be possible to use the entire set of events which constitute a semantic reaction as a dependent variable, what considerations enter into the choice of an event or events from the "set" of such events to serve as a dependent variable? There are several issues here. For example, a decision must be made on the OPTIMUM NUMBER of behavioral events to sample. The more behavioral events sampled, the higher the level of reliability of the dependent variable is likely to be. From this last statement it would seem reasonable to conclude that the more behavioral events sampled the better. This is, however, usually not the case. Beyond a certain number, which we will refer to as the optimum number, there isn't enough increase in the reliability level of the dependent variable to warrant the additional effort. The optimum number for a given study would vary as a function of several factors, including (1) the minimum reliability level required for the dependent variable and (2) the reliability level of the measures of the individual behavioral events which make up the dependent variable. The lower the minimum level of reliability required for the dependent variable and the higher the reliability level of the individual measures, the fewer behavioral events it would be necessary to sample to achieve the optimum number.

In addition to deciding on the number of behavioral events to sample, a decision must also be made concerning WHICH behavioral events of a "set" to sample. One consideration when making this decision could be referred to as "ease of quantification." Some behavioral events of a set may be easier to measure, or quantify, with a given degree of reliability than are others. If one is willing to assume that two or more behavioral events provide similar information concerning the degree of conditionality of the semantic reaction with similar levels of reliability, then the choice between them might be made on the basis of ease of quantification. For example, if an investigator wished to include a behavioral event in his dependent variable which was thought to be a physiological correlate of "anxiety," two possibilities might be heart rate and degree of pupil dilation. Since instrumentation for measuring heart rate is usually easier to come by than that for measuring degree of pupil dilation, heart rate might be chosen for this reason.

If ease of quantification is similar for two or more behavioral events, the choice between them might be made on the basis of the estimated level of reliability possible for the measure of each event. As you may recall from my discussion of the optimum number of behavioral events for the dependent variable, one consideration is the reliability level of the measures of the individual behavioral events. The higher the reliability levels of these measures, the fewer "events" that will be needed to achieve the optimum number.

Another consideration in the choice of behavioral events would be the degree of "validity" of the available measures of the events. If the validities of the possible measures of a particular behavioral event were either unknown or questionable while those of an alternative one were well established, the latter event would probably be selected for the dependent variable.

V

A second consideration in devising strategies for assessing the effects of training on the degree of conditionality of the semantic reaction is specification of the training program to be employed. A training program is made up of a "set of behaviors." Some of these behaviors are manifested by the student and some are manifested by the instructor. Each behavior in the "set" functions as an independent variable.

It is unlikely that all the behaviors included in a training program will influence the dependent variable – in this case, the degree of conditionality of the semantic reaction. For this reason, if a training program was judged to be "successful" (i.e., resulted in an increase in the degree of conditionality of the semantic reactions of the participants), it would be advantageous to be able to specify the sub-set of behaviors which was responsible for the effect of the program. This would permit the program to be simplified and thereby made more efficient. Obviously, a necessary condition for doing this is being able to specify the set of behaviors which constitutes a training program. While it is unlikely that one would ever be able to specify the complete set of such behaviors, the closer one could come to achieving this objective, the more likely it would be that one would be able to devise an efficient training program.

In addition to cataloging the behaviors which make up a training program, it is also necessary to specify the "co-ordinates" of each in space-time. A particular behavior by the instructor, for example, might have a different effect upon the dependent variable depending upon whether it preceded or followed a specific behavior by the student. The importance of specifying the temporal relationships between behaviors has been repeatedly demonstrated by learning theorists – particularly those with an operant orientation.[8]

Assuming that it is desirable to be able to specify the set of behaviors which constitute a training program, how would one proceed? Since a training program includes behaviors of both the student and the instructor and since some of the behaviors of both are of necessity contingent upon those of the other, it isn't possible to specify "COMPLETELY" the set of behaviors which constitute a training program prior to beginning it. Hence, in order to be as complete as possible it is necessary to add to the catalogue of such behaviors following the completion of the program. Audio or video tape recordings of the training sessions would be useful here. If recordings of the sessions are used, it would be advisable to have at least one observer in addition to the experimenter (instructor) review the tapes. The experimenter's catalogue of behaviors might be quite different from those of other observers since he has an additional "filter" influencing his abstracting process – the knowledge of what he was attempting to do.

It could be argued that specification of the behaviors which constitute a training program is too time-consuming to be practical. Undoubtedly, the time investment could be considerable especially if audio or video recordings of the training sessions were analyzed. Nevertheless, considering the time, energy, and financial investment of the training program itself, as well as those of future training programs similarly structured, the investment required to specify these behaviors seems relatively small. There is also a second consideration here: Without such an inventory a training program can not be communicated to others. Thus, the inventory would also be desirable for "time-binding" purposes.[9]

VI

In order to assess the immediate and long-term effects of a training program on the degree of conditionality of the semantic reaction, a temporal sampling strategy has to be developed. That is, a decision must be made concerning the co-ordinates, or points, in space-time at which to sample and measure each behavioral event to be included in the dependent variable.

There are several considerations which should enter into the choice of a temporal sampling strategy. First, is the requirement that the strategy selected result in a BASELINE which has adequate stability. Without such a baseline, it obviously wouldn't be possible to reliably infer the effects of a training program on the dependent variable(s).

There has been a great deal of discussion in the psychological literature concerning the characteristics of an "adequate baseline.[10] One point which has been made by a number of writers[11] is that a sample of a behavioral event at only ONE point in space-time is not usually adequately reliable to serve as a baseline for determining the effects of a training program. One reason for reaching this conclusion is that the performance of an individual at different points in space-time can be expected to vary as a function of such organismic variables as "alertness," "interest," "anxiety" level, etc. Hence, at any given point in space-time a person's performance could be "better" than usual, the "same" as usual, or "worse" than usual. If an experimenter wishes his baseline to represent the "usual," he will be more likely to achieve his objective if his baseline consists of a composite of measures (mean, median, etc.) from several points in space-time rather than a measure from only one point in space-time. Stated differently, rather than sampling each behavioral event only once prior to beginning a training program, each should be sampled several times. Hence, using Campbell and Stanley's notation scheme[12] in which X represents the beginning of a training program and O represents a sample of a behavioral event, the strategy which can be diagramed,

O O O X

would probably result in a more stable, more reliable, baseline than the strategy which can be diagramed as follows:

O X

Unfortunately, the latter strategy has frequently been incorporated in educational and psychotherapy outcome research into an experimental design which can be diagramed as follows:

O X O

The severe limitations of this "pretest-posttest" design are discussed in depth by Campbell and Stanley.[13] Hopefully, investigators of the semantic reaction will profit from the experience of researchers in psychology and education and avoid this design whenever possible.

A second consideration in the selection of a temporal sampling strategy is frequency of sampling once a training program has begun. An investigator must decide (1) whether he wishes to sample the behavioral events which serve as his dependent variable(s) only at the termination of his program or during the course of the program as well and (2) whether he wishes to continue to sample following the termination of his program and if he does, at what intervals. In other words, he has to decide (1) how he is going to assess the immediate effects of his program and (2) whether or not he is going to be interested in long-term effects.

Of what value is a temporal sampling strategy which includes periodic sampling of the dependent variable(s) throughout the course of a training program? I feel that this strategy has several advantages. In the first place, it permits a more reliable assessment of the immediate effects of a training program than does a strategy which only samples the dependent variable(s) at the termination of a program. The argument here is the same as the one outlined earlier for using a baseline consisting of a "composite" measure rather than a single measure.

Another advantage of a strategy which incorporates periodic sampling during the course of a training program is that it permits questions concerning "process" to be answered, such as the following:

1. If a training program results in an increase in the degree of conditionality of the semantic reactions of the participants, what was the course of the increase? Did most of the increase occur at the beginning of the program, at the end of the program, or did it occur gradually during the course of the program?

2. Could the program be shortened without "significantly" reducing its effectiveness? This might be possible if most of the "gains" are made early in the program.

3. Which behaviors of the set which makes up a training program are NOT responsible for an increase in the degree of conditionality of the semantic reactions of the participants? If most of the increase in conditionality produced by the program were observed prior to the introduction of certain parts of the program, these activities obviously could not have been responsible for this increase in conditionality.

In addition to devising a strategy for assessing the immediate effects of his training program, an investigator should also give some consideration to assessing long-term effects. It would be naive for him to assume that the degree of conditionality which he observes at the end of a training program will necessarily remain unchanged over a relatively long period of time. Based upon reports by clinical psychologists, psychiatrists, and educators concerning stability of behavior change, the "best" guess one could make would be that some reduction in the degree of conditionality of the semantic reaction would occur for most participants in a training program following the termination of the program. The degree of regression would vary: The effects of some training programs could be expected to hold up better than others. Since an objective of most general semantics training programs would be to produce a permanent increase in the degree of conditionality of the semantic reaction, and since training programs can be expected to vary with respect to the permanence of such an increase, it obviously would be desirable to use a temporal sampling strategy which would include some sampling of the dependent variable(s) following the termination of a program.

If an investigator wishes to incorporate some type of follow-up into his temporal sampling strategy – that is, if he wishes to continue to sample his dependent variable(s) following the termination of his training program – he must make at least three decisions: (1) which behaviors he will sample; (2) the interval(s) at which he will sample these behaviors; and (3) the length of time that he will continue to sample these behaviors.

A decision should be made on whether the temporal sampling strategy will include follow-up prior to selecting behaviors for the dependent variable. Some behaviors which might be considered would be difficult or impossible to sample following the termination of a training program, especially if the participants do not live in the community in which the program was conducted. Therefore, it may be necessary to include some behaviors in the dependent variable which can be sampled either through the use of the telephone or by means of a "paper and pencil" task.

The decision on how frequently to sample following the termination of a program is a difficult one for which to provide guide lines. If the temporal sampling strategy used included periodic sampling during the training program, it would probably be a good idea to plan to continue sampling following the termination of the program at the same interval which was sampled during it.

The final decision – the length of time to follow-up – is also a difficult one for which to provide guidelines. I know of no data on the long-term effects of general semantics training which would provide a basis for making this decision. Perhaps until such data are available it would be advisable to be "conservative" and adopt a five-year follow-up period as have many investigators in the fields of medicine and psychotherapy.

VII

Once the measures of the dependent variable(s) which are dictated by the temporal sampling strategy have been completed, how are these data manipulated to answer the question(s) asked? There are two general approaches: (1) combining the measures obtained from all the participants in a training program and (2) keeping the measures from each participant separate. The first would permit the questions to be answered for the "typical" or "average" participant and the second would permit the questions to be answered for individual participants.

The approach which has probably been used most frequently in "outcome" research is the first – combining the measures obtained from all participants. The methodology, which involves the application of statistical procedures such as the t-test and analysis of variance, is well developed. Perhaps the main limitation of this approach is that it only provides answers to questions about a relatively high order abstraction[14] – the "typical" group member. Since it isn't safe to assume that all or even most of the participants in a training program will behave similarly to the "typical" participant, the usefulness of this approach is questionable.

The second approach, which involves examination of the data from individual subjects, appears to be the more satisfactory of the two for assessing the effects of a training program on the degree of conditionality of the semantic reaction. It yields answers which are lower order abstractions than those provided by the first approach. For this reason, it is less likely to obscure individual differences in performance. The criterion measure with this approach would be the per cent or proportion of participants in a training program who behave in a particular way, e.g., increase the degree of conditionality of their semantic reactions.

VIII

I have attempted in this presentation to specify some considerations when devising strategies for assessing the effects of training on the degree of conditionality of the semantic reaction. Obviously, all such considerations were not discussed and a great deal more could have been said about those which were discussed. I have attempted to present a beginning – a first approximation – of a methodology for assessing the effects of general semantics training. Much research will be needed to refine this methodology.[15] Considering the central role played by the semantic reaction in such recent "happenings" as the ecology crisis (i.e., How can people be mobilized to take action?) and student riots (e.g., opposition to the extension of the War into Cambodia), research which would provide the methodology necessary to improve our ability to increase the conditionality of semantic reactions should be of highest priority.

REFERENCES

1. Alfred Korzybski, Science and Sanity: An Introduction to Non-Aristotelian Systems and General Semantics,sr, 4th ed. (Lakeville, 1958), p. 25.
2. Korzybski, p. 18.
3. Korzybski, p. 24.
4. For a discussion of organismic variables, see Allen L. Edwards and Lee J. Cronbach, "Experimental Design for Research in Psychotherapy," in The Investigation of Psychotherapy: Commentaries and Readings, eds. Arnold P. Goldstein and Stanford J. Dean (New York, 1966), p. 72.
5. For representative examples, see Donald T. Campbell and Julian C. Stanley, Experimental and Quasi-experimental Designs for Research (Chicago, 1963); The Investigation of Psychotherapy: Commentaries and Readings, eds. Arnold P. Goldstein and Stanford J. Dean (New York, 1966); Problems in Measuring Change, ed. Chester W. Harris (Madison, 1963).
6. Korzybski, p. 783.
7. Korzybski, p. 796.
8. For an introduction to operant learning theory, see James G. Holland and B. F. Skinner, The Analysis of Behavior (New York, 1961).
9. Korzybski, p. 797.
10. For a comprehensive discussion of baselines, see Murray Sidman, Tactics of Scientific Research: Evaluating Experimental Data in Psychology (New York, 1960).
11. E.g., Campbell and Stanley.
12. Campbell and Stanley, p. 6.
13. p. 7.
14. See Korzybski, pp. 386-411, for a definition of high order abstraction.
15. This study is part of a program of methodological research on the semantic reaction being conducted by the author. For a report of other research in this program, see Franklin H. Silverman, "Use of Psychological Scaling Methods for Quantifying Observable Aspects of Semantic Reactions", in Research Designs in General Semantics, ed. Kenneth Johnson (Washington, D.C.: Gordon and Breach, in press).

The Relationship of Systems Theory to Education:

The Problem of Internalization in Systems Education

BELA H. BANATHY

I have elected to discuss the relationship of systems theory to education within the context of systems education and to explore the problem of methodologies of internalization.

I shall develop my topic in the line of the following inquiry:

1. How can we conceptualize education as a system?
2. How can we design, develop, and implement it as a system?
3. What is the system problem of systems education?
4. And more specifically what is the systems space and system problem of internalization in systems education?

Using a well known general systems model we can state that the system problem of education is that of

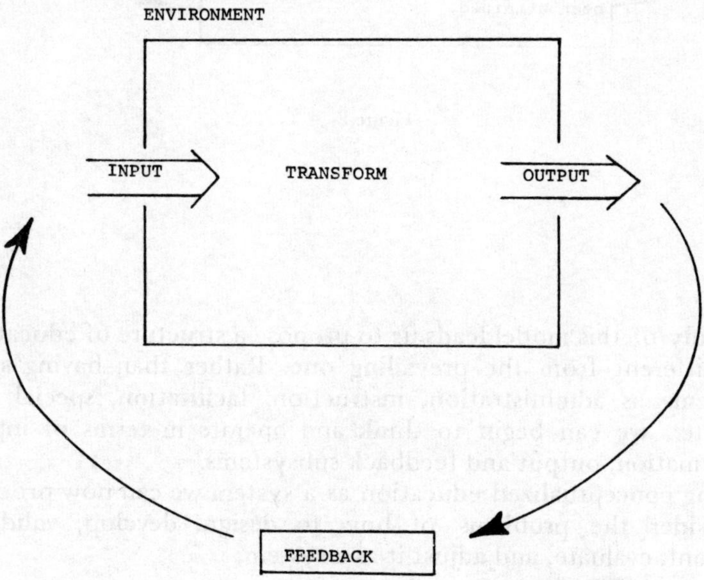

Figure 1

1. Defining the system space of education by establishing its boundaries, specifying the goals, resources, constraints and other input entities of education;

2. Transforming the learner and other entities of education into a state where they have attained goals;

3. Measuring the attainment of goals;

4. Feeding back information on this attainment and, based on it, design change to ensure the attainment of goals.

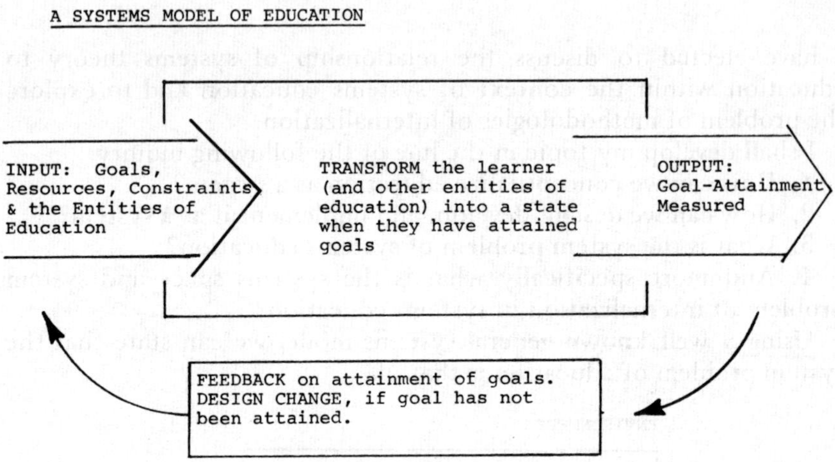

Figure 2

A study of this model leads us to propose a structure of education quite different from the prevailing one. Rather than having such subsystems as administration, instruction, facilitation, special services, etc., we can begin to think and operate in terms of input, transformation, output and feedback subsystems.

Having conceptualized education as a system we can now proceed to consider the problems of how to design, develop, validate, implement, evaluate, and adjust it as a system,

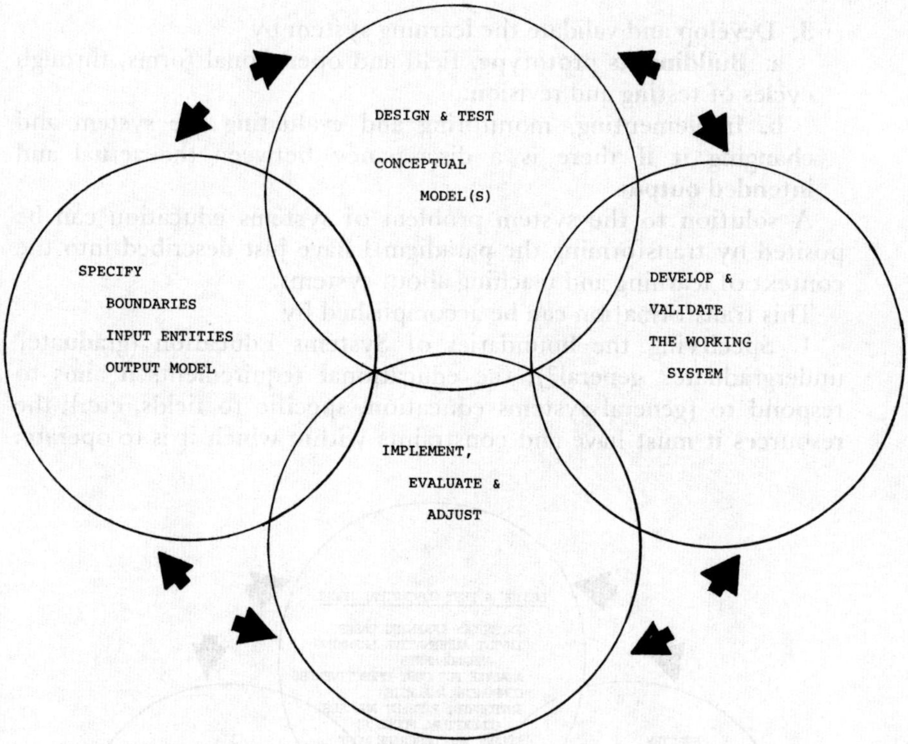

Figure 3

Transforming this giant-steps paradigm into a more specific plan of research for the development of educational products, we shall:

1. Identify the input entities of a specific educational program – such as, for example, systems education – and develop the performance specifications and criteria of its expected output;

2. Design and test the conceptual model. (The model is a descriptive representation or other kind of display of what the educational program might look like.) This model is developed by:

a. An inventory of the learning tasks which are to be mastered to enable the learner to attain the desired performance;

b. Considering alternatives in arrangements by which to facilitate the mastery of the learning task;

c. Selecting the most cost-effective alternatives;

d. Integrating and testing the conceptual model.

3. Develop and validate the learning system by

a. Building its prototype, field and operational forms, through cycles of testing and revision;

b. Implementing, monitoring and evaluating the system and changing it if there is a discrepancy between the actual and intended output.

A solution to the system problem of systems education can be posited by transforming the paradigm I have just described into the context of learning and teaching about systems.

This transformation can be accomplished by

1. Specifying the boundaries of Systems Education (graduate? undergraduate? general?); the educational requirements it aims to respond to (general systems education, specific to fields, etc.); the resources it must have and constraints within which it is to operate;

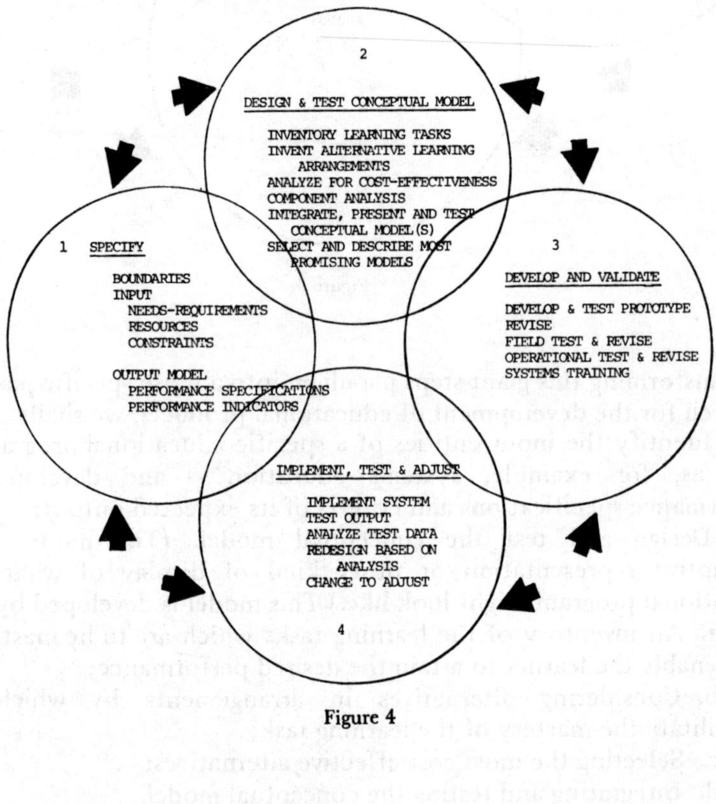

Figure 4

2. Developing the output model of whatever performance capabilities the learner should attain as an outcome of an effort in general systems education. Should he acquire the systems view; what theories, concepts and principles should he be able to work with? What operational capabilities should he have? Much of the success of systems education depends upon an in-depth exploration of these input-output relationships. Once these aspects are specified, we do NOW have the technologies and expertise to design, develop, validate and implement learning systems for general systems education.

But let me move on to the problem of internalization.

The system problem of internalization in systems education can be framed by two major inquiries:

WHAT to internalize, and

HOW.

The boundaries of the systems space of this problem can be projected in this model.

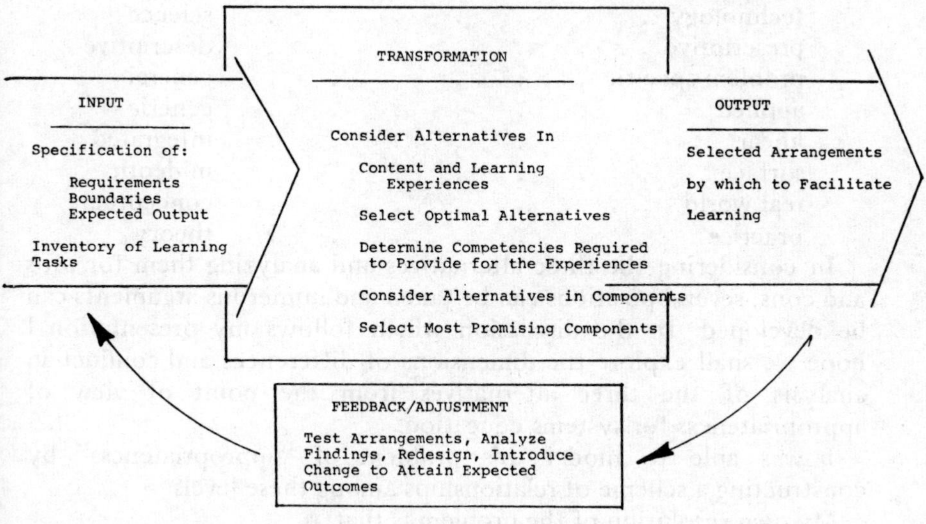

Figure 5

In the domain of transformation, the two strategies which my topic leads me to focus on are:

1. the consideration of alternatives in the content of systems education (or what to teach) and
2. alternatives in the method of internalization (or how to teach).

ALTERNATIVES IN CONTENT

At least three alternatives offer themselves for consideration as levels of content in systems education:

1. Systems strategies – steps of systems approaches
2. Specific-to-field of application models
3. General systems models.

There are many ways to characterize differences among these three alternative levels. Dimensions can be proposed for such characterization of differences as:

technology	science
prescriptive	descriptive
problem specific	general
applied	generic
linear	integrated
surface	in depth
real world	conceptual
practice	theory

In considering the three alternatives and analyzing them for pros and cons, several positions can be taken and numerous arguments can be developed. In the discussion which follows my presentation I hope we shall explore the dimensions of differences and conduct an analysis of the three alternatives from the point of view of appropriateness for systems education.

I was able to model this problem of "appropriateness" by constructing a scheme of relationships among these levels.

My own resolution of the problem is that:

1. The main thrust of systems education should be to help the learner internalize general systems concepts and principles, build and work with systems models;

2. Only when models are internalized and a systems view developed should we move toward fields of application, and activate strategies of systems approaches.

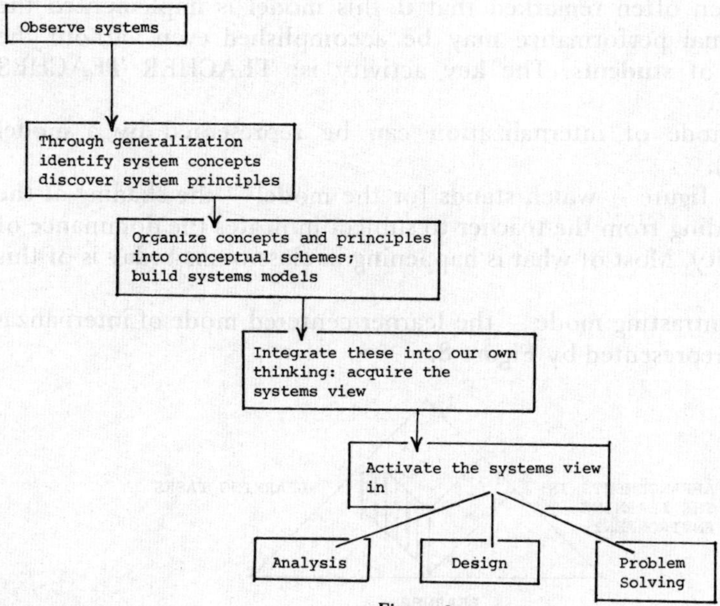

Figure 6

MODES FACILITATING INTERNALIZATION

Two contrasting methods can be projected:

1. The teaching focused or non-systemic, and
2. The learning focused or systemic.

In the teaching focused mode, the teacher appears to be the actor on the instructional scene. Arrangements are made in the instructional environment in order to optimize his performance. Teaching aids – if used – are introduced to enhance teaching, and students are the audience and are expected to pay attention to the performance of the teacher. The subject can be anything to which the teacher can address himself and which students may be required to study.

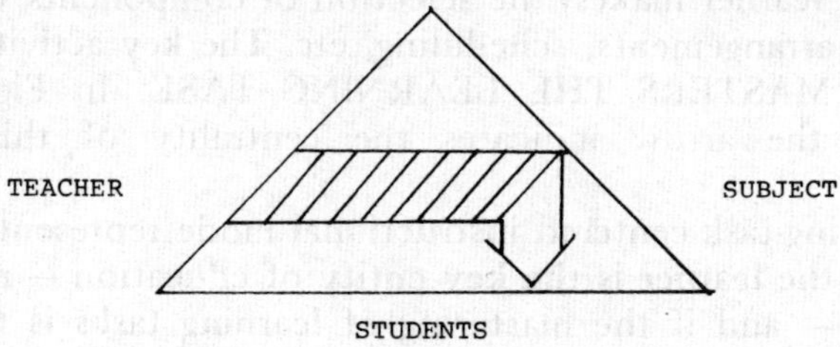

Figure 7

It has been often remarked that if this model is implemented the instructional performance may be accomplished even without the presence of students. The key activity is: TEACHER TEACHES SUBJECT.

This mode of internalization can be represented by a model (Figure 7).

On the figure – which stands for the model – the shading of the arrow leading from the teacher to subject indicates the dominance of this activity. Most of what is happening in classrooms today is of this kind.

The contrasting mode – the learner-centered mode of internalization – is represented by Figure 8.

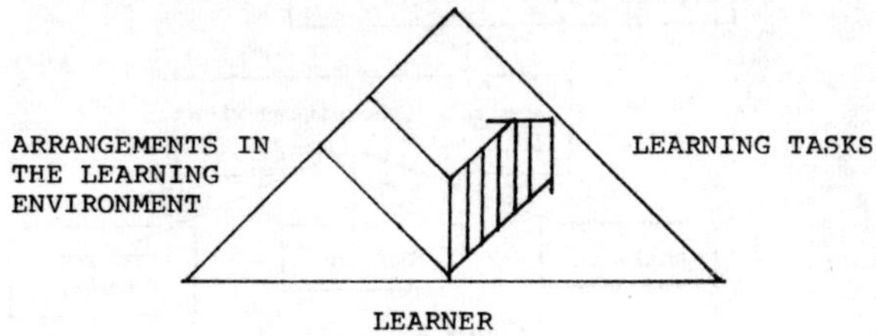

Figure 8

The figure reads: Arrangements are made in the environment of the learner through which he is confronted with the learning task so that he can explore and master the task. In this model it is the learner who is the actor on the instructional scene and arrangements are made around him in order to facilitate his mastering of the learning task. Arrangements as a term stands for a lot of things, such as the selection and organization of content which (best) represents the learning task; the selection and organization of learning experiences by which the learner is confronted with the content; the management and motivation of the learner; the assessment of the progress the learner makes; the selection of components which take part in the arrangements, scheduling, etc. The key activity is: THE LEARNER MASTERS THE LEARNING TASK. In Figure 8 the shading of the arrow indicates the centrality of this primary interaction.

The learning-task-centered instructional mode represents systemic thinking. If the learner is the key entity of education – rather than the teacher – and if the mastering of learning tasks is the central process of education, rather than instruction in itself, then the making of a choice between these two modes is not a difficult one. What may be of some difficulty, however, is for us to learn to operate in the learning-task-centered mode.

The success of systems education – or for that matter any education – depends a great deal upon our ability to break the educational mold patterned according to the instruction-centered model and create a fresh mode in which learning comes into focus and the learner becomes the key performer.

General Semantics, Research and General Systems Theory

ALVIN A. GOLDBERG

Despite the attempts by Elwood Murray, Irving J. Lee, Wendell Johnson, and others to encourage research in general semantics, relatively few studies that deal explicitly with hypotheses derived from general semantics have been reported. General semanticists have been more interested in spreading their gospel through teaching and writing than in studying its validity or applicability. The first section of this paper identifies the attitudes of some general semanticists toward research and attempts to determine why relatively few studies of general semantics have been reported. Various relationships between general semantics and general systems theory are then explored and it is suggested that some general semanticists may deal with general systems theory in a manner that is similar to their approach to research.

The failure of general semanticists to systematically investigate their methodology and its principles may conceivably suggest that general semanticists harbour a latent hostility toward scientific inquiry, but if they do, it is certainly not reflected in the literature. If anything, the general semanticist goes overboard in his enthusiasm for the scientific method. Science, according to Korzybski, is the road to sanity. We can avoid conflict and overcome our quandaries, Wendell Johnson tells us, through scientific living. General Semantics, according to Johnson, is the "...systematic attempt to formulate the general method of science in such a way that it might be applied . . . generally in daily life."[1]

But if the general semanticist is so committed to the scientific method as a general methodology, why has he been so sluggish in his use of the method to study general semantics? Perhaps one reason is that many enthusiasts who possess the scientific rigor to conduct general semantics research are identified with other disciplines as well, and the bulk of their research is done in these other areas. Just as general systems theorists are also biologists, economists, engineers, etc., general semanticists are often chemists, physicians, communication specialists, and the like.

Others factors may also be responsible for the paucity of research in General Semantics. The discipline is still young. Korzybski and his disciples devoted so much energy to teaching and proselytizing that they had little time to conduct or encourage research. Many general semanticists have literary or artistic backgrounds and are lacking in research skills. And few institutions have ever granted graduate degrees in general semantics. These and other reasons can be offered to explain why little general semantics research was done in the past. However, some of the events that occurred at the 1969 Conference on Research in General Semantics that was held on the Pennsylvania State University campus suggest that there may be more profound reasons why relatively little general semantics research is being conducted at the present time.

Many of the conferees at the Penn State University gathering were young and vigorous scholars with good backgrounds in research methods. They possessed the skills necessary to conduct first rate research in general semantics. The lack of total commitment to the goals of the conference that some participants occasionally expressed and that manifested itself as a subtle undercurrent of ambivalence, might be interpreted as an uncertainty about the ultimate value of science itself. In a sense, some of the participants seemed to be questioning the basic assumption that the scientific method can enhance the quality of our lives in any significant way. The participants were not opposed to scholarship, but a number of them seemed to be rather dubious about the necessity to test general semantics scientifically, or for that matter, anything else.

The disenchantment with scientific methods that some of us sensed at the Penn State Conference might have been part of a general revolt against objective inquiry. Robert Nisbet considers this growing rejection of objective, dispassionate approaches to knowledge on the part of some social scientists an "unbelievable thing". According to Nisbet, it represents an ". . .astonishing reversal of belief in the scientific, that is, the objective, the detached, the dispassionate character of the social sciences".[2] Since truly objective inquiry is not possible, the "radical" scientists assert that the goal of the social scientist should be to remake society, not merely to try to understand it.[3]

The ambivalence toward scientific investigation of General Semantics principles expressed at the Penn State Conference seems to have been part of the more general contemporary disenchantment with science referred to by Nisbet. However, it should be noted once again that general semanticists have seldom devoted much of their energy to the scientific study of general semantics. The general semanticist is committed to science as a methodology for enhancing individual and group effectiveness, but he has never really been very excited personally about science as a methodology for objectively and systematically investigating physical or social phenomena. By his behavior, the general semanticist seems to be saying that science is fine because it is individually and socially therapeutic, but it is not something he would care to dirty his hands with. So in some ways, the general semanticist, like the contemporary radical social scientist, sees science as a way to remake rather than understand society.

But what does science have to offer that is useful in dealing with our individual and collective problems? The general semanticist believes that it offers us an orientation. According to Wendell Johnson, a scientific orientation as opposed to a prescientific orientation recognizes the process character of reality and emphasizes differences between events rather than similarities. It encourages a willingness to change, to challenge traditional beliefs, to experiment, and it rejects rigid and conservative outlooks. A scientific orientation is more concerned with inquiry than advocacy; it does not rely on vested authority for answers to problems. The language of science is clear, direct, and factual. It avoids statements that are vague, meaningless, or value-laden. An individual with a scientific orientation recognizes that every statement tells us something about the speaker as well as the events the speaker is referring to. He evaluates statements and theories on the basis of their usefulness in making predictions.

How does a scentific orientation enable us to ameliorate our social and personal ills? Well, according to general semantics, science provides us with rules and principles for improving our daily lives. For example, science suggests that our language should accurately reflect the reality it represents. Maps that are not isomorphic with their territory will get us into trouble. We must recognize that events are more real than the symbols we use to talk about them, and our behavior should reflect this awareness. The word is not the thing, and meanings exist in people, not words. The proper order of evaluation consists of events first, then words. People who respond primarily to words rather than the reality they represent are intensionally oriented and this will cause them grief. But they need not despair. They can overcome their conflicts and confusions by becoming more extensional in their orientations. But even if we learn to prevent words from coloring our perceptions of events, we will never see those events as they "really are", because perception is a transaction between events and their perceiver. "The universe is a product of the observer as well as the observed."

An awareness of these matters will help us lead more effective and satisfying lives. But awareness in and of itself is not enough. Our grasp of map-territory relationships must be more than superficial, and we must change our language maps or habits so that they are more compatible with reality. The general semanticist is convinced of the validity of the Sapir-Whorf hypothesis, namely, that a person's language is a major determinant of his thoughts and actions. Hence, a person who changes his language habits so that they are more congruent with real events will begin to think and behave in a more satisfactory way.

Certain language habits are bad because they contradict what we know about the non-verbal world. Allness statements, ignorance of the abstracting process, a failure to report enough similarities or enough differences between events, two-valued orientations, reacting emotionally to words, and acting as if words are the things they represent, are language habits that get us into trouble and that we should get rid of. Bad language habits should be replaced by good ones. We should index the labels we use, date our assertions, use etc. to remind ourselves and others that one cannot say all about anything, and hyphenate words that refer to processes that can only be separated verbally and not in the real world. We are admonished by the general semanticist to avoid the is of identity, to recognize that words are multiordinal, to note that meanings are in people, and to realize that dictionaries merely tell us how words have been used in the past. We are told not to think of words as events that can be considered apart from the sentences in which they appear, that all events are unique and that absolute sameness is impossible. The general semanticist believes that if we follow these and other principles and prescriptions, we will obtain a better fit between our language behavior and reality, and this will enhance our individual and collective chances for survival. So in a sense, General Semantics provides us with a therapy, a life style, and an explanation of language behavior.

It has been pointed out that General Semantics is based upon a scientific orientation, and perhaps, what might be considered scientific insight into language-thought-behavior relationships, but it seems to be apparent that the general semanticist is primarily a change-agent rather than a scientist. And for whatever reason, he seems to prefer it that way. At least, he has devoted much more of his time to teaching, advice-giving, consulting, and counseling than he has to scientific inquiry. I suspect, therefore, that the general semanticist is likely to approach general systems theory in a similar manner. Von Bertalanffy has stated that " . . .system theory should prove an important means in the process of developing new branches of knowledge into exact science."[4] Hence, it is conceivable that general systems theory will enable general semantics to become more of a science. However, I have a feeling that the general semanticist will be more interested in the relevance of general systems theory for individual adjustment and survival. He will also want to know if it supports his own principles and formulations.

I think the general semanticist who investigates general systems theory is likely to be pleased with at least some of the things that he finds. General systems theory, for example, offers an explanation of the expression "the whole is more than the sum of its parts," and it is non-elementistic in its approach. General systems theory is process-oriented. It deals not just with static elements, but with their interrelationships and it recognizes that a change in one part of a system influences all of the other parts of the system. General systems theory sees relationships between disciplines and identifies the implications of findings in one discipline for all other disciplines. This concern for the general implications of scientific theory and research is something the general semanticist has in common with the general systems theorist. A primary interest of the general semanticist is the relevancy of scientific findings for improving human interaction and adjustment. A knowledge of general systems theory should be helpful to the general semanticist in making these applications and in differentiating between those that are meaningless and those that are meaningful. And general systems theory itself provides the general semanticist with formulations that have implications for human survival. Systems, for example, can be closed or open. A closed system has no input from outside the system and therefore, is subject to entropy. Human systems are open systems that have input and output of matter and energy. They seek and maintain homeostatic or steady states and they interact with their environments. The general semanticist who is searching for the practical implications of all of this might conclude that to operate at an optimum level, individuals and groups must be open enough to allow for an adequate exchange of matter and energy, steady enough to avoid entropy, but tense and active enough to maintain health.

General systems theorists have already discussed some of the implications of their work in the area of psychopathology. Mental disease, according to Von Bertalanffy, can best be described as a disturbance of system functions, since isolated symptoms such as illusions and hallucinations are characteristic of healthy people as well as schizophrenics. "An individual is mentally sound," Van Bertalanffy asserts, "if he has an integrated universe consistent within the given cultural framework."[5] This implies that any therapy must take many aspects of the individual as a system into consideration. Von Bertalanffy points out that occupational therapy might be appropriate for an active individual who needs a reorientation toward the future more than he needs insight into his past. Frankl's logotherapy, therapy at the symbolic level, might be most appropriate for someone suffering from the meaninglessness of life.

The methodological implications of other system concepts such as homeostasis, finality, feedback, and the like could also be pursued by the general semanticist. I do not think that the general semanticist OUGHT to restrict himself solely to the applications of general systems theory. I am merely suggesting that the general semanticist seems to be most interested in the practical methodological implications of theory and research and that he is likely to approach general systems theory from such a vantage point.

REFERENCES

1. Wendell Johnson, 'People in Quandaries,' New York, Harper and Row, 1946.
2. Robert Nisbet, 'The New York Times Book Review,' April 5, 1970, p. 1.
3. See Hugh Stretton, 'The Political Sciences,' London, Routledge and Kegan Paul, 1969.
4. Von Bertalanffy, 'General System Theory,' New York, Braziller, 1968.
5. Ibid., p. 219.

A Systems Approach to Human Communication

JAMES C. McFARLAND

> To know this, and know that however ugly the parts appear, the whole remains beautiful. A severed hand is an ugly thing, and man dissevered from the earth and stars and his history. . . for contemplation or in fact. . . often appears atrociously ugly. Integrity is wholeness, the greatest beauty is organic wholeness, the wholeness of life and things, the divine beauty of the universe. Love that, not man apart from that, or else you will show man's pitiful confusions, or drown in despair when his days darken.
>
> Robinson Jeffers

Each human being, whether he realizes it or not, has a communication theory. He has some organized way of perceiving and ordering the constant flow of information he encounters from his environment. This process of receiving, codifying and sending information is something we all do, yet few of us understand the process by which we do it. Information about how men communicate with other men is as old as the written literature of mankind. Every great thinker, as well as every common man, has devoted some attention to what happens when he encounters another human being. In order to survive, we, of necessity, have had to develop some knowledge of how to successfully interact with our fellow human beings. Yet, because communication is such a common every day event, we have failed, to a large extent, to make any kind of systematic study, or advance any kind of general communication theory to account for human interaction. Our communication studies have, for the most part, been restricted to formal situations. We have studied extensively how men share ideas through literature, poetry and formal rhetoric, but a general theory of communication is yet to emerge.

The present essay is not an attempt to advance a general communication theory; it is merely a suggestion that if a general theory were to emerge, it would be based on assumptions that can be broadly classified as systems theory. The nature of systems theory and the assumptions it makes will, therefore, be outlined. How these concepts apply to human communication will then be considered. Finally, systems theory will be compared to other prevalent approaches to the study of communicative acts.

A general definition of a system has been given by Hall and Fagen as, "a set of objects together with relationships between the objects and between their attributes."[1] Miller defines systems as "bounded regions in space-time, involving energy interchange among their parts, which are associated in functional relationships, and with their environments.. . . .A system is all of a thing."[2] McClelland adds, "A system may be defined as an assembly of components, having identifiable properties, among which relationships are perceived. To a person who has yet to comprehend some relationship between components or between their properties, there is no system."[3] And finally, Rapoport says a system is "a whole which functions as a whole by virtue of the interdependence of its parts."[4] From these definitions we may conclude that systems consist of an aggregate of dynamic elements that are in some way interconnected and interdependent, that they continue to operate together according to certain laws and in such a way as to produce some characteristic total effect. A system represents some kind of activity and preserves a kind of integration and unity in terms of balanced configurations, patterns and organized wholes. The systems may represent relationships that move to infinitely smaller or to larger systems. Systems levels move from molecular, cellular, organismic, interpersonal, social, cultural and cosmological systems. When we begin to deal with the laws of how systems operate, we begin to deal with a general systems theory.

Basically there are two general ways of classifying, symbolizing and ordering the world around us. One is to look for parts, for Aristotelian substances, where elements are isolated and put into classes. The second approach deals with Galileon variables, where events are propositional and patterned in field relationships. The weakness of the first is that it sometimes doesn't see the forest for the trees, and the weakness of the second is that it can at times cut down the trees to save the forest. General systems theory is an attempt to bridge the gap between the two. The systems theorist does not limit his investigation to the smallest unit or the largest whole, essentially they are both only a part. He looks for what Murphy has called the "structured organism-environment field," and the dynamic transactions between all levels.[5]

Philosophers and scientists have long recognized the unity and interrelatedness of various systems levels. Plato speaks from a systems viewpoint when he maintains that the greatest truth is the harmony of the whole. It is only in recent decades, however, that we have seen a "General Systems Theory" emerge. We are witnessing in the mid-twentieth century men trying totally new ways of seeking integral awareness between part and whole. General systems theory is one attempt in this direction. Perhaps as much as a theory, the systems concept represents an attitude, a way of going about ordering knowledge and discovering it. More than a "theory" made up of axioms, the systems approach is an epistemology. The systems theorist does not attempt to create a taxonomy or to become like a brick layer, assembling parts to make a whole. He seeks more to be the architect who endeavours to understand the functional relations of the whole and the mutual dependence of the parts in interaction. While systems theory is interdisciplinary, it is not mere eclecticism. Again, unlike a brick layer, the theorist does not collect units of information and assemble them to create a whole. R. L. Ackoff, a practitioner of the systems approach, describes it as follows:

> The tendency to study systems as an entity rather than as a conglomeration of parts is consistent with the tendency in contemporary science no longer to isolate phenomena in narrowly confined contexts, but rather to open interactions for examination and to examine larger and larger slices of nature.[6]

It was in the late forties when Ludwig von Bertalanffy, a biologist, made public his idea of a General Systems Theory. Since that time significant applications of the systems concept, not always in the name of general systems theory, have been made in both the physical and social sciences. Many developments have encouraged the increased use of the systems approach; it would be impossible to list all of them, but the following events seem to be among the forces most important to the emergence of systems theory: (1) The development of parallel and related movements such as Cybernetics, Information Theory, Game Theory, Decisions Theory and Factor Analysis cross fertilized systems approaches and added methodological insights. (2) Improved research methods and insights of the social and physical sciences have revealed man as being part of many systems. It is, therefore, hard to determine the boundaries of any environment and to isolate events. When a dog chases his tail, is he reacting to his environment or to himself? (3) Concepts of multiple causation based on fields of simultaneously acting forces have replaced mechanistic notions of simple cause and effect relationships. A person is composed of, and carries within him, the influences of many systems, both physical, cognitive and social. This produces ideas of multidimensionality, mutual dependency, transaction, feedback loops and self regulating systems. (4) Rapid and tremendous changes in technology organization and complexification on all systems levels have produced the need for interdependencies and synthesis. Where does the man leave off and the computer or machine begin? (5) For utilitarian as well as for theoretical purposes, there is a need to make predictive statements. Traditional methods were not always able to make such predictions or to investigate other areas. Due to the interconnection of systems, a broader theory was therefore needed. (6) The overly specialized academic disciplines, yet the need to investigate common concerns, has produced an interdisciplinary movement requiring a broader philosophy of science. Rapoport concludes that, "The systems approach to the study of man can be appreciated as an effort to restore meaning (in terms of intuitively grasped understanding of wholes) while adhering to the principles of disciplined generalizations and rigorous deduction. It is, in short, an attempt to make the study of man both scientific and meaningful."[7]

The difference between systems theory as a theoretical or epistemological approach, and systems theory as an applied science as represented by Systems Engineering, Operations Research, or Human Engineering should be made. While many concepts are shared by both approaches, they are not the same. General systems theory seeks to understand the nature of systems in general, while systems analysis seeks to control the operation of a specific system and is more closely connected with modern automation.

BASIC ASSUMPTIONS

The fundamental premise of general systems theory is that one event is interrelated to a whole series of events; any part of life is made up of smaller and larger parts, a series of systems within systems. Poetic, mystical, scientific and social expressions have been made of this concept. John Donne says, "No man is an island, no man stands alone". If a piece of earth falls from the cliffs of Dover into the sea, then all of England is diminished. In other words, a change in one system will result in a change somewhere else. The poet Yeats expresses the problem of knowing where one system ends and another one begins, "O body swayed to music, O quickening glance. How shall I tell the dancer from the dance?" One part of the system cannot be understood in isolation from the other parts, but only as it is affected and affects other systems.

We can make three main generalizations about almost all systems; they are complex, dynamic, and studies in communication. Each of these aspects will in turn be elaborated upon. Scientists today tell us that our planet is a sea of buzzing complexity – when looked at closely enough it seems to resemble a multitude of dancing fireflies. It is in constant flux, motion, flow, process, and it is forever changing. Yet, in spite of its dynamic, complex nature, there is form and unity. The universe contains, or perhaps more properly IS, a communication system, a series of interacting elements.

Man has not yet seen the smallest or the largest participant in the various systems levels. The noted paleontologist and theologian, Pierre Teilhard de Chardin, has described systems levels in three terms: "plurality", "unity". and "energy". The scaling down of the earth is "unlimited". Yet, Teilhard points out that the "more we split and pulverise matter artificially, the more insistently it proclaims its fundamental unity".[8] Unlike Norbert Wiener, who sees entropy (the running down of the universe) as the dominant trend of matter,[9] Teilhard holds that as matter becomes more diverse and complex, it becomes more united and stable. As Sir Julian Huxley points out in the Introduction to Teilhard's book, the energy forces that hold together and perpetuate the universe are "an anti-entropic process, running counter to the Second Law of Thermodynamics with its degradation of energy and its tendency to uniformity. With the aid of the sun's energy, biological evolution marches uphill, producing increased variety and higher degrees of organization". Teilhard concludes that the universe by reason of its unimpeachable wholeness is a "system", a "totum", and a "quantum": a system by its plurality, a totum by its unity, a quantum by its energy, all three within a boundless contour. Further, "It is impossible to cut into this network, to isolate a portion without it becoming frayed and unravelled at all its edges. All around us, as far as the eye can see, the universe holds together and only one way of considering it is really possible; that is, to take it as a whole, in one piece". Life can be seen as a system of geological and biological cooperation. Teilhard calls it the "biosphere", a network of mutually interdependent organisms of the most astounding subtlety and complexity. The balance of nature is a harmony of contained conflicts.

When we speak of the inner relatedness of the universe, or its systems functions, this is meant in a very literal sense. Scientists now know that except for nerve cells, the cells that made up our body a few years ago are not the same cells that compose it today. The same chemical elements that compose our body might just as well make up a cloud, or a rock, or a piece of iron. Indeed each moment oxygen comes into our body, gives it life, and returns again to the cosmos. A molecule of oxygen may be a cloud one time, water the next, part of our blood stream the next, and it could end up burning our house down. Because these are all different manifestations of the same thing, we think they are separate; but science now knows that they are not separate, they are all connected in systems relationships.

Today writers in different fields are not only recognizing the connectedness of the physical universe, but are relating this phenomenon to the asertion that man too is connected to the rest of the univers. We have a physical scientist, David Bohn, who speaks of quantum theory and says, "The world cannot be analyzed correctly into distinct parts; instead, it must be regarded as an indivisible unit. . . ."[10] The biophysicist, Erwin Schrodinger, adds, ". . . .you – and all other conscious beings as such – are all in all. Hence this life of yours which you are living is not merely a piece of the entire existence, but is in a certain sense the whole".[11] A philosopher of science and religion, Alan Watts, says that man "is an organism inseparable from his social and natural environment".[12] Teilhard sees physical matter, mankind, and the spirit as interconnected and bound by a "genesis" into what he calls "The Divine Milieu".[13] The physicist and philosopher Lancelot Law Whyte speaks merely on the "unity" of the universe.[14] The psychiatrist and neurologist Kurt Goldstein refers to what he calls the "sphere of immediacy" which bridges the gap between the material world and the human world.[15]

The foregoing analysis has been made to show that the universe is not only dynamic, flowing and changing, but that it is a series of systems relationships. The universe does not have a spring that will one day unwind, but it is self regulating and self maintaining. It is never the same and it never repeats itself, rather it is involved in a creative advance into more refined and unified states. All matter is in a continuous state of evolution – while there is flux and change, there is order, unity, pattern and form. Since everything is related to everything else, there is multiple causation and complex interrelationships of forces operating between systems.

A couple of examples of complexity and dynamic multiple causation in systems functions may be given. One could say that the most frequent cause of automobile deaths are cars running into telephone poles; and therefore, if we remove the poles from the roadsides, we will eliminate deaths. This is reacting to an event with multiple causes as though there were a single cause. It ignores what might have happened before the car hit the pole and that there are other objects behind the pole. In other words, there are different systems levels that must be accounted for before statements about causation or prediction can be made. The pollution of our environment is another example of how systems relationship have been ignored. It seems that modern man, in his short-sighted concern with parts, has ignored the whole. The result is that our ecosystems are being seriously pushed out of balance. For instance, the hippopotamuses along the river banks in South Africa were shot. This caused the silt, which had previously been kept in motion by the hippos, to build up. The rivers then flooded which favored a proliferation of schistosomiasis-carrying water snails. The result is a serious public health hazard. Similar situations in each local area have produced what is becoming a national crisis. The science of "ecology", a systems-oriented approach, is attempting to deal with the problem.

There are basically two types of systems, open and closed. Open systems maintain themselves by a communication process that involves transactions with other systems in the environment. Closed systems maintain themselves by a communication process that is limited to the system. Since closed systems are isolated from their environment, they are subject to the Second Law of Thermodynamics.[16] According to the Second Law, systems move toward entropy, which is a running down or disordered state. At the same time, there is a counterbalancing tendency for the system to maintain itself or create, through feedback mechanisms, a steady state of "homeostasis". The feedback loop is the mechanism that determines the flow of materials within the system, but not necessarily outside it. Modern servomechanisms, automation, Weiner's[17] cybernetics, as well as many phenomena in living organisms are based upon feedback systems. In a feedback system the emphasis is on control and maintenance of the system.

There are two types of feedback that can be distinguished, positive and negative. Positive feedback tends to trigger more of the same kind of action, while negative feedback is more flexible to adjust to changing conditions within the organism. An example of positive feedback might be two persons under an electric blanket, each having controls that are switched. A is hot, B is cold; so B turns up the heat which makes A even hotter and A turns down the heat which makes B colder. The result is a perpetuation or continuation of the existing condition. If A and B had their right control, they could properly adjust their environment to their needs, which would represent negative feedback.

Both positive and negative feedback are necessary and useful in maintaining a system; some conditions need to be perpetuated and repeated, and others need to be adjusted and changed. For instance in physiological systems, the hormone "FSH" stimulates the female ovaries to produce eggs (positive); however, when ovulation occurs, another hormone "estrogen" shuts off FSH (negative) and prepares the uterus to receive the egg. The process of positive and negative feedback continues to regulate and meet the needs of the body as the unfertilized egg is rejected or allowed to mature as a human embryo. This same process is found in computer systems, "closed loop" programs (positive) are desirable for repeating tasks that must be done over and over the same way. "Open loop" programs can sense an error and make the necessary adjustments (negative).

In contrast to the feedback system, the open system is concerned with dynamic interaction, not only within one system, but between multivariable systems. All living organisms are open systems. While cybernetics deals with the feedback-type system, general system theory includes both feedback and open systems functions. The added characteristic of the open system is that it allows matter (or information) to be introduced from the outside. As material is introduced into the system, it is processed or converted and emitted in a new form. This process is referred to as "anabolism". Open systems are not limited to the establishment of internal equilibrium, but they are engaged in a continuous exchange and flow of materials that result in ever-changing states. The goal of open systems functioning is spoken of as "equifinality," as opposed to the "equilibrium" or steady states produced by feedback systems. Equifinality is equated with growth, or actualization of the system and can be seen as "negative entropy" in that it displays characteristics that are paradoxical and even contradictory to the Second Law.

GENERAL SYSTEMS THEORY AS A COMMUNICATION THEORY

No theory of communication exists that accounts for every event that can be spoken of as "communication." The following is not an attempt to provide such a theory. It is rather a statement of how some of the assumptions of general systems theory have been applied to human communication. As such, it represents a theoretical approach to the study of communicative acts. In one sense, once we have outlined a general theory of how systems operate, we have outlined a communication theory since, like communication, a theory of systems is an account of the dynamic interdependencies between systems levels. Both general systems theory and communication theory must account for transmission of knowledge and information in terms of shared perceptions between different systems. As such, systems do not merely provide an analogy for communication, but common operations within the limits of some obvious differences. Such concepts as "equilibrium," "equifinality," "homeostasis," "anabolism" "feedback," etc. apply not only to physical but psychological and social phenomena. General systems theory and a general theory of communication both represent unifying disciplines. As Kenneth Boulding has pointed out, however, a theory that says something about almost everything, says very little about anything. [18] Like general systems theory a communication theory should be general enough to apply to the realities of the world we deal with, yet be specific enough to have some usable content. That such a theory is possible and has practical value will be subsequently illustrated. Two qualifications should be made, first no pretense of a "general" theory accounting for all communication is being made. Systems theory is presented as a perspective by which communication can be studied. As such it is more general and inclusive than other current communication theories. Secondly, our concerns are limited to human communication.

Almost all of the present and past communication models and theories hold a mirror up to communicative events in an attempt to reflect what they are like. On some levels the same can be said for systems theorists. Human communication has been compared to such things as cells, computers, machines, games, mathematical formulae, dramas and marketing systems: in other words almost everything except man himself. We have elaborate theories about how men play games with one another, put on shows for each other, and control and persuade one another, but few theories of how men can transact with one another in growth-promoting ways. The result has been the proliferation of structural models that describe what men do to each other. Examples of structural models would be the cybernetic or the dramatistic theories. The alternative is a functional model that accounts for how a system operates and improves itself, rather than how it looks. It is being suggested that we can come closer to understanding communication if we hold, not a mirror up to it, but a prism. A structural model can tell you something about a part function, and on that level is useful and valid. We need descriptions, but in the end a description can only help us to maintain equilibrium with the way things are, it can even create self-fulfilling prophecies to keep them that way. A functional model is more adequate to deal with the whole system since it more closely resembles an open system. Functional models are more difficult to deal with and tend to be more abstract. For one thing, it is hard to draw a picture or put parts in square boxes with straight lines between them. A communication theory based on the systems approach would move in the direction of using man as an open system for its model.

In order to understand a communicative event one must look at all of the systems relationships involved. This is the most fundamental assumption in a systems-oriented approach. In other words one cannot understand the individual by studying him only; the system he is a part of must be studied also. Structural models that have been reductionistic, for instance, have tended to stress the individual or the part rather than the whole. On the other hand, the social interactionists have looked at society, another part, and have likewise failed to see the whole. Allport has commented on this problem, "If we tackle a subsystem, we lose the whole; if we pursue the whole we overlook the part functioning." He concludes that we need a "complementary of investigators and theorists."[19] A systems approach would encourage this sort of cooperative effort. It would also encourage enlarging our outlook or attitudes and creating larger gestalts.

If we are to enlarge or change our "images"[20] of communicative events, we must be aware of the levels of theoretical discourse that are possible to us. Boulding has developed these mental models consisting of two general types, mechanical and biological.[21] These levels can well represent the various systems levels. The first level consists of a "static structure" and is called the "level of frameworks." This level deals with the geography and anatomy of the universe and can be represented by a pattern of atoms, a blueprint or an organizational chart. The next level is that of the simple dynamic system and is referred to as the level of "clockworks." This is the predetermined dynamic structure, the world of mechanics and machines. This is the level of action and reaction, cause and effect, stimulus and response. Systems on this level generally maintain a state of equilibrium. The last mechanical system is the control mechanism or cybernetic system, called the level of the "thermostat." This is a self-regulating system, where causes may produce effects that react upon their own causes. This allows the system to transmit and interpret information, not just react to it.

The fourth systems level moves to the biological level of living of "open systems" which Boulding calls the level of the "cell." The cell suggests life, growth and multiplication; they are open systems, feeding on the resources of their environment. The fifth systems level is the "genetic-societal" level and is typified by the "plant". Its distinguishing characteristic is the division of labor among cells to form a cell society and a self-expanding structure. We now move to the level of the animal kingdom on the self-moving system. With this greater mobility and expanded sensory capacity, the information input is greatly increased producing an "image" or knowledge structure that is significantly different than the additive accumulation of the input. We now arrive at the "human" level where self-reflexive man can be considered as a system. Here, through man's capacity to symbolize his experience, "he not only knows, but knows he knows." While it is hard to separate man from his social organizations, Boulding adds this level to the human dimension. Man in the presence of others acquires a new dimension. The final level is the ultimate "transcendental systems" level of cosmic consciousness. That man can transcend himself and identify with something higher is a common assumption in all societies. Boulding concludes that, "beyond the fourth level it may be doubted whether we have as yet even the rudiments of theoretical systems." This means that we function and seek to understand our affairs mostly on the machine and cell levels. A systems approach to communication would allow for a much broader approach.

Applications of systems concepts are being made by several communication theorists and practitioners. A brief review of them follows. Dean Barnlund has said of the systems approach to communication, "It may not be the best, but seems to be the most promising."[22] In other words, even though there seems to be no ultimate model within reach, the systems approach seems to provide concepts that lead in the right direction. Based on the following postulates, Barnlund has proposed a "transactional" model of communication. (1) Communication deals with transactions rather than interactions or reactions. Messages containing meaning are not transmitted from one person to another; only cues are given off. Thus meaning is in the person, not the message. (2) Communication is dynamic, not static; it occurs as a happening rather than exists as an entity. All behavior is communication; once born into this world, there is no way to stop communicating. (3) Communication is continuous; it starts at birth and doesn't stop until death. (4) Communication is circular which means a person can develop meanings by himself since he has his own internal systems as well as external systems. A message is a function of all that has gone before and continues to affect all messages after it. (5) Communication is unrepeatable; the same message cannot be given twice. Communication does not occur as a closed loop, feedback process, but is modified each time it is processed by the system. (6) Communication is irreversible; you cannot go back, only forward. (7) Communication is complex, it deals with a multitude of variables. Barnlund's model, which results from these considerations, consists of a circular process of two persons transacting with a variety of public and private cues as well as a variety of verbal and non-verbal cues.

Perhaps one of the most direct applications of systems theory to communication has been made by Watzlawick, Beavin and Jackson.[23] Based on von Bertalanffy's notion that the principles of systems theory apply to all systems in general, they have used the same terminology and find parallel occurrences in human interaction. Interaction itself is seen as a system that involves an environment and subsystems. Properties such as wholeness, nonsumnativity, feedback, and the equifinality of open systems are found operating. In their approach it is the pattern in communication, as represented by a repetition or redundancy of events, that is stressed. "So in the analysis of how people affect each other in their interaction, we will not consider the specifics of genesis or product to be nearly so important as the ongoing organization of interaction. . . . The system is then its own best explanation."[24] How this type of analysis is applied is illustrated by an in-depth analysis of the play, "Who's Afraid of Virginia Woolf?"

Virginia Satir, who has worked in conjunction with Watzlawick, Beavin and Jackson and their associates, has developed a system of family therapy based on the systems approach.[25] The family itself is seen as a system, and dysfunctional family relationships are seen as a breakdown in communication on one of the subsystems levels. One part of the family operates in conjunction with all other parts of the family. Because of this, it helps little to treat one person when it is the whole system that is dysfunctional. A pathology in one part of the system can be seen as the systems way of balancing out a pathology in some other part of the system. Thus it is the relationship and communicative patterns that must be understood and dealt with.

Gordon Allport has used the open system as a basis for his personality theory.[26] His analysis ends in a similar manner to Boulding's, i.e. we have not developed theories of man that fully account for him as an open system. In other words an open system must become more than what he already is, not merely maintain what he is. They emphasize "being rather than becoming." Like our theories of personalities, our communication theories have not yet accounted fully for human growth; they emphasize permanency rather than change, maintenance rather than progression, pleasure and success rather than meaning. Man needs to know the whys, wherefores, and nature of his existence. In other words, he needs to know where he fits into the system or pattern of that which surrounds him. When he is cut off from other systems' functions, he becomes frustrated, alienated or neurotic. As an open system, man is capable of giving and receiving in a transactional way; his relationships are reciprocal, mutual and democratic. When he has access to all systems levels, he is in a better position to transact and communicate meaningfully and totally.

One final area has made considerable use of systems concepts to deal with communication problems; this is the area of organizational psychology and communication. Seiler,[27] Bennis,[28] and Schein[29] have viewed the organization as a system consisting of multiple interacting subsystems, multiple purposes, dynamic environment and vague boundaries between systems. When viewed in such a way, it is obvious that a change in one systems level will result in a change somewhere else, and as with the family, you can't understand any part of the system without having some concept of the total systems functioning. To achieve good interpersonal relations in an organization, Bennis, et al. feel that the system must have the following characteristics: (1) It should be "open". (2) The Influence Structure should be balanced and characterized by interdependence. Basically this refers to transactional types of relationships rather than interactional or reactional types that have previously been expressed. (3) Decisions should be made through consensus. This characteristic expresses a transactional systems relationship. (4) The communication structure should maximize clarity. This is accomplished through open feedback-type relations between all systems levels. (5) Finally, the system must have adequate reality testing mechanisms. It must have some way of establishing 'truth" about its internal and external systems relationships and functioning.[30]

There are several general themes that seem to run through the foregoing examples of systems theory applied to human communication. By way of summation, these general principles can be highlighted: (1) Any communication event can be viewed as a system and displays systems characteristics. This applies to the intrapersonal, interpersonal, small group and societal levels. (2) When seen in terms of systems communication, it can be a transactional event in addition to an interactional or reactional event. The word "transaction" is being used to represent the human functioning on the highest human systems levels as outlined by Boulding. It has been defined by English as a "psychological event in which all the parts or aspects of the concrete event derive their existence and nature from active participation in the event."[31] Hormones can cause reactions, thermostatic feedback mechanisms can interact, but only a fully-functioning human being can transact in the fullest sense of that word. (3) A transaction is represented by the dynamic outcome of open system functioning. In other words, each time the system processes a message, the outcome is new and unique. This circularity does not imply repeatability or irreversibility. (4) The systems approach looks for the whole rather than the part. It thus accounts for communication occurring on all systems levels. (5) It is the form, pattern or relationship that is important, not isolated structural components. (6) The systems approach accounts for multivariable events occurring on any systems level. (7) Effective communication occurs when all systems levels are transacting in a mutually interdependent way.

How these principles operate in an actual communicative situation can be illustrated by a hypothetical interchange between person A and person B. For our purposes, communication can be defined as the sharing of perceptions. Person A is able to perceive and understand the systems function of person B and vice versa. It is the process of making a perception common to both A and B, of making them of one mind – not necessarily changing their minds, but creating mutual understanding. A sends his message to B in full awareness of what it is like in B's system; B responds with an awareness of A's system. This is the process by which humans maintain themselves internally and extend this inner function to external transactions. This process not only moves towards states of equilibrium, which often produce defensive behavior, but towards equifinality, which produces growth behavior. A perception is a systems function between an event, the sensory input, and the codification in the central nervous system. Perceptions are not transferred in toto from one mind to another. It is only through trial and error of open systems functioning that A knows when B has shared his perception.

The systems function between A and B may be thought of as relatively open or closed. If the system is open, there will be a free flow of information between A and B. This is an indispensable condition if perceptions are to be shared. B will be free to listen to A's message knowing that he will be free to respond to it in shared, honest and open transaction. In the achievement of optimum levels of communication through the use of an open system, several concepts are involved. First, a psychological contract must exist between A and B. A expects to be trusted by B and is willing to return the trust. If the contract that A and B agree to operate under is open, they will have a better chance of knowing when perceptions have been shared. If one of them breaks the contract, the system becomes closed and they have no way of double checking when a common mind has been achieved. A talks to X and B responds by referring to Y. The closed system violates the norms of reciprocity. What is reciprocated is not based on what is sent. Psychological contracts and norms of reciprocity imply not "interaction," which is one way, but "transaction" which is two way. Two way, open, transactional encounters involve feedback; that is, they are self reflexive and self regulating. They can adjust and adapt to each other. A one-way, closed system operates from a frame of reference of preconceived assumptions that may not fit the event and has no way of referring back to the event to verify whether it matches or not.

In interpersonal communication systems, both positive and negative feedback are found. In social interaction, however, there are not two situations the same, and one behavior can seldom be repeated in exactly the same way in any two situations or between any two persons ($event_1$ is not $event_2$). As a result, negative feedback is most frequently needed to meet the demands of changing situations. However, an open system can serve the need to perpetuate desirable conditions as well as shut them off when they are no longer appropriate. A closed system will perpetuate its errors, or fail to take advantage of favorable trends. So, if perceptions are to be shared, the interaction between A and B must be flexible and adaptive to meet changing needs.

Another difference between intrapersonal and interpersonal systems is that the former tend to be based more on stimulus-response relationships. Social systems, on the other hand, do not have hormones that flow and automatically regulate the balance between positive and negative feedback. Man must rely on his free agency, his free will and his conscious ability to choose. He is self reflexive and self regulating. Systems theory tends to deny stimulus-response, single cause relationships. If perceptions are to be shared and A is operating as an open system at an optimum level where there is a free flow of information, he will be sensitive to the system function of B. He will understand any event in terms of all of the systems that are interacting upon it; he will take into consideration human will, emotion and reason, physiological functioning, and social influences. Open systems are not used to make things happen, but to have the sensitivity and information to understand the multiple causes of any event. If A and B are to share perceptions, they must flow WITH each other, not do things TO each other; they must, in other words, reciprocate and transact. We cannot know all of the functions of a system, but if we have a sense that there are many functions as well as dysfunctions, we are in a position to be curious. Being curious produces inquiry, which stimulates transactions which creates understanding.

We have mentioned that the three basic steps in the communication process are import, processing, and export of messages. We will now analyze them more fully in terms of (1) what has been said about open systems, and (2) as a way of understanding the process of sharing perceptions or optimum communicative functioning.

We must cope with information and adapt to it in a non-repeatable, circular fashion. This process has been referred to as the "adaptive-coping cycle." The first step is to sense that some message has entered the system. A and B must be good diagnosticians. They must sense the changes in the internal and external environment. This requires a curious, inquiring attitude, an openness and willingness to discover and test reality as it relates to the current event. A and B must be able to understand their proper identity and their relationship – What are their roles? How is the interaction to be structured? What can be said? On what topics? What are the motives and assumptions behind the interaction? An open system looks for these things and is, therefore, more likely to share perceptions. A closed system presupposes them, or takes them for granted, or is threatened by them, and is, therefore, less likely to reach a state of commonness with those he interacts with.

The second step in the adaptive-coping cycle is to be able to process this information and translate it into the system. The incoming information must be codified in such a way that it will fit existing internal cognitive systems. There must be a stable, balanced, congruent level in the system. This above all requires an open attitude of flexibility and adaptability. Rigidity produces premature closure and causes messages to be ignored and distorted because they are seen as threats.

The third step involves the ability to act on or respond to the message. Until there is a discriminating response, meanings have not been transferred. When messages are fully listened to and heard, the response will be based on the message, not some tangential meaning. The exporting of new messages, expressions, and actions include, and are in line with, the original environmental event. Again, this is the difference between interactions which may bounce off each other or pass on by, and transactions which integrate and create entirely new meanings because they include something from A and B and the event they are discussing.

The fourth step is double checking through feedback mechanisms on the success or effect of the interaction. Is the message sent the same as the message received? Through trial and error and continued interaction this question may be answered. When the system includes a psychological contract of openness and trust, A and B will not be afraid to double check with each other. Distrust will create defensive behavior and close meaningful interaction. Therefore, the possibility for sharing perceptions will be diminished.

Carl Rogers has used the word "congruence" to refer to an "accurate matching of experiencing and awareness" as well as "communication." In other words, perceptions are likely to be shared when there is a congruent relationship on all systems levels, both internal and external. To illustrate how the adaptive-coping cycle operates more directly in actual interaction we refer to twelve statements Rogers makes about a hypothetical interpersonal interaction: [32]

1. Any communication of A to B has some degree of congruence.
2. The greater the congruence the clearer and less ambiguous A's message will be.
3. Because of this, the more B will respond with clarity.
4. The more congruent A is the less he will have to defend himself about. When he doesn't present a facade to be defended, he can listen more accurately to B's reply.
5. This empathy on the part of A makes B feel understood – A sees B as B sees himself.
6. Because B feels understood, he experiences positive regard towards A.
7. To the degree that B experiences A as congruent, positive and empathetic toward him, it will produce a helping, facilitating, mutually satisfying relationship.
8. To the extent that B experiences this kind of a shared relationship, he will have fewer barriers to communication, he will present himself more as he is, and his defensiveness will decrease.
9. B can now listen more accurately to A and will not distort A's communication. This is a repetition of Step 4, but now in terms of B.
10. To the degree that B is able to listen, A feels empathetically understood. The process now repeats itself only in terms of the other person. The result is that it becomes reciprocally satisfying.
11. Reciprocity produces greater unity, integration, less conflict, more energy utilizable for effective living, change in behavior in the direction of greater maturity, greater sharing of perceptions.
12. The introduction of threatening messages into the process, which touch on areas of incongruence, produce defenses and the whole process begins to occur in reverse.

In summation, we can see that the systems point of view in interpersonal relations encourages us to maximize the reciprocal, shared and interrelatedness of the relationship. The goal is to understand, not manipulate. Since life itself, or any event we single out, is composed of a series of systems, we cannot separate or divide parts of the system so that one seeks to dominate or eliminate another part. To do so creates anxiety, defensiveness and interpersonal alienation. An understanding of systems and systems functions encourages oneness, commonness and reciprocal functioning. It encourages us to see the fellow humans we interact with on interpersonal levels, not as parts or objects to be acted upon, persuaded, changed, controlled or told, but as fellow travellers with whom we can achieve understanding, share messages, and engage in mutual transactions as we accompany them through the various systems levels of interaction.

RELATIONSHIP BETWEEN SYSTEMS THEORY AND OTHER COMMUNICATION THEORIES

Whether it be implicit or explicit, every major philosopher, poet, rhetorician or novelist has a theory of communication. Certainly no attempt will be made here to relate all of them to systems theory. There are, however, a rather limited number of major movements that communication scholars have turned to in order to understand how men share ideas with one another. Systems theory will first be related to the two major methodological positions, historical-critical and experimental-empirical. It will then be related to each of the major theories of communication.

SYSTEMS THEORY AND METHODOLOGIES

While several methodologies can be applied to the concepts of a theory, most methodologies have built-in theoretical assumptions. To some extent the theory produced will reflect the method used to produce it. Methodologies can be lumped into two broad categories: critical-historical and experimental-empirical. In terms of theoretical assumptions, it is possible to make some generalizations about these two theories. The critical-historical method has tended to look at the big picture without careful attention to the parts that make it up. The experimental-empirical approach has tended to reduce variable to isolated parts without careful attention to where they fit into the big picture. As previously stated, one of the purposes of the development of general systems theory was the unification of this sort of fragmentation or what Joseph Royce has called "encapsulation."[33] When systems theory is viewed as a methodology, it allows for both approaches and encourages an interaction between them. The primary implication for research of systems theory would be the dynamic interrelationship between part and whole rather than emphasis on either one. Multiple variables and continuous relationships would become the focus of attention rather than separation and fragmentation.

The implications of systems theory for the historical-critical approach to communication can now be considered. The rhetorical critic can operate from a systems point of view. According to Hillbruner he looks at both the intrinsic as well as the extrinsic factors in a communicative situation.[34] He assesses the speaker and the audience in terms of values, beliefs, and philosophies; he looks at the current social conditions in terms of the intellectual milieu and the current and past stresses and strains of society. He can take into account aesthetic and subjective concerns. From a systems point of view, he would then look for the dynamic relations between these variables in terms of patterns and forms emerging, not simple cause and effect relationships. The rhetorical critic has not always operated in this way, however. As Hillbruner has pointed out, "Most of the so-called criticism today really is historical scholarship." A systems-oriented approach would move much beyond this limited type of investigation. In an interdisciplinary spirit he would not only be concerned with history, but philosophy, politics, psychology, social psychology, sociology and cultural anthropology. As Ness has pointed out, the study of communication should be, "interdisciplinary and pluralistic. Rhetorical criticism, for example, involves historical interpretation, for it must examine the event and all the actions and transactions of the 'moment'."[36] It involves sociological interpretation, for it must determine the value systems operative in the event. It involves philosophical interpretations, for it must be concerned with ethical standards. And it involves rhetorical interpretation, for it evaluates the accommodation of ideas to men and men to ideas. . . . Ideas and concepts, too, are not held in a vacuum."[37] Ideas do have consequences and effects beyond the mere expression of them as Richard Weaver has stressed.[38] The systems approach would provide a framework in which to consider ideas and their relationship to men. "To see only one facet of a diamond is not enough for total comprehension and appreciation of its beauty. So, too, with the examination of a speaker and his ideas."[39]

A systems orientation would allow the traditional rhetorical critic to move beyond a strictly speaker-audience type of concern. He might look for systems operations in terms of maintenance and growth as expressed in such phenomena as the rhetoric of silence, protest, urban blues, rock music as well as other linguistic and paralinguistic phenomena.

A systems approach would tend to unite the critic with the empiricist; they would tend to move toward each other, thus enhancing each other. The experimentalist need not leave his laboratory, but he may be forced to relate more directly his findings to other systems functions. Rather than the reduction of variables, he would have to see the mutual dependency of multiple variables. When pushed too far, his method of controlling variables runs counter to the systems viewpoint. If one does succeed in controlling all of the variables, he has disrupted the systems balance and can, therefore, only make statements about subsystems rather than the relationships between systems. As Rapoport has pointed out, the scientist can no longer answer questions like, "Why is the water in the kettle boiling?" with answers like, "Because the kettle was placed over a fire."[40] Complex human interaction is certainly much more complicated to deal with than such seemingly simple phenomena. In other words, the empiricist will have to reconsider simple cause and effect relationships in terms of multiple causality. This does not mean that science will be abandoned, but rather as Rapoport holds, a systems approach would be "built on recent developments which show promise of re-establishing holistic approaches to knowledge without abandoning scientific vigor. The analogues established or conjectured in systems theory are not 'mere' metaphors. They are rooted in actual isomorphisms or homomorphisms between systems or theories of systems."[41] For instance in systems-oriented research, questions such as, "What leadership styles are best?" or "who solves problems best, groups or individuals?" would be less likely than questions involving functional process.

SYSTEMS THEORY IN RELATIONS TO OTHER COMMUNICATION THEORIES

In the classical period some of the greatest minds of all time devoted themselves to the study of communication. Plato, Aristotle, Cicero and Quintilian all considered rhetoric important enough to devote major treatises to it. With the possible exception of Plato, their primary concern was with formal, public discourse as it occurred in political and legal deliberation. Due to this limited emphasis, the classical approach fails to emerge as a general theory of human communication.

The classical model can be represented by Aristotle; indeed it was not until centuries later that any major departures were made from it. The Aristotelian model consists of a speaker sending a message to an audience. Obviously this is the core of any communication model and all others since are merely elaborations of Aristotle's format. Aristotle's treatment of the speaker deals primarily with logical and political concerns; the emphasis on the audience is primarily ethical and psychological, and his treatment of the speech or message deals with literary or stylistic matters. He develops a rather elaborate classification system for each of these parts. The emphasis is placed more on the typology of the rhetoric than on the function of the rhetoric. Linear static dimensions are stressed rather than the dynamic circular elements of communication. It is difficult to relate Aristotle's model to Boulding's systems levels; however in general it falls within the second level of clockwork systems. Certainly his treatment of psychological and ethical considerations would place his approach more on human systems levels; but even there, Aristotle tends to see man operating as a rational being and his discourse following set rules of logic, according to cause and effect relationships. As such, Aristotle's 'Rhetoric' is more concerned with the proper practice of a speaker and his persuasive appeals than offering a general theory about how communication in general operates.

The first major departure from Aristotle's model is represented by the dramatistic approach to communication as formulated by Kenneth Burke and elaborated upon by Hugh Duncan.[42] The Burkean model, with its emphasis on "identification" rather than "persuasion," is more similar to the systems model than the Aristotelian model. Burke emphasizes the enactment of "roles" for "social appeal." As a result of this observation, he feels that the best metaphor for communicative events is a drama.[43]

Burke feels that social or interpersonal dramas do not occur in a vacuum, they must be studied in the context of their social setting as an organic event rather than as a static event. The concern is not with observing any two segments, such as the relationship between the act and scene, but the function of the act itself. Burke centers his attention on the interrelationships of the communication rather than on any one aspect such as speaker, message or receiver. All this is in the spirit of systems analysis.

We have spoken of feedback-type systems and open systems. The Burkean concept of "identification" can be closely related to open systems functioning. Men identify with one another to overcome the divisions that exist among them; there must be unity or equilibrium. Identification, however, produces "con-substantial" states, which is, "not just trying to tell people how things are. . . it is trying to move people." Thus it is not only equilibrium that identification produces, but growth and change, as in an open system.

As we look at some of the other aspects of the dramatistic theory such as hierarchial roles, mystification, victimage and strategies, we find that they seem to relate more to equilibrium of feedback systems than the growth of open systems. They operate to provide what Duncan calls "social control"; it is a description of how the system maintains itself more than a prescription of how the system improves itself.[44] The dramatist model offers valuable insight into the understanding of human communication as it operates on a social level. It is less certain that it contains the kind of dynamism that helps us to improve communication. Thus in terms of systems theory we can say that dramatism shows some of the ways social systems maintain themselves, a necessary systems function, but not necessarily how they change themselves.

Closely related to dramatism, if not including it, is the social interactionist or role theory approach to human interaction. This tradition begins with William James who stresses the "social me" as the most important aspect of self. Influenced by James, Cooley developed a concept of the "reflected" or "looking glass" self. Thus Cooley, along with James, defines self largely in terms of what is reflected from the interactions with others in one's environment. Following this tradition, Mead speaks of man as essentially a role-taking animal; mind and self are both products of social experience. Man's ability to communicate is measured by his capacity to assume the role of another person. Through this process of "role playing" and "role taking" with "significant symbols" and "generalized others," man develops a self.

Current statements of this general position can be found in the writings of Erving Goffman, who places his emphasis on situated social interaction. In a similar manner to Burke, he views man's interactions as performances in a social setting. To Goffman not only the nature of communication but the very nature of man himself is defined in terms of the situation. "Not, then, men and their moments, rather moments and their men."[45] "A self, then, virtually awaits the individual entering a position; he need only conform to the pressures on him and he will find a ME ready-made for him. In the language of Kenneth Burke 'doing is being'."[46] Without a stage there is no man. To Goffman the "proper study of interaction is not the individual and his psychology, but rather the syntactical relations among the acts of different persons mutually present to one another.[47] In other words, Goffman de-emphasizes the smaller part in favor of the larger part.

When Goffman's position is related to a systems approach we see that he has placed the emphasis with a subsystem and defined the whole in terms of that subsystem, rather than defining the whole in terms of the dynamic interplay between all systems levels. As Deutsch and Krauss conclude, "Although Goffman touches on the importance of the relationship between the 'performed self' and the 'phenomenal self' (the self one experiences), it is probable that he makes too little of it."[48]

On the other hand, there is the existential-phenomenal point of view of interaction, which looks at communication in terms of "being" not "doing." In this approach the individual person becomes the focus of attention. Abraham Maslow, who in part represents this position has said, "It is my vague impression that communication is studied too exclusively at the sociological level and not enough at the biological level."[49] While Maslow begins by stressing man's "natural and intrinsic inner nature," he goes on to point out how this inner self, through interaction with other people moves toward growth, maturity and actualization.[50]

In Carl Rogers' theory of personality we find a similar position. "Every individual exists in a continually changing world of experience of which he is the center."[51] The organism transacts as an organized whole to this phenomenal world in a goal-directed attempt to satisfy its needs as experienced and perceived in the field. In a similar vein Gardner Murphy has outlined a "field theory of communication."[52] He begins the examination of the interaction process on the biological level, but extends it via a "pattern-for-pattern isomorphism" to other systems levels, including the social Murphy has summed up the position well as follows.

> As one inspects the central nervous system or the endocrine system, one finds that contributions to any given symbolic activity, whether expressing understanding or overt communication, are bound together with all the other centers which control physiological activity. . . . Man is, then, a nodal region, an organized field within a larger field, a region of perpetual interaction, a reciprocity of outgoing and incoming energies. . . . The life history of the person is a history, not of events within the skin of the individual, but of transactions between the individual and his environment.

Another point of comparison, from the systems perspective, between the social interaction and the existential point of view centers around how a system maintains itself. Rogers speaks of "congruence" in the system when there is an "accurate matching of experiencing and awareness. He is one unified person all the way through, whether we tap his experience at the visceral level, the level of his awareness, or the level of communication."[54] In other words, all systems levels are connected in symmetrical ways. All parts of the system are open to exchange information with other parts as well as the environment around it and can therefore operate as an open system.

On the other hand Goffman offers the concept of "role distance" which is a "wedge" between doing and being. It allows a person to take a "double stance" to disidentify the real self from the role one is required to play. The performer denies, not the role, but some self that might be implied by the role. To Goffman,

> It is true that here and there we can pounce on a moment when an individual sits fully astride a single role, head erect, eyes front, but the next moment the picture is shattered into many pieces and the individual divides into different persons holding the ties of different spheres of life by his hands, by his teeth, and by his grimaces.[55]

Since we can seldom get with it fully, we must use the manipulations of performance, concealment and defense in our communicative behavior in order to maintain the system. This concept of "role distance" is not at all the same as "congruency." From the systems viewpoint we have already illustrated how perceptions can best be shared – in a situation where there is openness and transparency. In such a system, maintenance is achieved by creative growth and change. The system that uses performances, strategies and the defense mechanisms of role distance is more concerned with equilibrium in the system than with growth of the system. The systems point of view would recognize legitimate times to use these methods, but it would not stop there – it would go on to methods of growth.

In connecting all systems levels together in an organism-environment field in dynamic transaction and allowing for systems maintenance by the kind of goal-seeking growth activities we find in open systems, the Maslow, Rogers, Murphy position well represents a systems theory approach. Other, more traditional, existentialists may not typify this position to the same degree. For instance the European existentialists may tend to define man as separate from his environment and see only the organization within the individual organism. To the extent that they do this, they could not be considered as being consistent with the systems approach. Like the social interactionists, who emphasize the systems organization within the environment, they have not accounted for all systems levels. A systems-oriented approach would look for the cross-organization "a structured organism-environment field."[56]

Another major position that personality theories, learning theories and communication theories have developed around are the reinforcement or social learning theorists. There are three primary orientations to be considered here; behaviorism, which deals with stimulus-response reactions; associationism, dealing with integration of elementalistic units; and hedonism, which stresses the pleasure-pain principle. In terms of the systems levels developed by Boulding, these approaches seem to be using models of human behavior based on the lower animals and machines. Systems theory, as it has been developed by the writer, is a more humanistic approach to communication. There can be little question that some of these approaches explain how an organism learns information; it is less certain that they account fully for how incoming information is processed in an open system. Much of human communication goes beyond the principles of utility, stimulus-respose and conditioning. A behaviorist may be able to condition the phobias out of a man and otherwise repair the machine, but he may not be equipped to prevent the man from committing suicide before or after the repair job. When you are dealing with man in a mechanistic way, as a "thing", to be controlled and manipulated, it is easy to lose touch with moral and ethical considerations.

Behaviorists tend to see human behavior in terms of cause and effect, billiard ball, domino relationships. Systems theory, as a more general theory, accounts for stimulus-response phenomena but goes on to allow for open systems functioning in terms of dynamic transactions. Von Bertalanffy has summed up mechanistic views as follows;

> The organism is essentially considered to be a reactive system. Outside stimuli are answered by proper responses in such a way as to maintain the system. The feedback model is essentially the classical stimulus-response scheme, only the feedback loop being added. However, an overwhelming amount of facts show that primary organic behavior as, for example, the first movements of the fetus, are not reflex responses to external stimuli, but rather spontaneous mass activities of the whole embryo or larger areas.[57]

Another expression of the machine theory of human behavior is the utilitarian conception based on an economic outlook of human motivation. The organism is seen as a balance sheet, weighing pain against pleasure, profit against loss, reward against punishment. Out of this concept has developed the Game Theory approach to communication that has been dealt with by Homans[58] Thibant and Kelley[59] and Rapoport[60] as well as many current researchers. The fact that on some systems levels man operates this way is not to be disputed, certainly an organism seeks balance and control in these areas. This approach does not, however, account for the communication that is most uniquely human, the interactions produced by open systems and motivated by what Maslow calls "meta-needs".[61] In summary, we can say that reinforcement-oriented theories have not moved beyond the homeostasis principle.

Mention has been made of Information Theory as outlined by Shannon and Weaver[62] and the Cybernetics of Norbert Wiener.[63] Both theories have made important contributions to communication theory in that they moved away from purely mechanistic views by introducing feedback loops of self-regulating systems. To this extent these theories run parallel to systems theory. They do not, however, account fully for the dynamic, circular, transactional nature of human communication. Communication is seen as control and control can only produce equilibrium, not growth. As such these theories are subject to the Second Law of

Thermodynamics. Again, systems theory allows for open systems, not only feedback systems, and, therefore, is not fully subject to the Second Law. As Boulding concludes,

> Information Theory as it stands today is by no means adequate to bear the burden of providing the necessary abstractions (for a general theory). The information concept of Shannon is merely a statistic. . . .It is not a measure of any mysterious or impalpable quality of a message. . . .It is a convenient statistic for use in certain problems in solving the communication of messages over limited channels. In and of itself, it is no more than this.[64]

One final approach to communication will be considered – general semantics. While some approaches to language deal primarily with word usage, general semantics is not concerned with language as such; rather its concern is with how language affects us. In the limited sense, "semantics" deals with the accuracy of information in a manner similar to Information Theory. In the more general sense, semantics concerns itself with all phases of communication. In systems terms, general semantics can be said to stress the lowest levels or "event" level. Both Korzybski's 'Semantic Differential'[65] and Wendell Johnson's communication model,[66] stress what goes on inside a person's head as he abstracts from the outside world. This "extentional"level of abstraction seems to be stressed more than the social context of communication.

Korzybski's system is idealistic and prescriptive. While it attempts to be scientifically based, his great concern was to distinguish man from animal. Korzybski saw man as the "time binding" class of life and animals as the "space binding" class of life. He stresses the point that his system is "non-Aristotelian" in that it attempts to move away from static classes to more dynamic concepts. This idea encourages linguistic relativity or the "word is not the thing" idea; the "non-allness" idea; and the "self-reflexive" concept.

> In spirit, general semantics is for the most part compatible with general systems theory. From the perspective of at least one general semanticist, Samuel Bois (1966) the two movements virtually encompass each other. Bois has been influenced directly by Boulding, von Bertalanffy and other Systems writers. To the extent that Bois, and others of his mode of thought, continue in this approach to General Semantics, there is the prospect of a fruitful cross fertilization between the two movements.

We may conclude that there are few truly "general" theories of communication. While systems theory has not yet produced one, it seems to be working in the direction of providing the broadest possible outlook, yet still have some usable substance. Most theories have taken one of the parts and have built a theory around it. As a result, they accurately describe a certain aspect of communication; that is, they operate on one systems level, but they do not account for communication in general. A communication theory based on the systems approach would attempt to combine all systems levels, not as an assemblage of bricks that are added together to create a whole, but a combination of non-additive processes that create a dynamic unity. The theories reviewed in this essay mostly dealt with man on a limited systems level, i.e. man as a persuader, man as an actor, man as a machine, etc., rather than man as a fully functioning human being.

Ten years ago, Johnson and Klare reviewed the communication models developed during the previous decade.[67] One of their conclusions was that "most of the models are related to Shannon's general communication system".[68] In other words they are basically feedback systems rather than open systems. It is interesting to note that a current textbook on communication limits its discussion to feedback type, 1950's models.[69] The two models mentioned in this essay which most closely resemble open systems models are the Barnlund transactional model[70] and the Bois semantic reactor model.[71] At the present time neither of these models enjoy widespread recognition or acceptance. In addition to these models, the humanistic orientation of Rogers, Maslow and others is beginning to encourage thinking about communication in terms of open systems.

Man, in his most human sense, operates as an open system, moving toward growth. However, most current theory, research, and practice views man as operating on a feedback level, with his goal being systems control, maintenance, and stability. With a couple of exceptions this condition is reflected in the theories reviewed in this paper. We may conclude from this that communication theory which expresses man's full humanness is yet to emerge. It is predicted, however, that current trends will continue to move towards theories and models that reflect man's full humanness.

REFERENCES

1. Hall, A. D. Fagen, R. E., "Definition of Systems", 'General Systems', I, (1956), pp. 18-28.
2. Miller, James G., "Toward A General Theory for the Behavioral Sciences", 'The American Psychologist', X (1955), pp. 514-515.
3. McClelland, Charles A., "System and History in International Relations. Perspectives for Empirical Research and Theory", 'General Systems', III, (1958), pp. 236.
4. Rapoport, A., "Foreword", in 'Modern Systems Research for the Behavioral Scientist', ed. by Walter Buckley, Chicago: Aldine Publishing Co., 1968, p. xvii.
5. Murphy, Gardner, "Toward a Field Theory of Communication", 'Journal of Communication', XI 1961, 196-201, p. 199.
6. Barnlund, Dean C., 'Interpersonal Communication: Survey and Studies.' Boston: Houghton Mifflin Company, 1968, p. 11.
7. Rapoport, A., "Foreword", p. xxii.
8. Teilhard de Chardin, Pierre., 'The Phenomenon of Man', New York: Harper and Row, 1959, pp. 40-43.
9. Wiener, Norbert., 'The Human Use of Human Beings'. Garden City, New York: Doubleday and Company, Inc., 1954.
10. Bohn, David., 'Quantum Theory'. New Jersey: Prentice-Hall, 1958, pp. 161-162.
11. Schrodinger, Erwin, 'My View of the World'. Cambridge: University Press, 1964, pp. 21-22.
12. Watts, Alan W., 'The Joyous Cosmology'. New York: Vintage Books, 1962, p. 94.
13. Teilhard de Chardin, Pierre., 'The Divine Milieu'. New York; Harper and Row, 1965.
14. Whyte, Lancelot Law., 'The Next Development of Man'. London: Cresset, 1949.
15. Goldstein, Kurt., 'Human Nature in the Light of Psychopathology'. New York: Schockon Books, 1966.
16. Bertalanffy, Ludwig Von., "General System Theory – A Critical Review", in 'Modern Systems Research for the Behavioral Scientist', ed. by Walter Buckley, Chicago: Aldine Publishing Co., 1968.
17. 'The Human Use of Human Beings'.
18. Boulding, Kenneth E., "General Systems Theory – The Skeleton of Science", in 'Modern Systems Research for the Behavioral Scientist', ed. by Walter Buckley, Chicago: Aldine Publishing Co., 1968, p. 3.
19. Allport, Gordon W., "The Open System in Personality Theory". 'Journal of Abnormal and Social Psychology,' LX (1960), p. 309.
20. Boulding, Kenneth E., 'The Image'. Ann Arbor, Michigan: The University of Michigan Press, 1965.
21. "General Systems Theory – The Skeleton of Science", pp. 6-10.
22. Barnlund, Dean C., Class notes, "Communication Theory", San Francisco State College, 1967.
23. Watzlawick, Paul, Beavin, Janet, and Jackson, Don., 'Pragmatics of Human Communication'. New York: W. W. Norton & Co., Inc., 1967.
24. 'Ibid', pp. 127-129.

25. Satir, Virginia, 'Conjoint Family Therapy', Palo Alto: Science and Behavior Books, Inc., 1967.
26. "The Open System in Personality Theory."
27. Seiler, John A., 'Systems Analysis in Organizational Behavior'. Illinois: Richard D. Irwin, Inc., and The Dorsey Press, 1967.
28. Bennis, W, 'Changing Organizations', Boston: McGraw-Hill Book Co., 1966.
29. Schein, Edgar H., 'Organizational Psychology', New Jersey: Prentice-Hall, Inc., 1964.
30. Bennis, W., Schein, Edgar, Steele, F., Berlew, D., 'Interpersonal Dynamics', Homewood, Illinois: The Dorsey Press, 1968, pp. 662-668.
31. Patterson, C. H., 'Theories of Counseling and Psychotherapy,' New York: Harper and Row Publishers, 1966, p. 561.
32. Rogers, Carl R., 'On Becoming a Person'. Boston: Houghton Mifflin Co. 1961, pp. 342-344.
33. Royce, Joseph R., 'The Encapsulated Man'. Princeton, New Jersey: D. Van Nostrand Co., Inc. 1964.
34. Hillbruner, Anthony, 'Critical Dimensions: The Art of Public Address Criticism'. New York: Random House, 1966.
35. 'Op. Cit'., p. 6.
36. Ness, O. G. "Contributions of Historical and Critical Studies to Communication Research", 'Journal of Communication', XII (1962), 160.
37. 'Op. Cit'., p. 82.
38. Weaver, Richard, 'Ideas Have Consequences', Chicago: Phoenix Books, 1948.
39. 'Op. Cit'., p. 96.
40. "Foreword", p. xiv.
41. 'Ibid'. p. xxi.
42. Duncan, Hugh D., 'Communication and Social Order'. New York: Oxford University Press, 1962.
43. Burke, Kenneth, 'The Philosophy of Literary Form'. New York: Vintage Books, 1941, 'A Grammar of Motives'. Cleveland: The World Publishing Company, 1962, 'A Rhetoric of Motives'. Cleveland: The World Publishing Company, 1962.
44. 'Communication and Social Order'.
45. Goffman, Erving, 'Interaction Ritual', Garden City, New York: Doubleday Anchor Books, 1967, p. 3.
46. Goffman, Erving, 'Encounters', Indianapolis: The Bobbs-Merrill Co., Inc. 1961, pp. 87-88.
47. 'Interaction Ritual', p. 2.
48. Deutsch, Morton, Krauss, Robert M., 'Theories in Social Psychology'. New York: Basic Books, Inc. 1965, pp. 210-211.
49. Maslow, Abraham H., "Isomorphic Interrelationships Between Knower and Known," 'The Human Dialogue', ed. by Floyd W. Matson and Ashley Montague, New York: The Free Press, 1967, p. 197.
50. Maslow, Abraham H., 'Toward A Psychology of Being'. Princeton, New Jersey: D. Van Nostrand Co., Inc., 1962, p. 3.
51. Rogers, Carl R., 'Client Centered Therapy'. Boston: Houghton Mifflin Co., 1951, p. 483.
52. "Toward A Field Theory of Communication".

53. 'Ibid', pp. 197-198.
54. 'Client Centered Therapy', p. 339.
55. 'Encounters', p. 143.
56. Murphy, 'Op. Cit'., p. 199.
57. "General Systems Theory — A Critical Review", p. 19.
58. Homans, G. C., 'Social Behavior: Its Elementary Forms'. New York: Harcourt, Brace and World, 1961.
59. Thibaut, John W., Kelley, Harold H., 'The Social Psychology of Groups', New York: John Wiley & Sons, Inc. 1959.
60. Rapoport, A., 'Fights, Games, and Debates', Ann Arbor: University of Michigan Press, 1960.
61. "Isomorphic Interrelationships Between Knowers and Known".
62. Shannon, Claude E., Weaver, Warren, 'The Mathematical Theory of Communication'. Urbana: The University of Illinois Press, 1964.
63. 'The Human Use of Human Beings'.
64. 'The Image', p. 154.
65. Korzybski, Alfred, "Introduction to the Second Edition", 'Science and Sanity', 4th ed., Lakeville, Conn.: The International Non-Aristotelian Library Publishing Co., 1958.
66. Johnson, Wendell, "The Spoken Word and the Great Unsaid", 'The Quarterly Journal of Speech', XLVII (1951). 419-429.
67. Johnson, F., Klare, G. R., "General Models of Communication Research", 'Journal of Communication', XI (1961), 13-26.
68. "General Models of Communication Research", pp. 26, 45.
69. Keltner, John W., 'Interpersonal Speech-Communication'. Belmont, California: Wadsworth Publishing Co., 1970.
70. Class notes, "Communication Theory".

A General Systems Model of Communication

DENNIS R. SMITH and BOB R. HUNT

One of the most popular approaches to the study of human communication today is referred to as the "systems approach." While there has been much discussion on the analysis of communication from a systems viewpoint, there is little agreement on how such an analysis can be accomplished. The communication phenomenon remains a highly complex phenomenon which does not lend itself readily to simple analysis. This paper describes in detail one application of a general system to the analysis of human communication.

As with all terms which reach popular usage, the meaning and implications of the term system have been constantly changing. In discussing the application of systems theory to communication, it is necessary to begin with a perspective on the development of the concept of systems. Widespread interest in the analysis and synthesis of complex functional units arose from the attempts to develop machines which could handle the relatively simple tasks involved in the storage, retrieval, and transmission of large quantities of information. Under the influence of Shannon, Weaver, and Wiener, communication became defined as CONTROL of automata. Biological organisms as well as machines were grouped together as automata. Both were conceptualized as being controlled in essentially the same manner through the encoding and transmission of information, and both exhibited essentially the same statistical properties in the handling of information.

The approach of communications engineering to the analysis of systems inspired by cybernetics was primarily inductive. "Inductive" here denotes the process of collecting specific facts and relating them, by a suitable set of generalizations, in one coherent structure or framework.

Ashby has called the cybernetic approach to communication a "Black Box" approach.[1] We begin by assuming that the object of our investigations is a "Black Box" about which we know nothing. We proceed by presenting a series of stimuli to the "Black Box" and recording the responses. In Ashby's words, "the primary data of any investigation of a Black Box consists of a sequence of values of the vector with two components: (input state, output state). . . . .all knowledge obtainable from a Black Box (of given input and output) is such as can be obtained by re-coding the protocol (p. 89). A protocol is a complete record of the input values or states and the accompanying output values or states. From the cybernetic point of view, the only knowledge one can have about the object of study is obtainable from a statistical analysis of this record of stimuli and responses. We begin a cybernetic analysis with the written record of responses, draw statistical inferences from those responses, and eventually formulate high-level mathematical abstractions which form a "model" of the behavior of the Black Box. The model is designed to predict explicitly the response of the system in terms of the individual characteristics of the components. The inductive nature of this process is easy to perceive.

Shannon's mathematical theory of communication is probably the most elaborate (and successful) set of inferences obtained by recoding a protocol.[2] Shannon treated engineering systems as Black Box automata and analyzed certain statistical regularities in the input-output functions. These functions were grouped together and the classical sender-message-channel-receiver-noise model was created. Virtually all communication models since Shannon have been influenced by his statistical model.

There has been a growing dissatisfaction with these communication models and their application to human communication. Human communication is a natural biological phenomenon; it is not engineered. Human organisms are adaptive. They do not merely respond to stimuli, they change themselves in the process of response to stimuli. Not only humans but all biological organisms appear to be teleologic rather than static, emergent rather than entropic. Thus, any stimulus or input into a biological system alters the system in such a manner that the administration of the same stimulus occurs under different circumstances than the first. Furthermore, the alterations in the system are apparently self-determined and are not satisfactorily explained in terms of action-reaction to forces external to the system. Let us use an example to contrast the classical mechanistic models and the teleologic model. In classical mechanics, we should be able to specify how much force applied in a given direction through a specified period of time will change an acorn into an oak tree. The transformation from acorn to oak tree, however, cannot be explained as action-reaction nor in terms of forces applied through time. There is nothing in the physics of the acorn which will allow us to predict "oak tree" as its final state. The system is apparently teleologic or goal oriented as indeed are all biological systems.

There is dissatisfaction also with the methodological assumptions of the cybernetic approach. Neither the cybernetic nor the attendant information theory approaches to biological systems have produced satisfactory descriptions of the process behavior of such systems. The success of these two "classical" approaches to the study of speech and language has been the production of a large body of data on the statistical properties of English which is neither generalizable nor explanatory of human communication.

When we use the "Black Box" approach to study human communication, we are acting as the blind man studying the elephant. We are led by our method to ignore much that is obvious. We not only have thousands of years of insightful observations on man's speech from which to draw, but we come to the process of communication with experience. The real methodological weakness of the cybernetic and information approaches to the study of communication is that the "Black Box" under observation has a KNOWLEDGE OF ITS OWN INNER WORKINGS. This important consideration is consistently denied by the "Black Box" methodology.

To account for the more complex behavior of biological organisms, GENERAL SYSTEMS THEORY has begun to replace the more traditional or classical inductive approach of communication analysis. The development of systems theory is taking place in an abstract manner, independent of particular hardware technologies. The objective of such development is the derivation of a set of principles which may be stated in rigorous symbolic (mathematical) form which will be applicable to all complex organizational patterns which might be designated as systems. The fundamental distinction between the traditional approach and the systems theory technique can be basically understood as the difference between inductive and deductive schemes. As von Bertalanffy writes, "In rigorous development, general system theory would be of an axiomatic nature; that is, from the notion of "system" and a suitable set of axioms propositions expressing systems properties and principles would be deduced".[3] In contrast to the cybernetic and informational approaches, the "deductive scheme" of systems theory is the process of starting with a few basic postulates or axioms and, by the introduction of more definitions or facts and the application of logic, arriving at specific conclusions.

One technique of problem solution particularly applicable in dealing with complex systems is to take the problem and imbed it in a broad class of more general problems. By concentrating on the general class of problems, a means of treating the whole class, and hence, the initial problem, may be found. What this means in the study of complex systems is that a structure must be created which is sufficiently general and powerful to encompass any and all things that, now or in the future, can be considered as "systems".

To do this a deductive scheme must be employed. A deductive scheme begins with a few carefully chosen principles or axioms. The introduction of additional definitions and hypotheses and the application of logic and inference allow the derivation of specific conclusions.In the case of systems this means that a few basic axioms or definitions about systems must be discovered which are not restrictive in any sense, i.e. they must allow any "system" to be included within their space. Then one must start to explore additional hypotheses to see what can be specifically proved about this general class of systems. The total structure, basic principles, hypotheses, and conclusions becomes a systems theory.

The real challenge is to identify the basic principles necessary for the construction of a sufficiently powerful deductive structure within a specific area of interest. The principles must be related to the real world. But they must not be too narrow. Most objection about the applicability of axiomatic structures in such remote areas as human communication is really an argument that narrow axioms are of no value. The human organism is the product of eons of evolution and to hope to encompass all modes of human behavior in even a very specific area by a few narrow axioms is not prudent. To tackle such problems it is necessary to start with general axioms, principles and hypotheses, then expand the size of the deductive structure by inference. Obviously there is a critical judgment involved here in deciding upon the initial principles and how broad or narrow they actually are. In this case only the results can justify or dismiss a set of basic assumptions. The theoretical system which follows is a deductive framework developed by Wymore.[4] In order to illustrate the application of the systems theory technique to a mathematical definition of communication as a biological phenomenon, let us begin by presenting the general definition of a system as presented by Wymore and then apply this definition to the phenomenon of communication in biological systems.

Definitions A system is a set:

$$Z = \{S, P, F, M, T, \sigma\}$$

with the following properties:

1. S is a non-empty set. This is called the set of SYSTEM STATES. A system state has two important properties: first, the states of a system uniquely characterize the possible configurations that a system may possess or transform itself into, i.e. the complete collection of distinct system states is a complete summary of all the possible modes of behavior of the system. This set can be either finite or infinite, depending on the system's nature. Second, the states possess the property that the current state, along with the form of the incoming stimuli (and a knowledge of a special function σ discussed below) allow the prediction of the future state of the system. Knowing what state the system is currently occupying, the input or stimulus being applied, and how the system transforms or responds to stimuli allows prediction of the future state or states of the system. This sounds mechanistic, but observe that this does not rule out a probabilistic law of transformation and stimuli – in which case one gets a distribution of future states. State as defined here is a fundamental concept in a large number of disciplines at present, especially psychology, neurology, sociology, and speech.

2. P is an non-empty set of input values. Presumably, there is a class of different phenomena which constitute the possible stimuli to be presented to the system. This total collection of stimuli or input values is the set P. This set may be either finite or infinite. Note that for both P and S, the possibility of their being infinite is allowed. In such a case they are not specifically enumerated but described by a GENERAL RULE, allowing one to distinguish, by the rule, whether a particular item is a member of the set or not.

3. F is a set of ADMISSIBLE INPUT FUNCTIONS* whose domain is the set T and whose range is the set P.

The set T is a set of time values, a set to delineate within what frame or segment of time the system operates, e.g. 7 minutes or infinity. Given a time function which is designated as being in the class of input functions, it is possible to include a time function also in the class which is the same as the first but with the time origin shifted. Thus, translation may be accomplished as in the following figures:

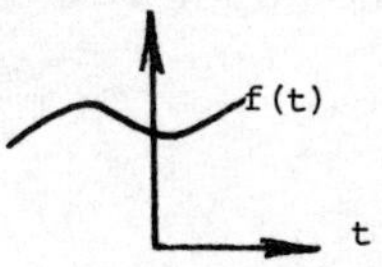

If this function is in the class

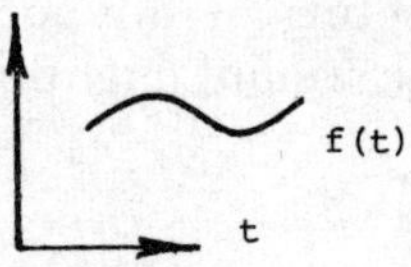

then this is also in the class, for any arbitrary repositioning or origin.

The reason for this shift being allowed is trivial: the function is unique, NOT the origin of time, e.g. the Greenwich Meridian could be set anywhere.

Another quality of the time functions in the input class is the ability to "shift" from one to another, thus building up complex functions. This quality is called segmentation. For example:

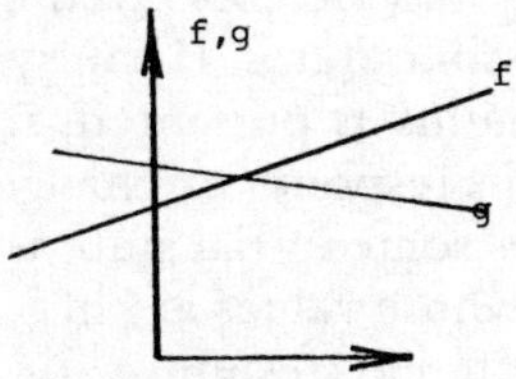

If f and g are both in the class of input functions . . . . . .

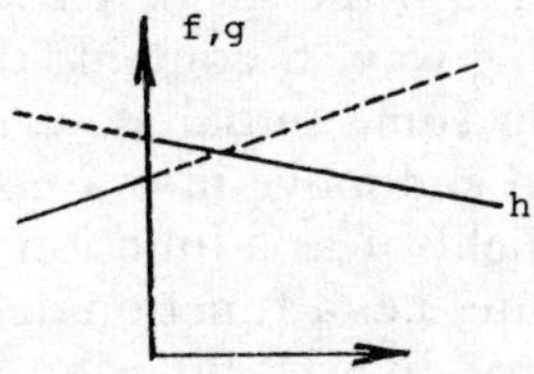

Then the function, h, composed by segments of f and g is also in the class.

The concept of functions does not apply only to numbers or curves. Any two sets can be placed into correspondence by a rule, and the rule becomes a function. Thus, the set of "admissible input functions" is the set of all possible functions, which may be generated by "translation" and "segmentation", which are defined on some set of time for the system and take on some set of stimulus or input values. NOTE THE DISTINCTION HERE: An input value P is a particular distinct level which may be assumed by a stimulus. But a particular stimulus is a function of time, i.e. it takes on different stimulus values in time to become a complete stimulus, or input function. Thus we have the following: If there are three stimulus values acting at any of three points in time, then two different complete stimuli might be:

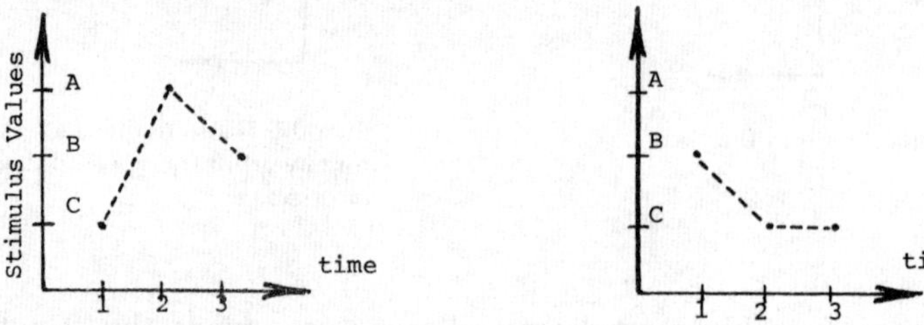

Special note should be taken here that a stimulus, or input function, is NOT just the stimulus value at a particular instant but is the WHOLE function. This is an important difference. To predict system behavior, one must know past stimuli as well as current ones.

4. M is a set of functions whose domain is the set S and whose range is also the set S. This set of functions may be conceived in the following way: S contains the set of all possible states. If the system starts in some initial state in S and a stimulus is presented, it will respond and move into a new state in S. This response or change can be thought of as a function which associates some initial state with a new state, i.e. a function defined on S but whose values are in S. But for every INPUT function (stimulus) a different response, or new state, may be seen even though the initial state was the same. So for one specific initial state, there is a different function on S into S corresponding to every different input function. The collection of all such functions, for every possible choice of input functions, is the set M. M is important conceptually because any possible change of states, due to some class of stimuli, is in this set. The nature of M is fundamental to the system behavior. Practically speaking though, M is so large for any reasonable set of function as to be useless. In such cases, we define M by saying: Let M be the range of the function σ. Note that M is a set whose elements are FUNCTIONS, not numbers or letters as we usually work with.

5. σ is a function of the two variables f and t, i.e. it is defined on the sets F and T for $f \in F$ and $t \in T$. The range of σ is the set M. For all possible initial states M contains all functions mapping S into itself, or responses, as they have been designated. Thus σ "operates", so to speak, on an initial state of the system and produces, for a given input function at a given time, the new state of the system. To

state it in equations: If the initial state of the system is x_0, and if we apply an input function f to the system, and then observe that at time t the state of the system is x_1, then by definition:

$$x_1 = \sigma(f,t)\,(x_0)$$

Thus by time t the input function has changed initial state x_0 into state x_1. This change is described by the function σ, which is what makes the system "go" since it details how the system actually responds. The set M contains ALL the possible changes. But if one actually knows a rule for σ then one does not bother with M. Because σ allows one to find the state at time t due to function f without having to work through the complete set of M (which is usually infinite).

Since the information sought is how the states change, σ is of prime importance, M is not. σ is the rule which tells how to find a new state given a current state and the input. Obviously, finding σ is the "gut" issue for any system because it tells one how to predict new states or a probability distribution among new states.

At this point the concepts outlined above may appear to be conceptually remote from the specific problem of analyzing speech communication of human beings. It should be recalled, however, that the system approach involves imbedding a specific problem in a general class of problems, and working deductively to apply a general system analysis of a specific system. In the case of human communication, the method of analysis involves moving from the general set of statements of a system outlined above to a more specific set of statements describing communication as a function within biological systems, and from there to specific deductions regarding speech communication of human beings. Let us illustrate this approach by describing a set of statements which might be used to describe communication as a phenomenon of biological systems, and from there to certain deductions about speech communication.

1. S is the state space. In applying this concept to biological systems, a multitude of system states may be introduced, each or any of which would satisfy the requirements of this set. The states of a system uniquely characterize the possible configurations that a system may possess or transform itself into, and the states possess the property that the current state, along with the specific inputs and σ, allow the prediction of the future states of the system. The biological concept which most closely fits the requirements for defining this set of functions is that of LEVELS OF ORGANIZATION. This concept encompasses Whitehead's concept of organism

but is somewhat more precise in that a measure of the level of organization is the relative complexity of the organism. Inherent in the use of the concept of levels of organization is the principle of emergent evolution and a certain inherent directionality toward increasing complexity. Each succeeding level would contain within it the principles of the lower levels of organization, but would be a more complex form embodying certain transformations in organization. Thus the principles of chemical communication between single cells is implicitly operative in the nervous system and consequently is embodied and operative in the more complex process of speech communication. In analyzing a communication system one of the system states to be defined would be the LEVEL of organization.

Before continuing, it is necessary to offer a brief explanation of the concepts mentioned above. Biologists have identified three basic processes which are necessary and sufficient for life. These are irritability – the ability to receive stimulation from an external source, conductivity – the ability to conduct energy, and integration. We have characterized integration as three interrelated "sub-processes." These processes are territoriality – the manifestation of a life-space by an organism, reproduction – the production of similar units of life, and communication – the process through which individual units transcend themselves to form social units and are simultaneously differentiated as parts within the total social unit. In this scheme, the individual unit and its processes constitute one level of organization which becomes differentiated according to the organization of the social unit, which constitutes a higher, more complex level of organization. Thus molecules have their intrinsic organization but function within cells which function within organisms, etc. with each succeeding level of organization providing constraint on every other level. The communication pattern of any part is the communication pattern of the unit of which it is a part, i.e. the communication pattern between two nerve cells is particular to the body in which they exist, and the communication pattern between two people is particular to the society or culture in which they exist.

The system states defined by levels are dependent upon the particular values of T, the time-scale on which one wishes to describe a particular biological phenomenon (in this case speech communication). This relationship will be described in more detail below. However, there is a second dimension to the system state which is relatively independent of T. This dimension operates within levels of organization and would be applicable only to those biological processes which are integrative in nature, e.g. communication, territoriality, and reproduction. This within levels' description of the system states assumes the form of a set of hyperbolic functions whose values are determined by the relationship between the facilitory and inhibitory dimensions of any specific communicative act.

Every communicative act has two dimensions – facilitation and inhibition. (1) Facilitation (x_f): Every act of communication acts to integrate a social unit. A communicative act of ant acts to integrate the colony; the gestures of a baboon act to integrate the group; a spoken utterance of a man acts to integrate a social unit whether it be family, state, society or culture. This direct integration associated with each specific communicative act is the FACILITORY dimension. (2) Inhibition (y_i): Every act of communication also acts to differentiate the communicator from the receiver of the communication, and thus to increase the complexity of the entire social unit. The communicative act of an ant distinguishes the sender as a worker, drone, queen, etc. The gestures of the baboon display his role in the dominance-subordinance hierarchy of the group as well as indicate his individual moods. Each communicative act of a man acts to form the self and the mind through the use of significant symbols. These dimensions of communication are a form of differentiation and are thus INHIBITORY to direct social integration.

In order for a biological system to preserve itself, certain limits are necessary between the facilitory and inhibitory dimensions of communication. If the system state is too facilitory, the system may become overloaded and destroy itself. For example, if a nerve is treated so that it receives and passes on all information, the entire system rapidly overloads and will literally burn itself out. The system prevents this by creating diversionary circuits and by inhibiting certain incoming stimuli, limiting the input into the system. These processes may continue to a point where the system becomes endangered by excessive inhibition (differentiation) at which time the system activates the facilitory dimensions once again. The system state representing the relation between the facilitory and inhibitory dimensions of communication may be described more fully as:

$$x_f = \frac{k}{y_i} \text{ or } x_f = \operatorname{csch}(y_i) \text{ for } x_f, y_i \; 0,$$

or further, as any function that has a graph of the form of

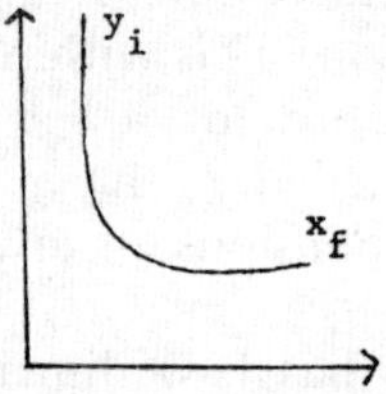

In describing the system state of a particular level of organization it is not necessary to know all values of the communication acts within the particular level. The range of such values is sufficient to describe the system state. The range would be defined as the maximum value an act of communication could obtain which was predominantly inhibitory. In human communication these values would be approximated by the most facilitory statements such as "Help!", "Look out!", or "Jump!" under conditions of extreme stress and perceived danger, and by the most inhibitory statements (statements which differentiate a sub-system from a larger social unit) such as "My name is John Doe" or "I am John" which are central to the formation of the self.

In conceptualizing the state space we have postulated three state variables: $\ell(t)$ which describes the current level of organization at a given time; $X_f(\ell(t))$, which describes the extent of facilitation within a given level ℓ at a given time, $Y_i(\ell(t))$, which describes the extent of inhibition within a given level ℓ at a given time. Two observations should be made regarding these relationships. First, even though we have identified three state variables, the latter two are related by some kind of function of a general hyperbolic form; if the functional relationship between them is $X_f = g(Y_i) => Y_i = g^{-1}(X_f)$ then knowing either state variable allows us to compute the other. If compactness of form is desired, we could reduce the dimension of the state space from three dimensions to two dimensions, using $\ell(t)$ and one of the two variables X_f or Y_i.

To emphasize the importance of the X_f and Y_i variables, however, it is desirable to retain both of them. Secondly, since the state variables X_f and Y_i are functions of the state variable $\ell(t)$, non-linear feedback is implicit in the system, since X_f and Y_i are non-linear in general.* If a useful human communication system model involves non-linear feedback, it is easy to see the advantage of the present model over the classical sender-receiver models of Shannon, etc. which are open loop and linear in nature.

2. P is a non-empty set of input values which constitute a class of different phenomena which constitute the possible stimuli to be presented to the system. In a systems analysis of communication as a biological function, for every organism there will be a finite set of input values. The elements of P may be grouped into subsets each of which may contain a number of values simultaneously, e.g. (P_{e1}, P_{e2}, . . . P_{en}). The various subsets of input values are:

(P_e) ENVIRONMENTAL INPUTS, especially ecological inputs related to the particular time and life space of the organism.

(P_o) BIOLOGICAL PREDISPOSITIONS OF THE ORGANISM. These input functions represent the biological "givens" in the communication system of any particular species of animal. These inputs would represent the particular medium of communication (e.g. chemical, visual, aural-oral, speech, etc.). The species predispositions toward signal or code formation would also be considered to be organismic inputs. In the analysis of human communication the biological principles underlying all speech and languages would necessarily be included. For example, one could use Lenneberg's concantenation, phonemitization and grammatization; Chomsky's structural grammar and kernel sentences, or Skinner's elaboration of operant conditioning as a set of organismic inputs.

(P_s) SOCIAL INPUTS. Every biological organism is social in nature. The communication of an organism arises as a function of the social unit in that communication serves to integrate the social unit. Therefore, in any communication system there are critical inputs from the social unit. Among these inputs are culture, general language systems utilized (i.e. English, French, German, etc.) and organism-specific variations within levels. In all animals there are variations of the general species signal system. Each colony of ants and hive of bees has its own particular sub-language system specific to that colony or hive. In man, local accents, dialects, and vocabularies serve the same function.

(P_n) NUMBER OF FUNCTIONING SUBUNITS which comprise the total biological system. This set of inputs is critical in all biological systems as it is very closely associated with σ. The number of units which make up the primary social system is highly influential in the analysis of speech communication particularly as certain inputs occur when the social system is a small group which do not occur when the social system is a triad or a dyad. Likewise, certain differentiational activity occurs in the communication patterns where several hundred neurons are present which is not possible with two or three isolated neurons.

(P_r) RANDOM INFLUENCES.

3. F is a set of admissible input functions whose domain is the set T and whose range is the set P. In determining the set F, certain limitations are imposed upon P by the system state. The subset of P which is admissible as input into the system is determined by the level or organization existing at any particular time in the system. For example, in human communication INDIVIDUALS very often become transformed into GROUPS or an ORGANIZATION. Each of these would represent a different system state. The admissible input functions would be determined by the particular system state. An organization would obviously have different admissible environmental, social, organismic, and random inputs (influences) than an individual. The same kind of analysis would hold true for other biological communication systems such as the neuron, the reflex arc, and the nervous system. Thus, the inputs P_e, P_s, P_o, P_n, and P_r are associated with the system state. The relationship between the system state $\ell(t)$ and the admissible input functions is another non-linear feedback mechanism of the form $P\alpha\ (\ell(t))$. This feedback is probably adaptive, corresponding to the evolution of the system to higher levels of organism and learning to respond to new inputs.

The set T is a set of time values, which delineate with what frame or segment of time the system operates. Each LEVEL of organization may function on a different time scale. An ant may have a life cycle of 6 months, the colony of which the ant is a member may have a life cycle of 30 years, and a cell within the ant may have a life cycle of 15 minutes. Likewise a human cell may have a life cycle of 24 hours, while the individual has a life cycle of 70 years and the life cycle of the society is still undetermined. The frame of time determines the level of organization observable. For example, the life cycle of human cells becomes virtually non-existent in geological time. With this in mind, it is possible to observe that the limitations on P which were ascribed above to the system state, are in actuality functions of the particular frame or segment of time observed. The general definition of admissible input functions is satisfied with the domain of F being T and the range of F remaining P.

4. M is conceptually very important for the application of a general systems analysis to human communication. The concept of a set M allows one to apply a general systems analysis to a phenomenon such as communication which integrates biological system. It is the nature of biological systems that they never return to the same state after a given stimulus. They are changed, having in some way adapted to that particular stimulus. The set M accounts for this adaptive process and becomes a set of all possible states generated by σ. However, the particular set M becomes so large as to be useless for practical analysis. A rule for determining a particular set is substituted rather than attempting to deal with the entire set M.

5. σ is the rule which operates on an initial state of the system and produces, for a given input function at a given time, the new state of the system. In all biological systems two basic principles appear to be operative in guilding the systems to new states. These two principles are applicable to all biological functions including human communication. We shall designate these two principles as σ_n and σ_p.

σ_n is the biological principle known as reduction of repetition. Reduction of repetition describes the transformation of organisms from one level of organization to more complex levels of organization, or from one system state to another system state. Reduction of repetition may be described in the following series of steps: aggregation of a number of similar units; fixing of topological differences in units; decrease in ratio of repetitious units; increase in local complexity; increase in total complexity; fixation and isolation of a more complex, totally new, greater unit. It can be demonstrated that this principle not only operates in the transformation of one level of organization into a higher, more complex level of organization, but also operates in the biological processes such as communication and perception.

σ_p is a general statement of the many theories of survival of the fittest. Without subscribing to any particular biological theory of survival, it is possible to derive a general principle of the operation of the preservation of the organism. In a general form the principle is: the variation allowed by a social organism is a reciprocal relation between the variation necessary for adaptive survival (V_a) and the variation beyond which survival of the organism is endangered (V_e). The variation in communication patterns between two members of the same super-organismic unit (species, colony, hive, family) is a reciprocal relationship between the variation in communication necessary for the super-organismic unit to adapt to its changing environment and that variation in communication pattern beyond which members of the same super-organismic unit can no longer integrate effectively through misunderstandings of the communication pattern.

CONCLUSIONS

The model which has been formulated above is a crude outline only. The real challenge is to now determine how the crude outline can be applied to specific cases; for example, how do we construct state transition functions, σ? What properties do such transformations have? The highly non-linear, feedback, adaptive nature of communication is explicit in the formulation of the elements S, P, F, M, T, σ of the model. It is obvious that communication is a non-linear feedback, adaptive process; but it is not obvious exactly where the non-linearities, feedback, or adaptation occur until a more specific application can be formulated. The advantage of the model presented here is that it leads us to such formulations.

Before continuing, it is important to take cognizance of two arguments that can be (and usually are) raised against the use of systems theory, or any axiomatic or mathematical structure, in the study of such topics as the development of human communication. The arguments are: (1) not all modes of behavior or quantities of interest may be describable within the scope of mathematics, i.e. some things just cannot be dealt with in such a concrete fashion. (2) Even assuming the quantifiability of something of interest, the systems theory approach starts in such an abstract fashion, from a very general set of axioms or principles, that it appears not only time-consuming but virtually hopeless to try to direct one's reasoning through all the specific conclusions to arrive at a point where information is provided about some area of interest. It appears exceedingly difficult to relate systems theory to communication behavior for example.

The first argument may be answered in two ways. First, mathematics is NOT just the manipulation of numbers or equations. The rigorous structure of axiom-definition-hypothesis-theorem-proof is more appropriately described as mathematics. Such a structure can be carried on independently of the conventional concerns of mathematics, i.e. numbers and equations, as systems of logic and symbolic logic, in particular, serve to demonstrate. In fact, symbolic logic can serve as a super-structure to use in describing the operations underway in more conventional areas of mathematics, as Whitehead and Russell's work implies. Mathematics can be thought of as the whole structure and process of deducing logical relationships among abstract entities. Secondly, a great many concepts are more quantifiable and treatable by proper symbolism than the casual observer might think. The application of the mathematical mode of thought forces the analyst to discard much of the chaff that may be obscuring the problem so as to achieve compactness in symbolism and expression. Value judgments as to the applicability and utility of the resulting model are obviously involved, but in the process the nature of the concept may become clearer with respect to its isolation and quantification.

The second argument may be answered by reference to another very active field of engineering – automatic control theory. The first forms of automatic control were developed by the inductive scheme of analysis. The individual components were isolated and described, their interactions determined, and the overall system behavior derived. A very large number of basic cases were solved and, in many cases, control engineering became the intelligent use of approximations and special charts and graphs. However, with the advent of more complex control needs, such as missile guidance and optimum trajectory control, the short-comings of such methods became more evident. The final result was to axiomatize control theory as a formal mathematical structure. Thus modern control theory today establishes concepts about stability of control, optimum control, etc. as mathematical entities by the use of definition, theorem, proof. The results have been very rewarding, giving new results not foreseen in the classical theory and still encompassing classical results as special techniques of modern control theory. In short, the employment of a general deductive scheme IN ONE SPECIFIC AREA brought results the inductive scheme never could have.

Obviously, human organisms are far more complex than control systems, and it would be over-optimistic to assume such relatively painless results could be achieved by applying a deductive scheme. The point to be made is that one does not have to pursue systems theory at the most abstract level imaginable. Reference to a specific problem, such as the integrative behavior of organisms manifest in communication should provide a set of basic principles or axioms general enough to allow a powerful deductive framework to be built, but specific enough to generate the hope that the results will have some direct relevance to the problem at hand.

The impact of general system theory may be highly useful in a specific area. For example, one aspect of the theory developed above deals with understanding interactions between systems. This treatment is abstract enough to be applicable to any kind of system. If one can develop a suitable system definition and description of individual organisms, the theory would allow the determination of the system behavior resulting from two organisms interacting with one another through communication. It is hoped that the general system elaborated above, and its application to the phenomenon of communication can provide a useful framework within which more explicit formulations and descriptions of communication may be developed.

REFERENCES

1. Ashby, W. Ross. 'An Introduction to Cybernetics.' New York: John Wiley & Sons, 1956, 1966.
2. Shannon, Claude and Warren Weaver. 'Mathematical Theory of Communication.' Urbana, Ill.: University of Illinois Press, 1948.
3. Bertalanffy, Ludwig V. 'General System Theory.' New York: Braziller, 1968.
4. Wymore, A. Wayne. 'A Mathematical Theory of Systems Engineering.' New York: John Wiley & Sons, 1967.

*The term "admissible input functions" is used by Wymore in a specialized sense that is too complex to develop profitably within the space allotted here.

*Feedback may be said to exist when two elements are related by a circularity of action which may be represented by

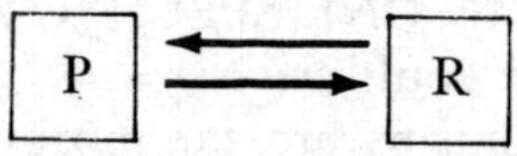

Feedback exists between two parts where each affects the other, as for instance in

$$x' = 2xy$$
$$y' = x - y^2$$

By contrast, feedback would not be said to be present in

$$x' = 2x$$
$$y' = x - y^2$$

for x's change does not depend on y's value; x dominates y in the latter case and the action is only one way.

A Simulated Communication Model of Community Action Organizations:

An Application of General Systems Theory and General Semantics

RUSSELL W. JENNINGS, JOE VINOVICH AND THOMAS J. PACE

In an effort to meet critical social needs apparent in poverty stricken localities in the United States, a national trend emerged in the 1960's focusing upon citizen involvement and participation of the poor in the social and political affairs of their respective communities. Although direct citizen involvement has been heralded as one of the hallmarks of participatory democracy,[1] lower income and ethnic minorities have been barred for all practical purposes from any meaningful political involvement and self-determination, except in a most superficial sense.[2]

> In today's American society majority rule might well be undemocratic. It seems . . . that in our very heterogeneous society we have what might be called . . . permanently outvoted minorities, people who always get outvoted by the various Establishments and the majority so that they can never achieve their goals. . . .[3]

A number of alternative approaches have been proposed that appeared to permit political participation by these excluded minority constituencies even though their inclusion ultimately proved to be counter productive. For example, at the turn of the 1960's one alternative was direct citizen participation through cohesive community-based citizen groups. But, the American Indian, Blacks, the Chicano, low-income whites, and in some localities, lower middle income whites were still excluded from the political process of their community affairs, especially in urban and rural-to-urban transition areas. The result of this exclusion was the emergence of social problems at essentially the community level that expanded to crisis proportions, such as taxation, real estate costs, land maintenance, environmental pollution, consumer protection, and criminality. Then, in 1964, the enacted Economic Opportunity Act posited that social problems at the community level were the product of economic deprivation and political alienation. The Act's most controversial features were contained in Title II-A, Section 202 authorizing the creation of Community Action Programs [CAP] "to provide stimulation and incentive for urban and rural communities to mobilize their resources, public and private, to combat poverty",[4] and which supposedly "were developed, conducted and administered with maximum feasible participation of residents of the areas and members of the groups served".[5]

In southern Illinois, the bicounty Jackson-Williamson Community Action Agency [JWCAA] was instituted to implement the approach established in the Economic Opportunity Act of 1964. The essential task confronting this, as well as other agencies of this type, was to develop and operationalize community-wide systems of information processing – communication models – that allowed target populations as well as those resources that were available to be identified and channeled towards the resolution of the target population's most pressing collective problems. Generally, this was an attempt to focus information about economic and social service resources toward the poorer sectors of the community. In southern Illinois, however, the result was a focusing of such information toward selected problems of a minor sub-sector of the potential target population – essentially that group ostensibly defined as "poor" by the local, political establishments in power at the moment. In some towns in the bi-county area, CAP services were not available because local officials declared that no poverty existed and that such services were unnecessary. In other places, extraordinary requirements were imposed upon potential recipients that exceeded those stipulated by state and/or federal agencies. Thus, uncoordinated and diversely functioning agencies attempted to deal with "the local problems" exclusive of a coordinated information mechanism that characterized the scope of those multiple problems.

A variety of plans and countless hours were committed by all parties concerned, both local, state, and federal, to overcome the most pressing effects of this lack of coordination. The efforts, however, were undertaken in lieu of dealing directly with local prejudices and political duplicity. The various plans initiated ranged from frequent reorganization of the agencies and replacement of their personnel to the development of "super-agencies" and "super-committees" that would coordinate various organizational functions and control the distribution of funds, with the major emphasis on the latter. The most frequent strategy, however, was to conduct "training workshops" for agency personnel focusing exclusively on the uses of management procedures in federally sponsored programs. The results realized by this approach were essentially the same each time – a limited understanding of selected organizational management procedures without an attendant understanding of information processing or information coordination principles. Thus after repeated failures, newer approaches were solicited that might deal directly with those problems associated with the manipulation, use, and control of information in loosely structured community oriented problem-solving organizations.

The purpose of this paper is to describe a generalized communication model of the JWCAA as a CAP agency upon which a "live simulation" is to be based and which incorporates General Systems Theory as a means of operationalizing the "organism-as-a-whole" approach espoused by General Semantics.[6]

> In General Systems Theory, the aim is to formulate and derive those principles of organization, of total overall of dynamic interaction. . . . There is a search for principles concerning "wholeness, growth, hierarchical order, dominance, control, competition." In the hierarchical levels within an organism there is a concern for the relationships and the coordinations of subsystems to the whole organism . . . the correspondence and parallelisms, or isomorphisms . . . which occur among the subsystems and levels.
>
> In General Semantics there is similar concern for wholes, in this case with particular emphasis around and upon the whole person in his environments. General Semantics brings a methodology for searching into, for evaluating, and coping with the relationships and interactions among the various orders of fact-phenomena [or systems and subsystems].[7]

This paper briefly outlines the procedures for the simulation of a functional model of communication as encompassed by the JWCAA in which the authors have been conducting research.[8]

The assumptions upon which this research project is based are:

1. That decision-making or administrative systems cannot be imposed arbitrarily on groups, small or large. The decision making and/or control mechanism MUST emerge from within a group, if that mechanism is to fulfil its function, since only an internally emergent control mechanism can be sensitive to the salient needs and demands of the group being affected.

2. The structural characteristics of groups, that is, the manner in which individuals self-reflectively integrate information for adaptive purposes, are invariant structures under transformation and, therefore, are evolutional.

3. The structural evolution in groups begins in a non-complex state and progresses non-cumulatively to complex states when the cognitive capabilities of individual group members allow, and when environmental conditions permit.

4. Underlying the above assumptions is the recognition that the human environment making such demands also provides those information resources necessary to cope with the limiting situational conditions. Therefore, in terms of information processing, the essential task is to cope with similarities in form but different in content.

> One problem is to determine the number of dimensional attributes processed in an environment, and if two or more dimensional attributes of information are perceived, the next problem is to determine the degrees of freedom involved in the rules of combination. [9]

The methodological problem becomes one of coping with processed information from the smallest fact-phenomena to that from which a person is able to abstract information with the aid of extended resources in the environment, to that from which they are able to abstract information within their cognitive limitations, and finally, to those relationships too large and complex to envisage through any one level of the system and target environment.

> These person-centered organizations must be organic in an almost biological sense because they must grow out of the human environment (as perceived and/or defined by human organisms) as it is today, not out of previous ideologies. They must be pluralistic . . . adaptive . . . organic . . . populist. [10]

As a systematic phenomena, not only must CAP programs maintain an inter-relatedness with their target environments, but they also must maintain that dynamic interaction through citizen participation to cope with socio-political change. For example, the JWCAA was placed on probation a number of times during its existence because of an apparent inability to maintain a viable relationship with all the critical sectors of its environment. Although it was able to establish a connection with the poor, it had only a TOKEN relationship with the middle class population, through the political sector, and no relationship with the upper class population. The agency, then, was confronted by attempting to directly serve part of the poor population, while being directly controlled by the middle-class through political intervention, and indirectly by the upper class population through the local, state, and federal political establishments.

The major hypothesis for this research project is that the existing cognitive levels of information processing of organizational group members, as components of an open system, impede the generation of situationally appropriate "map-territories", thereby limiting the evolution of the system. And it is posited that through live-simulated information processing models these organization members can be trained to better cope with the critical needs of the total target environment. A fundamental objective for using live simulation techniques to construct an interaction environment is to train its participants for adaptive, and sometimes, creative behaviors. This is the case for applying live simulation techniques to an organization such as the JWCAA. The basic dictum is: "in training, the question is not what an individual is taught, but what adaptive orientation he develops while he is learning these responses in a given training environment".[11]

The JWCAA, as a social subsystem, is organizationally assumed to be an open system existing in a socio-political environment.

> An open system can . . . be defined by its boundaries for the selective reception of inputs (a coding process) and for its typical transmission of outputs. It is further characterized by such properties as negentropy to counteract the tendency to maintain a steady state; by equifinality or the use of different patterns to produce the same effect; and by differentiation, the tendency toward elaboration of structure.[12]

The open system approach was adopted in this research primarily because of the input, energy transformation, and output characteristics of the organization itself, and secondarily, because the open-system approach holds the potential of accommodating an integration of diverse information.

> The functioning of any open system . . . consists of recurrent cycles of input, transformation, and output. Of these three basic systemic processes, input and output are transactions involving the system and some sectors of its immediate environment; transformation or throughput is a process contained within the system itself.[13]

In other words, the JWCAA and other CAP programs cannot be

closed systems since their resources, guidelines, and apparent social and political power are generated from outside. More important, the JWCAA's subunits must be focused on problems and people positioned outside of the system. Therefore, serious structural and functional information processing dysfunctions can occur if there is a lack of understanding and/or if the organization's structural and functional goals as well as its community relationships are not clearly delineated. This is particularly the case if those dysfunctions affect the decision-making and "helping strategies" to effectively combat poverty.[14] For example, the definition of NEIGHBORHOOD, as developed within the guidelines for CAP agencies, is predicated on predominantly urban community action experiences that denote not only a locality in which residents enjoy frequent "primary" or "face-to-face" contacts with each other, but also refers to distinctive territories within a geographic locale.[15] When programs and agency functions are initiated in a rural community, however, based on CAP guidelines, operations such as budgeting, travel, placement of Neighborhood Centers and Headstart Centers are found to be inappropriate and inadequate. This is primarily due to the differences in the defined size of the rural neighborhood's geography while its residents still maintain primary and face-to-face contacts. Sometimes twenty miles separated Headstart Centers, thus forcing each center to transport its clients much greater distances than the Urban Centers. Centers in both localities, however, were budgeted on the same basis.

The participating individuals and the formal organization are analogous as systems, each depending on "rules of the game" that serve as critical definitions of the structure in question. An individual's creative potential is approximate to the organization's adaptive limitations or its equifinality dimension.

> Organic life, in all its manifestations, from morphogenesis to symbolic thought, is governed by "rules of the game" which lend it coherence, order, and unity in variety; and that these rules . . . whether innate or acquired, are represented in coded form on various levels. . . . The rules are fixed, but there are endless variations to each game, their variability increasing in ascending order. . . . There is also an over-all rule of the game, which says that no rule is absolutely final; that under certain circumstances they may be altered and combined into a more sophisticated game, which provides a higher form of unity and yet increased variety; communication sequences are not anonymous units in a frequency distribution. . . .[16] Rather it

> is an inseparable characteristic of on-going processes whose order and interrelation occur over a period of time.[17]

The JWCAA and other CAP agencies, as social subsystems, acquire and accept definition not only from their own internal functions, but also from simultaneous interaction with the supra-system in which they exist. Included in this supra-system is not only the community it must serve, but also the various levels of government from which it derives its authority and financial support. Integration is made possible by delimiting interactions to essential or critical dimensions wherein each participant, as a system object, engages in information processing.

> (While the) objects may be individual humans, the attributes by which they are identified . . . are the communication behaviors (their systems of information processing). . . . The objects of interactional systems are best described not as individuals but as persons-communicating-with-other-persons.[18]

Thus, the open system approach as a model of information processing descriptively details important complex structural elements. This analysis of dynamic interaction in which the participating objects influence the operations of information processing events as they are taking place in and throughout the organization. For every act, there is a reaction, either counter to or in cooperation with the target act. These events as systemic phenomena can be delineated as time-space interactions to which defined limitations can be assigned for analysis. The JWCAA is a system in which time-space interactions occur as events, is defined operationally as a mediational device within the community. As such, it attempts to assimilate and place pertinent information about social and political problems in its environment in viable relationships with those resources potentially available for initiating an appropriate resolution of those specific problems.

As an information processing system, the agency is capable of acquiring and utilizing alternative meanings of the same stimulus and of constructing and employing different patterns and interrelationships within the same set of meanings.[19] As an example, the Emergency Food and Medical Service procedures that the agency had developed were generally applied throughout its target area, however, in some instances it was necessary to modify these procedures to deal with individual cases as they emerged in specific communities. In one community, the county welfare supervisor was unwilling to develop or implement emergency procedures for those cases that demanded immediate action. Thus, if a client needed emergency medical treatment, he was forced to wait from one to two weeks before his case was processed and the necessary authorization could be obtained. The JWCAA by-passed the welfare office and made direct contact with a sympathetic supervisor in the State Welfare Office to expedite the authorization. This, of course, solved the problem of the agency's client, but it also created an alternative to be used under special conditions with other clients. Unfortunately, it also disturbed the relationship between the specific county office and the agency. In such cases, the use of simulation would permit exploring not only the approach used but also other possible approaches to determine which one could get the emergency authorization without severely disturbing the relationship between the agency and the county office. The agency thereby fulfills a COPING FUNCTION within its specified environment, that being the attempted solving of problems which the system encounters.[20]

> In social systems . . . structures tend to become more elaborated rather than less differentiated. The open system does not run down, because it can import energy from the world around it. Thus, the operation of entropy is counteracted by the importation of energy and the living system is characterized by negative rather than positive entropy.[21]

In this role, the JWCAA is construed to be principally an adaptive system which is charged with "anticipation of changing forces which may affect the society, . . .and (moving) toward adjustments which place the (society) in a position of advantage (to cope with those changing forces)."[22] Additionally, this adaptive system is construed to be open and functionally negentropic since its continual existence is exclusively dependent upon its importation of information-as-energy from its environment. It is open "internally" as well as externally since the interactional relationships between its own respective components can result in significant changes in the nature of the components themselves and have important consequences for the system as a total entity. Its negentropic function is basically concerned with ingesting new information and contacting potential information sources that will preclude its own internal deterioration to critical functional limits. Although the agency as a system is stabilized by a set of "rules of the game" the components and their interrelationships can be significantly altered by changing the basic rules. Likewise, the rules can be changed by altering the characteristics of the system's component attribute values and the subsequent interrelationships (or decision rules) between them. The most important factors are the system's patterns and critical paths of internal external interchanges. These are maintained throughout by the system's information processing and organizing. Its feedback loops help in the self-regulation and self-direction function of adapting to changes in the environment. The internal and external interchanges, as this system's form of energy transformation are predicated on the assumptions that: (1) the target society is a potentially changing environment that varies within specifiable limits; and (2) that an organizational system's persistence and elaboration to more complex levels depends upon its ability to provide for itself appropriate verbal "mappings" of the environmental variations as well as functional constraints for inputs on a semi-regulated basis. Such specialized programs as Vista, Headstart, Neighborhood Centers, and Emergency Food and Medical Services[23] exist within the agency to process specific information under special constraints, as outlined in the guidelines, administrative directives and policies of the sponsoring federal, state, and local agencies. At the same time, these specialized programs attempt to fulfill particular adaptive tasks that secure the agency's survival and viability in the changing environment. As with all adaptive systems. the JWCAA:

> Must manifest (1) some degree of PLASTICITY and IRRITABILITY vis-a-vis its environment such that it carries on a constant interchange with environmental events, acting on and reacting to it; (2) some source or mechanism for VARIETY, to act as a potential pool of adaptive variability to meet the problem of mapping new or more detailed variety and constraints in a changeable environment; (3) a set of selective criteria or mechanisms against which the "variety pool" may be sifted into those variations in the organization or system that more closely map the environment and those that do not; and (4) an arrangement for PRESERVING AND/OR PROPAGATING these "successful" mappings.[24]

Each specialized program operates as an open subsystem as specified by its interrelated functioning with its respective participants as entities within the agency's structure. Structure, as used here, refers to:

> An interrelated set of events which return upon themselves to complete and renew a cycle of activities. It is event rather than things which are structured, so that social structure is a dynamic rather than static concept.[25]

The basic structure is manifested at all levels within the agency as an elaborated information processing system predicated on functional, as opposed to arbitrarily imputed, component attribute values and decision rules. Attribute values refer to the common and/or consensually held goals and expectations regarding how to achieve systemic objectives.[26] Thus, most resulting cooperative activities would grow more from shared values than from imperative demands of an objective task. Decision rules, on the other hand, refer to the coded directives specifying relationships between systemic objects.[27] The position that one occupies within the organization as well as the types of tasks that he is expected to carry out, and the manner specified for carrying out those tasks all imply valuative and specifiable relationships that exist between various component entities within the organization. These rules are predicted upon component values as well as the relationships between those valued components and cannot functionally exist without such valuation.

> The relationships to be considered in the context of a given set of objects depend on the problem at hand, important or interesting relationships being included, trivial or unessential relationships excluded. The decision as to which relationships are important and which are trivial is up to the person dealing with the problems, that is, the question of triviality turns out to be relative to one's interest.[28]

When members of the Agency use language inappropriately in interviewing or problem-solving, or information dissemination, their proficiency in systematically analyzing and planning "helping strategies" becomes seriously impaired. The fundamental requisite, therefore, for any analysis of the Agency resides in determining the type of organization-qua-task requirements and needs, which in this case are adaptive, as well as determining attribute values and decision rules upon which the Agency is based and which direct and regulate its function.

In the context of an adaptive organization with task requirements and needs as well as attribute values and decision rules, the criterion for participant performance involves three basic requirements. First, it requires an expanding awareness of subtle and dynamic environmental changes.[29] This essentially demands that the participant develop skills in abstracting the most obvious aspects of environmental change, and more importantly, to become sensitive to and capable of defining those more covert features of the environment's dynamic function. Second, it requires the evolution of new or inventive interrelated means of problem-solving based on the natural order of evaluation to produce viable solutions.[30] This requirement demands that once a participant is able to acquire appropriate data from the changing environment, he then be able to adapt or create evaluative systems and criteria that are appropriate for that data. For example, the application of traditionally white middle-class standards of problem identification and solution to those problems that reside uniquely in the province of lower class Blacks can yield little beyond divisiveness, dissension, and failure, as many of H.E.W. and H.U.D.'s programs of the past five years clearly indicate. Finally, the criterion requires the acquisition of multivalued perceptions toward diverse environmental information from which can be generated multiple coping alternatives in decision-making.[31] This requirement demands that participants constantly expand their own perspectives as well as seek out the viewpoints of others and then explore alternative approaches to resolve the problem as perceived. By instituting a "reality-bound" simulation, it is expected that the various components of the Agency can interpose and interrelate with socio-political situations within the environment. Further, it is expected that problems can be examined and handled so that the most appropriate solutions can be identified and initiated. Most important, problem identification and solution construction can be accomplished prior to their actual implementation in the "open community", thereby providing more operational stability for the Agency in its environment.

A systemic communication model is both the manifestation of organization structure and an operationalized "blue print" of information processing and organizing of the system generally.

> (Furthermore) once we accept all behavior as communication, we will not be dealing with a monophonic message unit, but rather with a fluid and multifaceted compound of many behavioral modes – verbal, tonal, postural, contextual, etc., – all of which qualify the meaning of all the others. The various elements of this compound (considered as a whole) are capable of highly varied and complex permutations, ranging from the congruent to the incongruent and paradoxical. A communication not only conveys information, but at the same time it imposes behavior.[32]

In terms of information processing and organizing, there exists integratively complex levels of communication, from precisioned efficiency to unrestricted noise production. Without functional decision rules and component values, little or no systemic operation or attendant meaning is possible. Communication is considered as the systemic process that defines, through the application of distinctive decision rules and discernable component values, the relationships between critical environmental information abstracted and the density of information available for abstraction. Component values are prescribed by the degree of criticality that any attribute of the system plays in any situational event or field. The decision rules are functional directives assigned by the system to meet specific situational objectives. Attributively, component values can exist without operationalized decision rules, but decision rules cannot exist exclusive of component values in order to maintain a functional level of organizational validity and reliability. Reliability of the "social" system refers to the amount and frequency of "positive" feedback from the objective or target environment – i.e. if it works, society usually maintains it; if not, it is terminated.[33] For example, if the JWCAA members lack functional understanding of the organization's goals regarding the social and political sectors of its environment as well as community-agency relationships, then serious dysfunctions, must occur in the promotion and application of helping strategies. Similarly, such misunderstandings are reflected in intraorganizational dysfunctions, especially in the manner that communication channels affect decision-making processes in efficiently coping with randomly emerging problems in the agency's environment. Basically, systemic communication is attributively information acquisition, processing and organizing which generate behavioral changes within the organizing and in the target environment.

> On the level of the symbol-based sociocultural adaptive system, where the more or less patterned actions of persons and groups are as crucial a part of the environment of other persons and groups as the non-social environment, the gestural activity and its more or less normatively defined constraints is encoded, transmitted, and decoded at the receiving end by way of the various familiar channels with varying degrees of fidelity. Over time, and again by a selective process – now much more complex, tentative, and less easily specified – there is a selective elaboration and more or less temporary preservation of some of this complex social as well as non-social constrained variety in the form of "culture", "social organization", and "personality structure".[34]

Traditionally, systemic communication has been viewed as an exchange of information, specifically the transmission of meaning, but as indicated, more is involved than purely adaptive and manipulative activities. Information processing and organizing is significantly a symbolic interaction of controls that must include all behavior. The degrees of control reciprocally give as well as gain impetus from the ratio of the information acquired and the levels of density at which the information is being acquired. Thus, systemic communication as described above must be perceived not only as a process occurring between a sender and receiver, but also as an integral characteristic of the organizational system in which it occurs, particularly in regard to the functional attributes of that system. [35]

Primarily, three categories of systemic communication can be identified in an open system. First, there are EXTRA-SYSTEMIC forms which are those non-task related functions carried out by organizational components, but that lack the necessary regulation and direction to accomplish specified organizational objectives. These encompass situations such as a secretary telephoning for a beauty appointment during working hours; the private discussion of a supervisor's performance by an organizational worker with a friend who is not associated with the organization; or the planning and preparation of a week-end party involving both selected members of the organization and people not affiliated with the organization. Second, there are SUB-SYSTEMIC forms which are the manifested functional alternatives for action within a system that usually operate through the integration of the multiple levels within the organization. These encompass situations such as a VISTA worker informally notifying a Headstart teacher of a potential client; a Neighborhood Center referring a client to Emergency Food and Medical Services for assistance; or the cooperative conducting of a joint information gathering survey between a number of the specialized programs, i.e., Emergency Food and Medical, Headstart, and the Neighborhood Centers. Third, there are SUPRA-SYSTEMIC forms which are the collective cueings of expected types of "normative" organizational behavior, exhibiting the prescribed constraints and restrictions of the assigned decision rules as perceived necessary to meet the organizational objectives. These include situations such as the VISTA worker formally submitting the required H.E.W. forms to transmit a case to the Headstart program; or the distribution of an official memorandum from the agency Director announcing changes in policy made by the Board of Directors.

The latter two forms are considered to be essential to task-relevant orientations within the system and differ primarily in the amount and form of intrinsic systemic control.

> There may well be identifiable factors intrinsic to the communication process – that is, aside from motivation and simple habit – that serve to bind and perpetuate a relationship . . . . (These might be subsumed) under the notion of the limiting effect of communication, by noting that IN A COMMUNICATIONAL SEQUENCE, EVERY EXCHANGE OF MESSAGES NARROW DOWN THE NUMBER OF POSSIBLE NEXT MOVES.[36]

Both the sub-systemic and supra-systemic forms function for system maintenance, the transformation of energy, and administrative control. This does not preclude the transposition by the sub-systemic and supra-systemic forms of their intrinsic systemic control capabilities. Often, for the sake of expediency this transposition does and should occur.

The fundamental problem that emerges is that the fewer communication alternatives available to and adopted by an open system, the more the tendency toward the equilibration of its methods to deal with environmental problems, and perhaps, the more the tendency toward organizational entrophy. The more alternative communication forms the system adapts to and maintains within controllable constraints, the greater its potential for regulating and directing decision rules and component values at all organizational levels.

Based upon its existing systemic properties, a CAP organization determines both the density and type of information to be acquired from its target environment, and how that information is to be transformed. An apparent limitation resides in the diverse functions and dynamics of the levels affecting the agency's productive system, its maintenance system, and its adaptive system.[37] Each level acquires specialized information, which then processes and organizes that information differently depending upon the perceived criticality assigned to the acquired information by that and other levels.

> In every communication the participants offer to each other definitions of their relationship . . . each seek(ing) to determine the nature of the relationship. Similarily, each responds with his definition of the relationship, which may confirm, reject, or modify that of the other. . . . If the process did not stabilize the wide variations and unwieldiness, not to mention the ineffi-

ciency of redefining the relationship with every exchange, (it) would lead to run-away and dissolution of the relationship.[38]

Structurally, the stabilizing of a relationship's definition "has been called the RULE of the relationship, (that is) it is a statement of the redundancies observed at the relationship level, even over a diverse range of content areas".[39] By necessity, this stabilizing process is more than critically dependent upon the abilities of each individual confronting the information field of the target situational environment. It serves as the definitional base of the organization-as-a-whole.

Summarily, a component analysis based on the above implications of systemic communication must be guided by the following hypotheses: (1) that an integratively complex open system, as in a community action agency, exhibits more differentiation and integration as it organizes and processes information; (2) that communication manifests invariance between the system's components, including the relationships between subsystems and between their attributes; and (3) without communication neither variance nor invariance is manifested at any level.

Using the preceding models, a multidimensional, multileveled investigation of the Jackson-Williamson County Action Agency has been initiated. The methodological strategy for this study is contingent upon and isomorphically related to the processes as incipiently morphogenic communication structures of the organization being analyzed. The basal unit of analysis is the SITUATIONAL EVENT which consists of the patterned activities of a number of individuals functioning in a common interact field. As such, a situational event is viewed as a time-space manifested interaction "within which relevant self-other patterns arise as dynamic relationships in response to interpersonal tension".[40] They are complementary or interdependent with respect to some common output or outcome, and are repeated, relatively enduring, and bound in space and time.[41] The patterned activities are those communicative interchanges, including both their FORM and CONTENT, that manifest the situational definitions held by the participants. These situational definitions are the "expected 'other' responses aroused in a given (participant) by a given situation".[42] Extraction of these definitions from the target communication interchanges will be accomplished by applying a propositional analytic technique,[43] that permits the formulation of verbal descriptions of the situation as it is perceived by the participants. Each communicative event will be examined to determine the information content, especially the emergent transformational patterns. Once these patterns are delineated, the functional value will be determined for each interchange and the rules that seemingly specify the mode of operation. Functional VALUE, based on given available alternatives for interrelationships within a system and their possible outcomes, becomes defined as a function of: (1) the probabilities of choice associated with courses of action; (2) the efficiencies of the courses of action for each objective; and (3) the value of the outcomes to the participant, as part of the system, which the participants might define themselves, as expected return, expected gain, or expected loss.[44] Value, then, becomes the determining factor regarding which course an interrelationship can assume while still maintaining its effectiveness and efficiency.

At this juncture, the preliminary information processing functions have been tentatively identified and will be controlled through a "live simulation" procedure. Simulation refers to the construction and manipulation of an OPERATING model, that model being a physical or symbolic representation of all or some aspects of a social or psychological process.[45] As used here, the "live simulation" will be a reconstructed stimulus environment in which the participants must functionally "play out" the interaction events under consideration. The environment will contain all those physical and interpersonal features that exist in the real environment, however, the simulated environment will be controlled by a programmer who resides outside of the main interact field. The described operational models are intended to designate the processes in action, and the simulation of these models is intended to provide further information about variable, component, and relationship changes as they occur within the organization over time.[46]

As a result of initial analysis of the organization, the following participant information processing capabilities have been identified to be factors basic to the development of the simulation: (1) the use of multi-valued logic for asymmetrical relationships; (2) the evaluating and planning of "helping" strategies; (3) the abstracting of organizational goals and non-additive community-agency relationships; and (4) the perception and management of intra-organizational dysfunction of communication channels affecting decision-making.

The ultimate intent of this investigation is to prepare measures of preventative maintenance that are appropriate to the organization and the multiple situations that emerge within it and between it and other organizations and sectors of the community. Because such an intent demands that the study be developed concomitantly with the emergence of the situational peculiarities of the organization, which themselves demand "tailored" analytic techniques, the current status of this investigation is at the completion of Phase I, the formulation of appropriate basic investigation strategies. Phase II has been initiated which is the development of a live simulation design that can be implemented in Phase III as a constructed live simulation model.

REFERENCES

1. Bennis, Warren G., and Slater, Philip E. "Democracy is Inevitable," 'The Temporary Society', eds. W. G. Bennis and P. E. Slater. New York: Harper and Row, 1968, pp. 1-19.
2. Clark, Kenneth B., and Hopkins, Jeannette. 'A Relevant War Against Poverty'. New York: Harper and Row, 1969, p. ix.
3. Glasgow, R. W., "The Ayn Rand Syndrome – A Conversation With Herbert J. Gans," 'Psychology Today', 3, 1970, 61.
4. Johnson, Lyndon B. "The President's Message to Congress," March 16, 1964. See also Clark and Hopkins for an in-depth analysis of CAP programs across the country.
5. Kramer, R. W. 'Participation of the Poor'. Englewood Cliffs, N.J.: Prentice-Hall, 1969, p. 1.
6. Murray, Elwood, "Streamlining the Speech Edifice I: General Systems Theory, General Semantics, Semantics," a paper presented at the Western Speech Association, Salt Lake City, Utah, November 28, 1958.
7. 'Ibid.'
8. Jennings, Russell W., "The Live Simulation of a Community Action Organization's Communication System," a research proposal submitted to the Office of Research and Projects, Southern Illinois University, January, 1970.
9. Schroder, Harold M., Driver, Michael J., and Streufert, Siegfried. 'Human Information Processing'. New York: Holt, Rinehart and Winston, 1967, p. 14.
10. Harris, George T., "Warren Bennis, A Conversation," 'Psychology Today', 3, 1970, 48.
11. Schrolder, Driver and Streufert, p. 12.
12. Katz, Daniel and Kahn, Robert L., 'The Social Psychology of Organizations'. New York: John Wiley and Sons, 1966, p. 453.
13. 'Ibid.', p. 454.
14. Jennings, 1-2.
15. Gould, Julius and Kolb, W. L., 'A Dictionary of the Social Sciences'. New York: The Free Press, 1964, p. 464.
16. Koestler, Arthur. 'The Act of Creation'. New York: Macmillan, 1964, p. 631.

17. Frank, Lawrence K., "The Prospects of Genetic Psychology," 'American Journal of Orthopsychiatry'. 21, 1951, 506-522.
18. Watzlawick, Paul, Beavin, J. H., and Jackson, D. D. 'Pragmatics of Human Communication'. New York: W. W. Norton, 1967, p. 120.
19. Schroeder, Driver, and Streufert, 15.
20. Miller, J. G., "Information Input, Overload, and Psychopathology," 'American Journal of Psychiatry', 116, 1960, 695.
21. 'Ibid.', p. 19.
22. Katz and Kahn, pp. 456-457.
23. "Community Action Agency, Plans and Priorities," Jackson-Williamson Community Action Agency, Grant No. Illinois CG-4118, May 1, 1970.
24. Buckley, Walter, "Society As A Complex Adaptive System," 'Modern Systems Research for the Behavioral Scientist', ed. Walter Buckley. Chicago: Aldine Publishing, 1968, p. 491.
25. Katz and Kahn, p. 20-21.
26. 'Ibid.', p. 36.
27. Hall, A. D. and Fagen, R. E., "Definition of System," 'Modern Systems Research for the Behavioral Scientist', ed. Walter Buckley. Chicago: Aldine Publishing, 1968, p. 82.
28. 'Ibid.'
29. Schroder, Driver, and Streufert, p. 12.
30. 'Ibid.'
31. 'Ibid.'
32. Watzlawick, Beavin, and Jackson, p. 50-51.
33. Bruyn, Severyn. 'The Human Perspective in Sociology'. Englewood Cliff, N.J.: Prentice-Hall, 1966, pp. 180 and 225.
34. Buckley, p. 492.
35. Katz and Kahn, pp. 223-229.
36. Watzlawick, Beavin, and Jackson, p. 131.
37. Katz and Kahn, pp. 112-113.
38. Jackson, Don D., "Family Rules: The Marital Quid Pro Quo," 'Archives of General Psychiatry', 12, 1965, 591.
39. 'Ibid.', p. 593.
40. Fearing, Franklin, "Toward a Psychological Theory of Human Communication," 'Interpersonal Communication: Surveys and Studies', ed. D. C. Barnlund. Boston: Houghton-Mifflin, 1968, pp. 32-34.
41. Katz and Kahn, p. 17.
42. Cottrell, L. S., "The Analysis of Situational Fields in Social Psychology," 'Interpersonal Communication: Surveys and Studies', ed. D. C. Barnlund. Boston: Houghton-Mifflin, 1968, p. 387.
43. Jennings, Russell W., Pace, Thomas J., and Winters, Dennis E., "The Situational Analysis of Urban Communication: An Extended Case Study of Racial Tension," 'Research Designs in General Semantics', ed. Kenneth Johnson. New York: Gordon and Breach, 1970; Russell W. Jennings, "Field Studies in the Urban Situation: Chicago, 1967-1968," 'Proceedings of the S. A. A. Conference on Research and Action', ed. James Roever, New York: Speech Association of America, 1969; Russell W. Jennings, "A Community Action Group's Definition of Urban Tension: A Situational Analysis Through Public Address in Cicero, Illinois, in 1967-1968," unpublished dissertation, Southern Illinois University,

1968; and Dennis E. Winters, "A Community Action Group's Definition of Urban Tension: A Situational Analysis Through Public Address in Chicago, Illinois, in 1967-1968," unpublished dissertation, Southern Illinois University, 1968.
44. Ackoff, Russell L., "Towards a Behavioral Theory of Communication," 'Modern System Research for the Behavioral Scientist', ed. Walter Buckley, Chicago: Aldine Publishing, 1968, p. 213.
45. Dawson, Richard E., "Simulation in the Social Sciences," 'Simulation in Social Science', ed. Harold Guetzkow. Englewood Cliffs, N.J.: Prentice-Hall, 1962, p. 3.
46. 'Ibid.', p. 5.

Systems Concepts in Literary Analysis

DONALD E. WASHBURN

A systems approach to the study of literature presupposes that poems, like atoms, crystals, cells, and societies, can be thought of as complex systems. It also assumes that to some extent words conform to principles that are familiar in the natural world, that is, to physical, biological, and social systems. In the plastic arts it is not unusual for artists to turn to nature for models. Frank Lloyd Wright, for example, derived many of his ideas from biological structures. Michelangelo and Leonardo da Vinci were also alive to the importance of design based on meticulous observation of natural forms. But the extent to which imaginative literature is ordered in accordance with principles common to a wide range of non-verbal systems has seldom been appreciated.

People who study verbal systems have an especially difficult task. Their most important concern is with meaning, and meaning is something difficult to define. To resort to an analogy, it is somewhat like the process that takes place when a bulb is switched on. Part of the house lights up. But it would be incorrect to say that the light is to be found in the bulb itself. Like a word in isolation, a bulb unscrewed is not very illuminating. The bulb is a special kind of conductor, part of a vast system of transformers and channels that will not function unless every part is in order. There is no light if a fuse is blown or if the power plant fails. Likewise, a word is merely ink on the page if no connections can be made in the network of systems and relationships that surround and inform the interpretative act.

To expand the analogy, an entire poem may be visualized as a kind of circuit drawing its energy and information from a semantic "grid." The grid is as vast as the cultural matrix itself, ending where human knowledge and experience end. Just as the Greek word LOGOS means both word and Word, a poem is both concrete and universal, carrying the potential perhaps of an entire world view or way life. But although the poem draws its life and meaning from the system as a whole, it also maintains its own special integrity and capacity. And since it establishes new connections, it may, in fact, subtly alter the nature of the semantic environment upon which it depends.

The enormous scope and complexity of this semantic environment poses a problem for the critic, whose job is to discover what connections are relevant. Naturally he wants to take into account everything he can, or at least everything that increases the amount of order (negentropy) that the poem seems capable of generating. But where does he stop? Strictly speaking, everything is connected with everything else. A full unfolding of the word "man" would by implication take in all the statements that have ever been made about anything.

In practice, of course, the critic, like the scientist, delimits beforehand the scope of his concern so as to make his task manageable. Sometimes he specializes in certain kinds of connections to which he has become particularly sensitive. There are schools of criticism, for example, which concentrate on exploring the relation between the sort of work the writer has written and the sort of life the writer has lived. Others propose to de-emphasize the biographical implications in favor of an account of how the work itself has been organized and how the implicit may be derived from the explicit. With some approaches, the categories of psychology, sociology, metaphysics, and mythology are brought to bear. Others concentrate on the linguistic and semantic schemes that determine the design of the text itself. Each approach engages some aspect of the extrinsic or intrinsic context. Each promises to enlarge the network of associations that may be validly "plugged into" the verbal system.

Practically all systems can be studied either as parts or as wholes made up of parts. To return to the electronic analogy, imagine a lighting system set up for the purposes of display. Someone wishing to understand as much as possible about how it works might decide to investigate the supersystems that make the display possible: the power plant, the generators, transformers, resistors, etc. Or he could ignore the outside agencies and concentrate on what is happening in the syntax of the lighting system itself. He could study the circuitry, the fuses, the switches, the lamps, the rheostats, the stroboscopes, the effects produced by these arrangements, and, in short, whatever may be considered interior to the system.

But with verbal systems there is an important complication. Since words are representations, the distinction between interior and exterior relations becomes ambiguous. Imagine that the lighting system already described was set up to display a series of mirrors aimed to reflect, among other things, aspects of the greater world that surrounds the system. The viewer is in the peculiar position of looking into the system and out of it at the same time. The things he sees in the mirror may be significant because they are part of the supersystem surrounding the lighting system or because they are an important part of the display which IS the lighting system. Whether he looks over his shoulder or into the mirror, he will be made aware of the same event. Literature, like art in general, aims to make us forget the distinction between inside and outside.

Because of what general semanticists call "self-reflexiveness," the lighting system may even display itself. And depending on how the mirrors are arranged and on the point of view, a reflection may contain a reflection of a reflection, or a reflection of a reflection of a reflection. In this sense a part may become interior to itself, much as a person may become a character in his own story. There are also warps in the surface of the mirror and the possibility for double exposures so that two or more images may be superimposed or taken in at a glance. Some things are given emphasis by the shape of the mirrors or the intensity of the light. Other things are rendered shadowy and indistinct. The critic who wishes to understand how the system works must know not only what is relevant but also what one kind of relevance is doing to another kind of relevance. A veritable hall of mirrors! It is hard to imagine a system with more functional variables or variable functions than language approaching its full potential as a communication medium.

But where there is order there is the possibility for understanding. Like the lighting system, a work of literature exists in a context. It is related to the larger systems which brought it into existence and which respond to its influence. Also, like the display, it has an interior design, interlocking levels of organization and patterning, that constitute its structure. And parallel with these, it contains a kind of UMWELT, the fictional world or vision which points beyond itself to some intuited order in the nature of things. These interlocking systems turn out to be aspects of the act of communication itself. This, of course, is an indivisible process, but for the sake of exposition it is possible to suppose that certain orders of relationship have a sufficient degree of autonomy and distinctness to warrant the name "system." If this is granted, then the following preliminary scheme may prove useful in suggesting the scope of literary studies and the variety of its concerns.

TRADITIONAL CONCERNS OF LITERARY STUDIES THAT MAY BE DESCRIBED AS SYSTEMS

1. SOURCE SYSTEMS

A. World-writer: systems of knowing and experiencing
B. Psyche-writer: systems of inspiration and imagination
C. Writer-writer: systems of influence and tradition
D. Writer-work: systems of invention and craft

II. MESSAGE SYSTEMS

A. Graphemic systems: graphemes, fonts, typographical forms, etc.
B. Phonological systems: speech sounds, rhythms, pitch, juncture, etc.
C. Lexical systems: vocabularies, connotations, idioms, etc.
D. Syntactic systems: constructions, tagmemes, transformations, etc.
E. Formal systems: patterns, distributions, motifs, symmetries, etc.
F. Epistemic systems: images, concepts, schema, etc.
G. Semantic systems: figures, symbols, metaphors, allusions, etc.
H. Affective systems: moods, attitudes, judgments, etc.
I. Presentational systems: accounts, episodes, settings, etc.
J. Interpersonal systems: witnesses, encounters, roles, etc.
K. Thematic systems: doctrines, philosophies, values, etc.
L. Mythopoetic systems: archetypes, mythologems, deities, etc.

III. TRANSACTION SYSTEMS

A. Work-reader: systems of comprehension and appreciation
B. Work-literature: systems of authentification and classification
C. Work-criticism: systems of interpretation and evaluation
D. Criticism-criticism: systems of systems and criteria

The hypothesis of this paper is that concepts useful in the study of systems in general can be useful in the study of literature, i.e., verbal systems. These concepts seem to provide a promising scheme for relating and integrating the diverse categories presently available for literary analysis, many of which have emerged as a consequence of historical accretion rather than as part of a unified and systematically developed theory of symbol formulation. It is promising because there is evidence that these concepts have in the past been utilized by some critics, but mostly on an intuitive basis. Usually the concepts appear in a context that makes them difficult to recognize. An act of translation, or at least transposition, is usually required to make them explicit.

There are three variables important to this process which define the parameters within which the present discussion will proceed. The first is the magnitude of the system, which can be indicated in a rough way by reference to the categories listed above. The second is the systems concept itself, which can be referenced in the literature of cybernetics, general semantics, general systems theory, etc. The third is the critical term or principle as it is used in theoretical discussions and in the analysis of literary works. A high degree of congruence between the systems concept and the literary principle regarded structurally, will be understood to lend support to the hypothesis and to confirm the impression that a systematic study of literary criticism from this point of view would produce interesting results.

Congruence cannot be precisely measured, but for the sake of this paper it is indicated whenever a key term can be assimilated without violence to a systems concept at a particular order of magnitude. For example, the concept of levels of meaning, which recognizes the stratified organization of literary forms, can easily be assimilated to the broader principle of Hierarchy (super-ordination/subordination) familiar to students of general systems theory. This, in turn, can be seen as a variant of the concept POLARITY: Centralization/ Differentiation, which is useful in describing complex organizations from zygotes to communes.

However, it must be made clear that the search for relationships has been conducted within certain limitations. Naturally iit is impossible here to treat in any comprehensive way the full range of general systems concepts, nor, for that matter, those that have been applied to literary studies. Therefore, a selection has been necessary based on what appears a priori to be most fundamental or interesting. Special attention has been given to twelve basic principles or concepts that are relevant to a wide variety of systems. Each concept is represented in capital letters and explained in terms of a contrast. Similar or parallel formulations in general systems theory and general semantics are also indicated.

CONCEPTS BASED ON FORMULATIONS FUNDAMENTAL TO GENERAL SYSTEMS THEORY AND GENERAL SEMANTICS

CONCEPT 1 Pattern: Order/Disorder
G. Sys. State: negentropy/entropy
G. Sem. Predictability: sanity/unsanity
CONCEPT 2 Integrity: Stability/Instability
G. Sys. Condition: equilibrium/disequilibrium
G. Sem. Irritability: steady state/disturbance
CONCEPT 3 Articulation: Continuity/Dissociation
G. Sys. Boundedness: field/closure
G. Sem. Relations: non-elementalism/elementalism
CONCEPT 4 Complexity: Aggregation/Supersummation
G. Sys. System: summative relations/constituitive relations
G. Sem. Structure: additivity/non-additivity
CONCEPT 5 Process: Invarience/Transformation
G. Sys. Isomorphism: constant/variable
G. Sem. Abstracting: similarity/difference
CONCEPT 6 Periodicity: Cycle/Frequency
G. Sys. Rhythm: activity/recurrence
G. Sem. Rhythm: function/rate
CONCEPT 7 Polarity: Centralization/Differentiation
G. Sys. Hierarchy: superordination/subordination
G. Sem. Multiordinality: intensionality/extensionality
CONCEPT 8 Perspective: Formulation/Bisociation
G. Sys. Information: input/codification
G. Sem. Awareness: orientation/illusion
CONCEPT 9 Proportion: Congruousness/Distortion
G. Sys. Allometry: compatibility/incompatibility
G. Sem. Abstracting: correspondence/false-to-fact relation
CONCEPT 10 Latitude: Necessity/Option
G. Sys. Degree of freedom: matrix/strategy
G. Sem. Flexibility: conditioned reaction/conditional reaction
CONCEPT 11 Regulation: Inhibition/Reinforcement
G. Sys. Feedback: symmetrical/asymmetrical

G. Sem. Reflexiveness: negative/positive
CONCEPT 12 Sufficiency: Optimization/Degeneration
G. Sys. Growth: evolution/decay
G. Sem. Evaluation: consciousness of abstracting and time binding/pathological identification and social disorders

Although many of these terms may seem ultramodern because they have become current in the scientific communities of our own time, the concepts or intuitions to which they refer are quite ancient. Even a cursory review of the history of literary criticism shows that ideas recognizable as systems concepts have played an important part in the development of literary theory.

The attempt to understand symbolic processes began in the West with Aristotle. In his 'Poetics' he provided us with our earliest description of the structure of literature. Unlike Plato he seems to have brought to the task a biological rather than a geometrical conception of form. It is not surprising then, that some of his most important ideas can be understood in a systems framework. For example, the evolution of Greek tragedy from the early rituals in honor of Dionysus to the highly sophisticated masterpieces of Sophocles reminded Aristotle of the process of biological development by which the young of a species arrive at maturity. Just as a developing animal undergoes a succession of transformations until it achieves a relatively steady state in the fully realized adult, so a literary genre fulfills the organic form implicit in its own nature as it is shaped by generations of practitioners.

The Aristotelian concept of "magnitude," a size and proportion appropriate to the occasion and purpose for which the work was conceived, is also both fundamental and familiar as a concept useful in considering many kinds of systems. In biology, for example, there is the principle of allometry, which describes the regularities in patterns of growth that determine the relative size of one organ in relation to others or to the organism as a whole. Because of this, the size of the head in relation to the body is not something that may assume any proportion, at least under conditions that presently prevail on the planet. These relatively constant ratios testify to the interdependence of biological processes. Aristotle was the first to suggest that a play is also governed by laws of proportion and that the ratio of its parts may be altered only within certain limits if it is to be communicationally viable.

This is easily explained in the case of a comedy. A comic plot expands by multiplying complications and opportunities for unlikely encounters. The number of mix-ups, coincidences, incongruities, reversals, etc. is potentially unlimited. That is to say, the author could go on adding new ingredients and recombining the old ingredients indefinitely, in which case the complications would increase at a geometric rate. In practice, of course, the system is stabilized at some level below that which would fatigue and finally bore the audience. After a certain point of complexity is reached, the plot must "unwind" and simplify or else become another kind of experience, one that may shatter the presentational frame that organizes the communication taking place.

Aristotle also invoked a systems principle to explain how an audience might witness distressing events, as portrayed, for example, in a tragedy, and still leave the theater with a sense of satisfaction and relief. The doctrine of catharsis is based on an analogy between systems in which energy is generated and discharged and the dynamics of semantic reactions. The images of the play are "conductors" which raise the level of psychoneurological activity. They excite, in Aristotle's terms, the emotions of pity and awe. Were these tensions to remain unresolved, the experience would be a painful one. But the structure of the play also provides for the release of negative emotions, some antecedent to the aesthetic experience. Hence, the play has a positive social value in that it serves to purge the audience of potentially disruptive energies. This principle of surcharge and discharge is familiar in many kinds of systems from electric batteries to the human bladder.

The classical idea of narrative climax also suggests many kinds of parallels. The climax is the moment of greatest tension before the central conflict or problem is resolved. Presumably catharsis begins to take place following the moment of climax. But catharsis refers to changes in the audience. The climax is part of the design of the play and it depends on the order in which events are presented. Any system which can be represented as having a rhythm or a rising and falling action can be considered climactic provided that something decisive has happened in the system that alters its character, direction, or magnitude. The range of possible application is enormous and includes climaxes which are musical, meteorological, botanical, sexual, and even architectural.

Aristotle was one of the first to identify the super-summative character of poetic meaning. He noted that in a play the whole is greater than the parts taken separately. The playwright has the task of creating a design in which everything contributes to the desired effect and everything is in exactly the right proportion. He must take into account the psychological limitations of his audience, the extent to which it is able to assimilate the details of time, place, and dramatic action. He must select only what is necessary to fulfill his purpose. When he is successful, nothing can be added or deleted without disturbing to some degree the unity he has created. The importance of unity has been stressed by countless critics down to the present time. Super-summative relations are also, of course, the rule in biological and social systems.

Judging from the 'Poetics,' Aristotle was especially interested in tragedy. This is a form that portrays the dramatic action that results when the human order becomes disastrously dissociated from the transcendental orders which sustain it. The central character is sometimes guilty of hubris, the overweening arrogance that provokes the enmity of the gods. In general systems terms, it is a case of a hypertropic sub-system that threatens the welfare of the whole. In his blindness or willfulness, the tragic hero violates some part of his ecosystem and sets in motion compensatory events that eventually lead to his own suffering or death. The concept of entropy is also apropos here, since the plot, as it is subject to various narrative complications, assumes a state of increased disorder relative to the purposes of the hero. However, since he becomes aware of what had been obscure before, his suffering may lead to a wisdom of a higher order or a new sense of the whole.

The special virtue of the tragic spectacle is that members of the audience may be made aware of super-ordinate realities and, by imaginatively participating in the hero's disaster, forestall in themselves corresponding tendencies. The experience provides a kind of negative feedback which is beneficial in preventing the community from exceeding limits which would threaten the systems upon which it depends. The religious origins of tragedy are well known, and perhaps it is safe to say that one of the functions of religion is to remind people of these limits. True religious piety is a hedge against tragic hubris since it assumes a cosmic frame of reference based on ultimate wholes. Therefore, it approaches reality as a kind of mystery — like the parabola of Korzybski's structural differential, which represents what cannot be directly known or exhausted.

Comedy, on the other hand, attacks the false piety that literalizes the mythopoetic and sustains a social order that has become inflexible and unresponsive to vital needs. The limits, in other words, have become outmoded. They are legalisms, semantic fossils, inappropriate to prevailing conditions. Some sort of adjustment is necessary in the calibration of the system to bring it back into a fruitful relationship with the larger universe. Unlike tragedy, comedy ends by celebrating the ability of the system to cope with its own dissociations. Human ingenuity is shown to be equal to the task of restoring the community to a state of equilibrium. The audience, in recognizing the folly of one-sided or outmoded behavior, becomes sympathetic to standards that supercede the arbitrary and self-serving piety of the comic figures.

The pioneering efforts of Aristotle were extremely influential and became the point of reference in discussions of literary form for over a thousand years. Much subsequent commentary turns out to be a rehash or elaboration of his ideas. The Romans were less interested in theory than in practice, so that much of their criticism is advice about how to write successfully. But even Horace emphasizes the importance of balance and proportion – Aristotelian ideals. In the Middle Ages allegory became important and with it a view of language which stressed its many levels of meaning. This is very apparent in works like the 'Romance of the Rose' and Dante's 'Divine Comedy'. An episode such as Dante's encounter with the three beasts (lion, leopard, she-wolf) in the 'Inferno' may be interpreted, for example, on the literal, the moral, the political, and the anagogic (spiritual) levels.

In some ways this is an extension of Aristotle's organismic model. The patterns of polysemy that are characteristic of complex literary structures are analogous to the hierarchical features and overlapping functions of biological organizations. Just as the human mouth may be simultaneously regarded as an instrument of ingestion, respiration, and communication, so a passage or scene from a work of literature may have numerous senses and accomplish a variety of artistic purposes. The medieval tradition of symbolic pluralism foreshadows the modern interest in intrinsic relationships (as opposed to extrinsic approaches which stress the relevance of biographical, sociological, psychological, etc. contexts). This trend has reached its apogee in the close textual analysis of the so-called new critics, who are primarily interested in the structure of the work rather than the conditions which gave rise to it.

It is interesting, however, the extent to which discussions of intrinsic relations have been colored by a vocabulary drawn from accounts of non-verbal systems of all sorts – biological, physical, even mechanical. Critics write about "levels" of meaning, or about the "architectonics" of a work. They discuss "texture" or "surface" meaning. There are "clusters" of images, which may be in a state of ironic "tension" or "polarized." A style may be described as "dynamic" or "decadent," "masculine" or "feminine," "obscure" or "lucid." Terms with spacial or visual origins are legion. Even the word "structure" is likely to suggest a three dimensional figure rather than a unified impression mediated by a complex of symbols. The study of the subtle effects of these metaphors on literary theory would make a project in itself.

But whether or not these metaphors oversimplify or misrepresent the nature of symbolism, they seem often to be a rudimentary effort to understand works of literature as systems. This is an especially difficult task because many relationships are "submerged," recoverable only by inferring the sequence of decisions that the writer has made in shaping his medium. The critic is, to some extent, in the same position as the biologist who is trying to reconstruct the genetic code that organized the animal he is studying. The code acts to suppress lines of development that are not in conformity with the overall plan. This is, of course, physio-chemical in ontogeny. In poetry or fiction it seems to be related to a feedback process in which output becomes input, the writer his own reader. In both cases the predispositions unfold themselves in a hierarchical fashion.

One of the best descriptions of the hierarchical nature of literary composition is found in the essay by Edgar Allen Poe in which he describes how he wrote his well-known poem, "The Raven." The series of decisions which he discusses begins with those most basic and ramifies in an increasingly self-limiting way until the basic plan of the entire composition has been realized. He begins, for example, by making a decision about length. The poem shall be short. Then he determines the kind of effect he wishes, in this case, beauty. The tone shall be one of sadness. The principal artistic device will be a single word refrain ("Nevermore") spoken by a non-human (the raven). The subject which best realizes these preliminary requirements is the death of a beautiful woman. He subsequently makes decisions about the protagonist, the moment of climax, the stanza form, the setting, the changes of tone, and the narrative sequence.

Whether Poe is telling the truth about how the poem was composed or whether the order might be varied in some ways (but not in others), the important thing to notice is that his fundamental strategy is based on the recognition of a hierarchy of decisions, which, like the rules of generative grammar, are applied in a definite order and in such a way that each decision places limits on what may be subsequently decided. Although he does not extend the method to the level of the sentence (nor to the stanza for that matter), potentially it could be done. The final result would be a statement about the structure of the poem and would show the relationship of parts to wholes, and of fixed rules to flexible strategies. For an appreciation of what literary hierarchies have in common with hierarchies in other kinds of systems see Arthur Koestler's 'The Ghost in the Machine,' especially Parts One and Two.

Many systems concepts are apparent in the works of contemporary critics. For example, I. A. Richards' concepts of "synaesthesis" describes the way in which literature functions to restore competing psychological impulses or attitudes to a state of equilibrium. The equilibrium, of course, is psychological, but it implies a semantic equilibrium dependent on the way that the work of literature itself is conceived and organized. If the work is ironic, that is, if it expresses experiences that are normally thought to be discordant or contradictory, it sets up a kind of verbal polarity which must be in some way transcended. This forces the reader to reorder his habitual categories of awareness and to find unity in diversity. In this way the poem assists in the creation of a new semantic equilibrium.

The number of instances in which systems concepts can be shown to be relevant in literary theory is enormous. They are applicable at every level of magnitude and in a wide variey of connections. A complete survey would be impossible. But it is possible at least to suggest the range of applications by searching for at least one instance of each concept at each level of magnitude. It is assumed that wherever relations are congruent, a concept may be said to be a component of some structure at that level of magnitude. For example, if cadence : caesura :: continuity : dissociation, then the concept of Articulation is shown to be relevant to the literary structures to which the terms refer.

What most of the concepts describe is a contrast expressed as a continuum in semantic space. The contrast is defined by its limiting conditions. Concept 1, i.e. Pattern finds its limits in absolute entropy or disorder on the one hand, and absolute negentropy or order on the other. Between are the various degrees of orderliness or disorderliness, depending on one's viewpoint. It is somewhat like the glass of water that may be called either half-full or half-empty. A great number of critical terms have the concept of pattern as one of its components. Some, like "orthography," draw attention to a high degree of order relative to a norm, while the contrasting term "cacography" points to an increase in noise or entropy.

Concept 2, Integrity, is a measure of equilibrium and disequilibrium. A stable system maintains itself within certain limits; an unstable system falls apart or becomes something else. When a system approaches its limits of viability, processes are set in motion to restore the necessary balance. Although symbolic systems are not dynamic in the same sense as physical or biological systems, they do participate in certain norms and in the illusion of flux created by the isomorphism between time and linearity. The notion of "decadence" in literature, for example, presupposes a condition of deterioration with respect to some ideal of balance. Likewise, certain plots gain their power from the suspense created by situations in which psychological or social equilibria have been disturbed.

The principle of Articulation (Concept 3) refers to states of continuity or dissociation. The extreme of dissociation or boundedness is the hypothetical, completely isolated system. The extreme case of continuity is the completely open system, which would be no system at all. Between, there are various degrees of permeability. The symbolic equivalent of boundedness is a distinction or abstraction which, in Whorf's language, "segments" reality. To some extent, these reflect the discontinuities of experience. Any classificatory system is based on what are understood to be bounding conditions. It also creates the possibility for logical inclusion, which mirrors the relation between whole and part, system and subsystem.

Concept 4, Complexity, denotes the degree of interrelatedness that prevails among the parts of a system. One limit is absolute aggregation or separation of parts; the other is a state of supersummation in which every part is multiordinally related to every other part such that the whole is more than the sum of its parts. In language systems this is usually expressed in a kind of indeterminacy or ambiguity which is not resolvable by reference to the immediate constituents. An idiom, for example, cannot be understood by putting together the words that make it up. It is a truism of criticism that literary works are supersummative.

Any process is simultaneously something that changes and something that remains constant. Process (Concept 5), therefore, can be understood as a relation between transformation and invarience. The familiar idea of isomorphism, of course, implies both, since similarity of form presupposes both important differences (transformations) and important constancies (invariences). Change/constancy seems to be one of the fundamental characteristics of the universe, and there is probably no system that is not process. In verbal systems this feature is expressed both in the transformations that take place in the abstracting process and in the processes implied in the representation of the space-time continuum.

Periodicity (Concept 6) is also very basic and may be thought of as a special kind of process in which a cycle or pattern is repeated. The cycle itself must first be perceived as a configuration. Usually at least one occurrence is necessary for this. Subsequent occurrences establish the frequency of the pattern. The gradient of periodicity extends from the single occurrence at one extreme to numbers of recurrences at the other. When the latter is measured with reference to a particular unit (i.e. time), we have the rate of periodicity. Periodicity is extremely important in verbal systems and seems to operate at every level of organization, from the phonemic regularities of poetry to the regularities familiar as the archetypes of mythology and religion.

The concept of hierarchy is useful in many kinds of systems. Probably the most dramatic example is the human nervous system itself. In its simplest manifestation, it can be seen in the polarity of the magnet or cell. In fact, Polarity (Concept 7) is fundamental to any system which depends for its functioning on the establishment of a gradient. Unlike a simple gradient (positive and negative), complex biological, psychological, and sociological gradients must usually be understood multi-dimensionally, in terms of centralization and differentiation. Our models of these most often resemble the branching networks familiar in the study of neurology, computer programming, and syntactical patterns. Structures of polarity in verbal systems range from the two valued contrasts familiar in antynomy and meter to the multi-leveled differentiations of generative grammar and literary analysis.

When information is formulated in more than one way, the possibility for Perspective (Concept 8) may arise. This is a kind of symbolic pluralism or bisociation in which one conceptual matrix is imposed on another. It leads to effects which are familiar in humor, metaphor, parody, allegory, irony, etc. These exploit the psychological dynamisms that are set in motion when one orientation is forced to give way suddenly to another. The juxtaposition reveals relationships formerly obscured and in some cases creates new gestalten and new attitudes. It is, therefore, a powerful heuristic device, one which aids in the transcendence of habitual categories of thought and perception.

Concept 9, Proportion, takes into account the limits that are placed on the parts of a system in relation to the whole. These limits may restrict the size of the part or its function. For example, the human nose normally assumes a certain size in relation to the rest of the body. A larger than normal nose (i.e. Jimmy Durante's) is perceived as incongruous. The same would be true of a nose that assumed functional dominance. A man who sniffed everything in sight would also be comic. The important thing is that a part should make sense structurally and functionally in the context of the total system. The concept of proportion ranges from absolute harmony and economy of means (beauty?) to some ultimate condition of distortion or incompatibility. As we have seen, there are laws of allometry that apply in literary systems as well as in biology.

Latitude (Concept 10) refers to the degrees of freedom available to a system or function. At one extreme is absolute necessity, i.e., no choice. At the other is absolute freedom, an unlimited number of options. The latter is, of course, hypothetical since practically all human decisions are made within limits. It is usually a question of determining what options are available within the set of conditions that hedge a given decision. Many of the most important limits are to be found in the structure of the system itself. Languages have a high degree of determinism at phonological levels, less at morphological and syntactic levels, and very little at the level of discourse. Works of literature are frequently designed in accordance with certain conventions or rules that increase the amount of redundancy present.

Most sophisticated systems are controlled by patterns of positive and negative feedback. Concept 11, Regulation, designates these important processes, whose function is to provide more of the same (reinforcement) or less of the same (inhibition). The relation of the writer to his text in the act of composing is dependent upon complex feedback loops. The text itself becomes a control in determining the limits within which the reader may respond. And the reader may initiate self-corrective activities as he becomes more deeply involved in the implications of what he is reading.

Concepts such as "optimization" and "degeneration" imply that a system may be judged by standards that assess its long-range health or progress toward some equifinal condition of self-realization. In the opposite direction lies an ultimate of degeneration and disintegration. This continuum, referred to here as Sufficiency (Concept 12), is implicit in diagnostic statements of every kind and in literary judgments. Much can be learned from "bad" literature since pathology presumably occurs because a principle of organization at some level has been violated. Likewise, the "best" literature usually marks a new level of achievement in the evolution of literary form and serves as a touch-stone for generations of writers.

It should be pointed out that these twelve concepts are neither mutually exclusive nor exhaustive. They are, however, representative ideas that have proven their usefulness in describing how systems work. The distinction represented by each set of literary or critical terms is only one of many possible applications. In some cases, a term may be referable to a number of concepts. Therefore, the survey must be regarded as illustrative, an indication of the extent to which the conventional language of criticism can be assimilated to concepts familiar in systems theory. A preliminary analysis suggests that a great many literary terms either suggest systems concepts or include them as components.

Even allowing for a certain amount of controversy over the degree of congruence for some of the entries, the chart does show that there are vast areas of overlap. Therefore, the hypothesis has in the main been confirmed. But the value of this sort of charting lies also in the relationships it brings to light. The scheme, like an analogy, creates fresh possibilities for organizing information. It may also suggest new avenues for investigation and call attention to neglected aspects of symbolic form. For example, assuming that the principle of Proportion is important at a number of levels, it is necessary to show exactly how laws of proportion apply in verbal systems. Perhaps this can be done by systematically altering various structural ratios and determining the consequences. Finally, there is the promise that a systems frame of reference will bring greater coherence to the study of literature and will assist in the eventual assimilation of literary studies into the slowly emerging program for the integration of human knowledge.

I. SOURCE SYSTEMS

		A World-Writer (Experience)	B Psyche-Writer (Inspiration)	C Writer-Writer (Tradition)	D Writer-Work (Craft)
1	Pattern:	knowledge	creativity	tradition	competence
		error	sterility	neglect	failure
2	Integrity:	adjustment	balance	vitality	concentration
		maladjustment	madness	decadence	distraction
3	Articulation:	sensibility	being	collaboration	design
		attention	ego	independence	fragment
4	Complexity:	abstraction	consciousness	plagiarism	itemization
		orientation	intuition	influence	synthesis
5	Process:	gestalt	archetype	norm	style
		ground	imagination	experiment	language
6	Periodicity:	event	impulse	practice	volume
		recurrence	obsession	trend	sequel
7	Polarity:	objectivity	earthiness	conventionality	medium
		subjectivity	spirituality	individuality	meaning
8	Perspective:	happening	dream	doctrine	strategy
		significance	interpretation	zeitgeist	correlative
9	Proportion:	veridicality	revelation	decorum	balance
		illusion	chimera	eccentricity	caprice
10	Latitude:	necessity	character	convention	rule
		agency	will	invention	improvisation
11	Regulation:	revision	repression	reserve	suppression
		confirmation	fantasy	participation	expression
12	Sufficiency:	intelligence	wisdom	erudition	perfection
		stupidity	superstition	pedantry	imperfection

II. MESSAGE SYSTEMS

		A Graphemic	B Phonological	C Lexical	D Syntactic
1	Pattern:	orthography	euphony	bon mot	grammaticality
		cacography	dissonance	malapropism	solecism
2	Integrity:	text	sense	standard	sentence
		corruption	echolalia	obsolete	ellipsis
3	Articulation:	elision	cadence	lexeme	construction
		hiatus	caesura	syncope	constituent
4	Complexity:	font	sounds	denotation	constituents
		style	onomatopoeia	connotation	idiom
5	Process:	grapheme	iatacism	etymon	pattern
		allograph	vowel	cognate	parallelism
6	Periodicity:	syllable	foot	name	line
		reduplication	meter	synonym	refrain
7	Polarity:	Roman	ictus	root	kernal
		italic	breve	derivations	expansion
8	Perspective:	typography	locution	word	proposition
		concretism	homonym	pun	ambiguity
9	Proportion:	script	pronunciation	designation	aphorism
		palimpsest	drawl	euphemism	pleonasm
10	Latitude:	spelling	pronunciation	word	grammar
		variant	intonation	neologism	usage

11	Regulation:	fine print	susurration	amelioration	curtailment
		bold face	exclamation	pejoration	amplification
12	Sufficiency:	calligraphy	mellifluence	euonym	fluency
		cacography	harshness	caconym	barbarism

		E Formal	F Epistemic	G Semantic	H Affective
1	Pattern:	structure	clarity	felicity	sympathy
		formlessness	obscurity	lameness	enmity
2	Integrity:	symmetry	insight	transcendence	equanimity
		asymmetry	riddle	paradox	ambivalence
3	Articulation:	passage	context	metonymy	mood
		parenthesis	allusion	distinction	transition
4	Complexity:	motif	description	conception	intimation
		effect	atmosphere	symbol	tone
5	Process:	pattern	representation	tenor	feeling
		variation	summary	vehicle	displacement
6	Periodicity:	stanza	image	citation	expression
		repetition	cluster	definition	reprise
7	Polarity:	concentration	perception	universality	satire
		diffusion	synaesthesia	concreteness	romance
8	Perspective:	expectation	impression	narrative	comment
		surprise	ambiguity	allegory	irony
9	Proportion:	form	rationale	simile	sentiment
		parody	absurdity	conceit	sentimentality

10	Latitude:	convention	experience	cliche	participation
		option	interpretation	wit	self-assertion
11	Regulation:	counterpoint	contradiction	understatement	approval
		amplification	corroboration	exaggeration	disapproval
12	Sufficiency:	realization	reliability	eloquence	power
		formlessness	fallibility	cant	flatness

		I Presentational	J Interpersonal	K Thematic	L Mythopoetic
1	Pattern:	contingency	communion	purpose	piety
		absurdity	alienation	arbitrariness	sacrilege
2	Integrity:	motivation	identity	synthesis	pollution
		deflation	anonymity	contradiction	atonement
3	Articulation:	plot	identification	development	adventure
		episode	distance	digression	exploit
4	Complexity:	narrative	action	statement	manifestation
		parable	enactment	implication	mystery
5	Process:	incident	character	theme	suppliant
		version	development	elaboration	purification
6	Periodicity:	event	performance	thesis	figure
		repetition	role	restatement	archetype
7	Polarity:	intention	protagonist	synopsis	deity
		complication	foil	explication	hierarchy
8	Perspective:	exposition	identity	assertion	substance
		reversal	mask	irony	epiphany

9	Proportion:	realism	portrait	rationale	piety
		farce	caricature	obsession	hybris
10	Latitude:	fortune	automatism	dogma	fate
		act	intention	dialectic	freedom
11	Regulation:	furtherance	prohibition	denial	repentance
		hinderance	sanction	confirmation	temptation
12	Sufficiency:	success	normalcy	profundity	paradise
		failure	license	superficiality	hell

III. TRANSACTION SYSTEMS

		A Work-Reader (Transaction)	B Work-Literature (Characterization)	C Work-Criticism (Evaluation)	D Criticism-Criticism (Theorization)
1	Pattern	comprehension	familiarity	judiciousness	consistency
		incomprehension	anomaly	rancor	inconsistency
2	Integrity:	synaesthesis	classicism	receptivity	tradition
		disorientation	romanticism	prejudice	revision
3	Articulation:	impression	type	orientation	program
		aspect	example	judgment	dogma
4	Complexity:	theme	composition	opinion	formulation
		meaning	situation	value	intuition
5	Process:	text	relevance	meaning	issue
		translation	allusion	paraphrase	dialectic
6	Periodicity:	reading	fashion	opinion	critique
		rereading	revival	consensus	school
7	Polarity:	passage	recognition	criterion	scholarship
		implication	association	evaluation	vision
8	Perspective:	message	genre	response	agreement
		ambiguity	mode	ambivalence	equivocation
9	Proportion:	interpretation	manner	sentiment	position
		misreading	burlesque	sentimentality	polemic
10	Latitude:	stock response	convention	bias	orthodoxy
		construction	experiment	receptivity	heterodoxy
11	Regulation:	distraction	censorship	rejection	reservation
		engrossment	publication	appreciation	endorsement
12	Sufficiency:	clarity	maturity	excellence	relevance
		obscurity	decadence	worthlessness	irrelevance

Epistemology and the Biosocial Crisis: The Difference that Makes a Difference (1)

ANTHONY WILDEN

> The things themselves, which only the limited brains of men and animals believe fixed and stationary, have no real existence at all. They are the flashing and sparks of drawn swords, the glow of victory in the conflict of opposing qualities.
> NIETZSCHE: 'Philosophy in the Tragic Age of the Greeks' (1873)

1. THE COUNTER-ADAPTIVE EFFECT OF ADAPTIVE CHANGE

Whether or not the present ecological and social crisis is actually the terminal state of an advanced industrial culture, there is an absolute limit to the possibilities of adaptive change in any ecosystem where adaptive changes in one part of the system set off irreversible, retrogressive changes in its other parts, such as its environment. Moreover, all major adaptive changes in any ecosystem tend to trade short term survival value for long term counter-adaptive value so long as the adaptive continues to operate within the original "phase space" (i.e. a spectrum of multiple and interdependent variables) which allowed the original change. As long as the adaptation is an extreme and not a mean, that is, as long as the phase space in which it occurred has not been shifted or RENORMALIZED, the adaptive and behavioral flexibility of the system is reduced.

For example, the selection of pre-existing genes which produced black moths to fit trees blackened by air pollution in England has original survival value in respect of the "selectors of incompatible differences" in the ecosystem (the birds who eat black moths on grey trees and grey moths on black trees). But the black moth's territory and behavior is now restricted by his dependence on the factory system, and if air pollution finally destroys the trees, his adaptation will have been valueless. Not only will the whole ecosystem be hard up against one set of its parameters and unable to renormalize sufficiently to survive, but the subsystem "black moth" will have an equally reduced flexibility. He should have emigrated, for by refusing the grey moth a visa to the Black Country, the State Department of nature, in consultation with British industry, has revoked the black moth's passport.

The same principle of counter-adaptivity applies to cultural and technological adaptation, but here the effects are qualitatively of a different order from the problems of the black moth, whose effects upon his environment are mimimal. The short range survival value of the domestication of animals – a typical instance of the short range extension of a culture's life span through technology – is in many instances countered by the destruction of the ecosystem in which it occurs, through overgrazing (cf. the effects of modern medicine on the cattle population of the Masai lands in East Africa). Those cultures that survived their own counter-adaptive changes did so in general by switching ecosystems: they moved on and messed around with somebody else's environment.

It seems to be true, as Philip Wylie has suggested, that, because of the irreversible effects of technological innovation on the biosocial ecosystem, the life span of any culture is in inverse ratio to its per capita "technological index." In distinction from other cultures lasting thousands of years, the advanced industrial culture of modern man seems to be up against the wall after only three hundred.

In another vocabulary, this particular ecological statement would be known as the "intensification of the contradictions."

Rome did not fall because of homosexuality, high living, or marauding "barbarians." The Roman Empire fell because its technology could not renormalize when the counter-adaptivity of its original innovations (e.g. colonialism) made itself felt as inflexibility. The Empire became a suitable "someone else's ecosystem" for the more flexible "barbarians" to move in on. "Someone else's ecosystem" is a metaphor of the EPISTEMOLOGICAL ERROR which lies behind the present biosocial crisis. It is an error whose effects may already be irreversibly damaging.

2. POSITIVE FEEDBACK

It was this epistemological error which was essential to the massive technological advances of the modern industrial system. But its short range survival value is now countered by its long range counter-adaptive value. The system has traditionally depended on an "ethic of disposability" for which natural resources, other people's ecosystems, "other" human beings in general, and the disposable beercan have had roughly the same (exchange) value. Having taught that all it defines as environment is disposable, modern industrial society has only just begun to learn that THE SYSTEM WHICH DISPOSES OF ITS ENVIRONMENT DISPOSES OF ITSELF.

It is common knowledge that the relationship between technological innovation and additional innovation, as well as that between population increase and further increase, generates exponentially increasing one-way processes. Increased population tends to increase the need for and the possibilities of new technology, and increased technology increases the population. The biosocial hubris of growth, efficiency, selective affluence, progress, diversification, and the maximization of profit, coupled with an almost religious fanaticism about technology, further increases the biosocial deviations of the ecosystem which economic factors are already amplifying. But technology, requires human and natural resources. And our global dependence on technology, along with the continuous increase of the technological index of every individual, has brought us to a situation which neither the human nor the natural resources will long stand for.

Each individual makes ever-increasing demands on technology because of the increasing demands technology makes on him. His demands include the psychological gratification and "escape potential" he seeks in an alienating, objectifying, plastic culture: more horsepower, more mobility, more gadgets, more "free time" more material status. It is now estimated, for instance, that every man, woman, and child in the United States depends upon an average of 500 "energy slaves" in order to go about his daily business, each "slave" expending the energy and producing the waste of an average human being (Los Angeles Times, June 21, 1970).

This increase in the technological index is not deleterious in itself (it could depend upon the recycling of energy and waste, for instance, whereas at present every discarded beercan increases the entropy of our culture). But in the context of an epistemology which refuses renormalization, the technological index becomes an index of decadence. The rapacious building up of quantitative change is leading us inevitably to a qualitative rupture, which may include our own destruction. Whereas in all previous cases of environmental and psychosocial desolation and exploitation, the victims were too localized or too weak to react effectively, mankind may ALREADY be facing the kind of massive natural retaliation which could be the genetic result of our turning on to DDT.

In cybernetic thinking, the present relationship of the industrial system and those who control it (state or private capitalism) to the biosocial environment (to me, to you, to nature, to its "resources") is known as a positive feedback or runaway relationship: the more you have, the more you get. Unlike the primary control system of nature, negative feedback which seeks out difference and neutralizes or transforms it, positive feedback increases the deviations between input and output in the communication between the subsystems of the ecosystem. In the short run, this is fine for those who invest their money at compound interest or who draw their profits from underdeveloped countries, but in nature, all runaway systems (such as a forest fire or a supernova) are INEXORABLY controlled, in the long run, by negative feedback at a second level. Second order negative feedback always takes the form of THE EMERGENCE OF A METASYSTEM (the elaboration of new structures, morphogenesis) or THE DESTRUCTION OF THE ECOSYSTEM INVOLVED. In social systems, the first of these responses to positive feedback is known as revolution. The second is extinction.

3. THE PARANOIA OF SYMMETRY

Because it seeks to ignore feedback and relationship, the prevailing epistemology is at every level an epistemology which leads inevitably to positive feedback. It is an epistemology of BIOSOCIAL IMPERIALISM. In Bateson's terms, it is an epistemology of lineal causation or "force" or "power"; for the general systems theorist it involves the imposition of closed-system thinking on those aspects of reality which are open systems (see below). It denies the relationship between energy and information by splitting symbiotic wholes (ecosystems) into supposedly independent "things".

The same epistemological error obtains whatever the ecosystem or the level of ecosystem we are concerned with: biological, psychological, socio-economic. Deterministic thinking in biology, instincts and intrapsychic conflicts in psychology, the free rational subject in economics, and uncritical attempts to apply the experimental method in the social sciences are particularly obvious examples of the error. In its ideological manifestations (for every ideology is dependent upon a theory of knowledge and vice versa), the same error feeds pollution, racism, alienation, exploitation, oppression, and ALL OTHER FORMS OF PATHOLOGICAL COMMUNICATION.

It is a necessary function of pathological communication to deny its own pathology at some level while admitting and using it at other levels. Thus, since no system can actually ignore feedback relationships our culture simply converts them into power relationships. More precisely, we note that any rational argument pinpointing the pathology at any one level will generally be met with a stereotyped set of defense mechanisms: a shift of level (e.g. pragmatic considerations in response to moral arguments), a refusal of content through an attack on form (e.g. complaints about the use of proper channels), a refusal to understand ("I don't know what you're talking about"), or a naked challenge ("So what?").

Each defensive response is a further manifestation of the primary epistemological error. The error justifies and encourages an overriding commitment to SYMMETRICAL relationships – such as arms races, corporate mergers, simplistic anti-communism, racism – within which COMPLEMENTARY relationships at another level – spheres of influence, exploitation of underdeveloped countries, domestic colonialism, and the like – provide the material for the overall commitment to competitive oppositions.

The danger, as I explain in more detail below, does not arise from the nature of man or from the nature of symmetry and complementarity as such, but rather from the general tendency of all components of the system to specialize in EITHER symmetry OR complementarity with any given set of other components.

It is an extraordinarily curious business. What happens is that the epistemological error allows our culture to deny the complementary relationships on which it depends, and consequently to justify them as symmetrical. Real relationships of dominance and subservience between unequal partners are defined as free competition in an open marketplace between free and equal legal, psychological, and socioeconomic subjects. The most obvious example of this pseudo-symmetry lies in our admirable eighteeth-century commitment to "freedom of speech" (rather than to freedom of communication). Within limits we are all free to say what we think; it is only the channels to the audience that cost money. Similarly, every man has the right to become a conscientious objector – provided he has the education (and the necessary psychological distance from a culture that equates killing with virility) to argue the philosophical and religious premises of his case.

It is in these ways that the conventional epistemology takes on an ideological PARANOIA OF SYMMETRY (Freud's "narcissism of minor differences"), which generates further pathological symmetry. To paraphrase a schizophrenic statement: "If the Russians (Chinese, blacks, hippies, students, etc.) aren't what I think they are, I'll prove it."

4. COGITO, ERGO SUM

For the sake of illustration, let us take Descartes' 'Discourse on Method' (1637) as a historical metaphor of this epistemology. Once Descartes had made the crucial epistemological error of "cogito ergo sum" (to say nothing of its psychological and social deficiencies),[2] he had made a COMMENTARY on the TEXT of the relationships of seventeenth century society – a statement in a meta-language about the (overdetermined) relationships in the "language of real life" (Marx), the object language of the existing material relationships. Commentary and the text are mutually interdependent, in the same way that the psychological symptom is a commentary on a text which affects the text and vice versa. Descartes' statement was a CRITICAL commentary on some aspects of a past epistemology (it was a partial metaepistemology), at the same time as, in keeping with Descartes' representative moral conservatism, it was a JUSTIFICATION of the present state of the text. The relationship between commentary and text is a necessary one. There are no accidental relationships.

In keeping with philosophy's historical function of responding to and elaborating on developments in science, the 'Discourse on Method' provided the Galilean revolution with its philosophical paradigm. But, as befits the foundation stone of a new philosophy which sought to overcome the epistemological obstacle of Aristotelian "final causes" in the physical sciences, the "cogito ergo sum" went much further. It corresponded to the requirements of the existing socioeconomic system that there be an epistemology from which an ethic of personal freedom, individual autonomy, equality, rationality, the universaliy of bon sens, and, above all, an ethic of SEPARABLE AND INDIVIDUAL RESPONSIBILITY could be derived. It was an ethic of symmetry, a doctrine of separate but equal.

The system had begun its counter-adaptive adaptation. It was no accident that Descartes' assertion of the existence of CLEAR AND DISTINCT IDEAS AND CLEAR AND DISTINCT PEOPLE should have paralleled and preceded the introduction of the technics of efficiency, substitutability, and organization in the culture that was soon to dominate the world. The rise of individualism, the invention of perspective, Protestantism, the discovery of alphabetical order for dictionaries, the ordering of practical and abstract knowledge in the 'Encyclopedie', the invention of interchangeable machine parts, the standardization of weights and measures, Linnaeus' ordering of the species, the substitution of legal codes and constitutions for spontaneous common law, and the creation of standardized tasks in the factory system[3] – these and other events were symptoms of the imposition of the "clear and distinct" on the complex, explicitly oppressive and hierarchical communitas of the Middle Ages. And if Descartes' commentary on the text of an ongoing adaptive change was to strike a blow for the ethic of personal freedom in the Age of Reason, it was also a blow on behalf of an economic system which would lead to the most complex and irrational organization of unfreedom mankind has ever seen.

Descartes' statement justified the ethics of symmetry by clearly stating the atomistic epistemology upon which we still depend. It is an epistemology related to the forces and energy pushing and pulling bodies about in what was soon to become a Newtonian universe of attraction and repulsion in mechanical equilibrium. Additive, closed-system energy models of ALL reality – the conception of reality as an aggregate[4] of individual bodies possessing forces, humors, sympathies, affinities, instincts, and the like, in an entropic universe – excluded those minority voices who, like Pascal or

Rousseau, to some degree or other protested on behalf of the multiplicative and fractionative communicational relationships of the biosocial open systems (systems involving life or mind, or simulating life or mind, ecosystems), which physics necessarily ignored.[5] The "social physics" which followed, the sociology of the energy relationships of clear and distinct billiard balls,[6] demonstrated the utility and the danger of the Cartesian epistemology. Every cogito was free to sell his disposable energy at the best price.

Following on some aspects of classical ontology and epistemology, of which it is only a variant, the Cartesian "revolution" made the crucial absolutist and analytical error of unjustifiably conferring a privileged ontological status on ENTITIES ("substance") as opposed to RELATIONSHIPS ("attributes", "accidents").[7] In spite of Aristotle, Hegel, and Marx, the truth that it is not entities which create relationships BUT RELATIONSHIPS WHICH CREATE ENTITIES was (and still remains) generally obscured. Moreover, the privileged ontological status of entities in the system encouraged the reification of whatever relationships it did recognize. Gravitation, energy, matter, people, and so forth, became things. (Cf. Marx on atomism and the general equivalent in exchange, Capital, Lawrence and Wishart, London, 1961, I, 92).[8]

The second error, compounded by Locke[9] and other anti-Cartesians,[10] followed from the first. By imposing analytical logical categories (such as the supposed ontological primacy of the simple over the complex) on the processes of biological, historical, and psychosocial reality, both the Cartesian and the anti-Cartesian epistemology matched the wrong "organism" or level of the organism to the wrong "environment" or level of the environment.

As Pascal, the dialectical thinker who so vehemently opposed Descartes, knew only too well, the epistemological error was to impose an "either/or" (closed-system) logic on a "both/and" (open-system) reality. The error has a venerable pedigree, for nowhere is it more clearly demonstrated historically than in the Socratic elenchus, which employs the tried and true "principle of assent" of the encyclopaedia salesman (i.e. he begins by asking questions which can only be answered "yes") against interlocutors forced by their cupidity for knowledge to choose between artificial oppositions.

5. THE BIOSOCIAL UNIT OF SURVIVAL

Without a new epistemology of communication, relationship, exchange – a holistic and symbiotic epistemology – mankind will not survive.

The Cartesian error appears in many forms. With our present hindsight, it seems almost incredible that Descartes could not have realized that the "cogito ergo sum" was not a "clear and distinct idea" but rather a MESSAGE between a sender and a receiver, and, moreover, a WISH FULFILMENT, like a dream. Today we would replace it with another slogan: "loquor, ergo summus."

It is no accident that the Cartesian epistemology entails solipsism. (There are only two kinds of solipsists: those who want to be consciously and those who want to be unconsciously Solipsism; what we erroneously call schizophrenia, is a "security operation.")[11]Of all the ways in which it is possible to focus on this error, the most important is to describe it as that perception of the biosocial universe WHICH DESCRIBES THE UNIT OF SURVIVAL AS THE INDIVIDUAL (the individual organism, species, family line, system, and so on). It is the error of the "survival of the fittest," which is ideologically quite different from the principle of natural selection, for nature selects the survival of the ecosystem, not the survival of the individual (subsystem).

There is no need to elaborate on the often remarked relationship of Darwinism and Social Darwinism to nineteenth century liberal capitalism. And we are not about to oppose this error with the usual moral arguments about the difference between "natural" and "human" selection, nor with the argument that Darwin meant the species and not the individual, nor with the point that Spencer was really thinking about the cooperative relationships of the parts of the organism rather than about the conflict between organisms in a world of restricted possibilities and the "law of nature."

In their traditional form, these arguments compound the original error by being variations of it. Since such applications of the axioms of the prevailing epistemology are dead wrong, they cannot be tinkered with or corrected; they can only be put back in their place.

The first step is to recognize that THE UNIT OF SURVIVAL IS THE MESSAGE IN CIRCUIT IN THE ECOSYSTEM (Bateson), whether the ecosystem in question be methodologically defined at the biological, the sociocultural, the psychological, or at some other level. Unlike energy, information (messages) can be both created and destroyed, primarily because the very possibility of information depends upon a code which is shared by both sender and receiver. (By "sender" and "receiver" I mean the heuristic device which enables us to talk about the message in circuit). The code, in fact, as Bateson points out, IS this relationship. Without the reciprocity of the code, the message is received as "noise." And when the possibility of information is destroyed by the breakdown of the sender-receiver relationship, the ecosystem perishes.

This is only another way of repeating Schroedinger's maxim that life feeds on negative entropy (information, order, improbability, organization). That is to say, when there is a breakdown of the information flow which maintains organization in living systems (homeostasis, reproduction) and which, carried by energy borrowed from an entropic universe, may also create localized packets of negative entropy (morphogenesis in the sense of the elaboration of new structures, evolution), positive entropy (disorganization, randomness) takes back its own.

In order to feed on negentropy, life must have an adequate supply of free energy, that is, it must have "a supply of uncommitted potentiality for change, i.e. flexibility" (Bateson). It is precisely the inflexibility of our present unrenormalized "phase space" of potentialities which makes our epistemology counter-adaptive.

Epistemology is a question of where you draw the line, and there are only a restricted number of loci through which to draw it (i.e. everything in this paper has somehow been said before). The line drawn between "organism" and "environment" by our conventional model of reality is such a line, and, like ALL such lines it is a mental fiction. Unfortunately, we think that it is real.

Psychologically, this conventional epistemological boundary is a refusal or denial (verneinung, verleugnung) of the interpersonal ecosystem; ideologically, it is a (schizoid) justification of the way material relationships are programmed in our culture (biosocial imperialism); epistemologically, it is an arbitrary punctuation of the discourse between sender and receiver, whose supposedly real status in the world is itself defined, not by "reality," but by the original epistemological error. Some epistemological boundary is necessary (boundaries are the conditions of thought), but it is always arbitrary in the sense of being methodological rather than real.[13] The problem here is that the conventional line serves inappropriate ideological ends.

For the sake of this illustration, imagine the message in circuit in an ecosystem which is maintaining homeostasis as a spiral in which information – the transformation of difference – flows in one direction. (Leave for the present the complex problem of the relationship between the synchronic and the diachronic aspects of this circuit. This relationship is another aspect of the problem of models – i.e. a question of our consciously chosen point of view. As much as I am suspicious of the geometric metaphor I am using, it is worth noting that, as Robert Nyberg has pointed out to me, projecting a helix on to a plane produces at one transformation a circle – synchrony – at another, a zig-zag ladder – diachrony.

Imagine that there passes through the longitudinal axis of the spiral a plane representing the epistemological line which divides the circuit into two "sides." These "sides" are arbitrarily defined; the barrier between them is imaginary both in the sense of its not being real (the flow of information ignores it) and in the Lacanian sense of the Imaginary (see note 22). This fiction does define the boundaries of sets of real, material differences in order to allow us to talk about senders and receivers. The problem is that our epistemology reifies the methodological necessity: by reifying the material differences between various sets of components in a symbiotic whole, it turns them into Imaginary OPPOSITIONS. And any culture that systematically turns difference into unmediated opposition is eventually doomed.

Our culture labels one "side" of the helix: organism, self, reason, we, ego, men, man, and so on; the other half it correspondingly labels: environment, other, emotion, them, superego, women, nature, and so forth.[12] We further compound the error of opposition by mixing up the correspondences: we match a biological "self" to a sociocultural "other," for example, thus confusing the various LEVELS OF organization within the model. The boundaries of my skinbound biological individuality are quite different from those of my psychic "self" or my social "role." There is not only an environment specific to every system but also to every level of the system; a symbiotic epistemology will always seek to remain aware of these various levels of correspondence at the same time as it remains aware of its own epistemological act of repeatedly homesteading a wilderness – "Nature hath no outline, but Imagination has" (William Blake).

Language itself (naturally) assists in the confusion, and at a level of categorization more profound than the relativity of conceptual categorization described by the Whorfian hypothesis. At both the phonological and the conceptual level, language (which, like mathematics, is neither superstructure nor base) is constructed on difference (Jakobson's "binary oppositions"). More significant at this point is the actual function of words like "I," "me," "you," "him" (Postgate's "subjective elements"), which we commonly take to refer to entities when in fact they are simply "shifters." The shifter ONLY designates the locus of the sender of the message; it refers to a relation, but it does not of itself supply the context (indicate the code) without which it can have no referent other than itself. Shifters indicate the ATTITUDE of the sender in every sense of the word (most uses of the subjunctive mood, for example, qualify as shifters) – that is, his locus in a context but not the context itself. "Me" and "here" are not entities; they are punctuation marks.

(Punctuation of difference in real relationships involves questions of power, responsibility, dominance, pseudo-dominance, and, especially, the power to be responsible, but lack of space prevents me from dealing with this aspect of the master-slave dialectic in this paper. The ethic of viability allows for opposition in matters of life and death).

Any act of categorization is a punctuation, and the unit of survival has been wrongly punctuated. The unit of survival is not EITHER this organism (species, etc.) OR that, nor EITHER organism OR environment, nor EITHER this side of the line OR that. It is both-and.

Nor is the unit of survival "in" entities – "in" the organism or "in" the environment – the unit of survival is in their relationship, WHICH IS NOWHERE. It is nowhere because it is information, and information (improbability, order, potential for organization, "news," pattern, etc.) is simply DIFFERENCE. Information is the difference which makes a difference (Bateson), that is to say, the difference, within any given set of variables, which makes the further transformation of difference (e.g. reproduction) possible.[14]

Thus, although the epistemological line has traditionally been drawn in the wrong place – in the place where it creates a (solipsistic) barrier between sender and receiver – its placing corresponds paradoxically with a real, material locus of difference (cf. note 13). The error, I repeat, is to turn the difference between the loci of a symbiotic relationship into an opposition of entities. In our historical example, the error corresponds to Descartes' confusion of the linguistic shifter "I" with the skinbound biological individual (the organism) which provided the bioenergetic base for the sociocultural subsystem (the man) which emitted it. At this point he chose a difference (between levels of organization and complexity of information flow), matched it with the wrong environment (matched a biological subsystem to a sociocultural environment) and called it IDENTITY. The referent for "I", which designates the sender, is not the biological organism, separate and different from his fellows, but rather a locus in the system (similar to what G. H. Mead would call a role).

And once you have consequently unconsciously correlated "Self" with "Body" (immanence of the self) in order to establish the transcendence or the non-corporeality of "Mind," (which you thought was "self"), you have a mind-body problem to which no satisfactory answers can be given within the epistemology which produced it.[15] The Cartesian axioms are perfectly consistent – but they generate Godelian sentences which we know to be true, but which cannot be "proved" from those axioms. The axioms generate statements of a logical type different from all other statements within the system, and all attempts to deal with those sentences without transcending the epistemology which produced them result in paradoxes. The "self" becomes one pole of a double-bound oscillation between "self" and "other." In reality, they are together (mitsein), as Heidegger would put it.[16]

To put it more precisely, the self is neither transcendent to nor immanent in the subsystem which thinks it. It is immanent in the UNIT OF MIND (Bateson), which, isomorphic with the unit of survival, includes the rest of the system in which the appropriate transformations of difference (information) are transmitted. The unit of mind is not either sender or receiver: it is both-and.

There is an environment specific to every organism or subsystem, and when either the environment or the organism adapts or changes – i.e. manipulates difference over time – the other must change also. In the terms of viability or survival, their overall relationship is SYMBIOTIC; within that relationship, the specific behavior of any two or more subsystems may be COMPLEMENTARY or SYMMETRICAL, or both.

Either symmetrical relationships – the more of the same, the more of the same – or complementary relationships – the more of the same, the more of the opposite – may lead to runaway. In Bateson's terminology (Naven, 1936), this progressive differentiation is called schismogenesis, which is beneficial only if controlled.

The possibility of runaway follows from the fact that every piece of information is a difference or a deviation. Every initial message, before feedback, moves the system towards the amplification of deviations. Lack of feedback, misconstrued feedback, or positive feedback can then lead to pathological communication. Unless controlled be deviation-reducing comparisons between input and output that are meaningful at some level or other to all components of the system, both symmetry (e.g. rivalry between father and son) and complementarity (e.g., domination and dependence between mother and child) may lead to the destruction of one component. If the circumstances are sufficiently dangerous, or if there are no other substitute relationships available (as in an arms race or a predator relation), the destruction of the one entails the necessary destruction of the other.

Every transmitted difference is simultaneously a report on a situation and a command to do something about it (McCulloch). When the reports and the commands get mixed up so that deviations between sender and receiver increase, there must eventually occur a situation in which no further "matching" or "understanding" – symmetrical imitating or complementary fitting – is possible, and controlled communication ceases. The message in circuit had been converted from "signal" to "noise" which, for the system, is equivalent to the destruction of the information. In some circles this phenomenon is known as the generation gap.

Thus neither symmetry nor complementarity are guarantors against destructive schismogenesis. And yet in nature, either destruction or emergence always rules. What guarantees emergence rather than destruction is not either symmetry or complementarity, then, but both: THE SWITCH FROM ONE MODE TO THE OTHER (cf. Bateson's NAVEN).

The supposed (entity-like) "inhibitory response mechanism" (epistemologically equivalent to the soporific power resident in opium) which prevents an attacking wolf from rending the exposed jugular of his defeated adversary provides an example, as Bateson has pointed out to me. The vanquished canine switches from symmetry (matching aggression) to complementarity (fitting submission), and it is this substitution of one kind of energy-triggering difference for another which sends the victor off to transfer his symmetrical relationship to the surrounding foliage. The same information in the new mode is the trigger to a different gun.

6. DIALECTICS AND SCHISMOGENESIS

Only in this century, with the beginnings of a communicational and relational epistemology, have the life and social sciences discovered how to free themselves from the deadly tutelage of physics. Symptoms of and contributions to the change can be found in many different writers (especially now that one knows what to look for), in the principle of indeterminacy, in the statistical theory of entropy, ecological biology, the Prague School of linguistics, information theory, French structural anthropology, topology, phonology, group and set theory, French psychoanalytical theory, existential psychoanalysis, conjoint psychotherapy, and cybernetics. The demand in France for a semiology, a science of signs, has been matched in the United States by the development of a general theory of systems.

All open systems (which include the theory elaborated here) are goalseeking, adaptive transducers of difference; a semiology or semiotics is the science of the transformation of oppositions: a dialectic. Truth is a process (bewegung), and every ecology must be an ecology of presentations and representations (vorstellungen).

Of all the representative quotations one might introduce at this point, perhaps the most apposite would be one from the man whose discovery (inbenientio) of the principle of the DIFFERENTIAL ELEMENT in language generated a new paradigm which enabled philology to become a science, a quotation from Saussure's 'Cours de Linguistique Generale' (1916):

> In language there are only differences. More than that: a difference generally presupposes some sort of positive terms between which it is established, but in language there are only differences without positive terms. . . . [In language] the value of a term can be modified, without its sense or its sounds being affected, simply by the fact that some other neighboring term has undergone a modification (p. 168).

and:

> Concepts. . .are purely differential. They are not defined positively by their content, but negatively by their relationship with the other terms of the system. Their most exact characteristic is to be what others are not (p. 162).

Symmetry and complementarity are manifestations in behavior, in ways yet to be fully defined, of the linguistic opposition between metaphor and metonymy, similarity and contiguity, code and message, text and context, substitution and combination, the paradigmatic and the syntagmatic.[17]

They seem to be related to what Bateson calls selective and progressional integration in decision making as well as to schismogenesis. Both processes are essential to communication; all behavior IS communication. And just as the choice of symmetry over complementarity (and vice versa) may lead to the destruction of the message in circuit, so the choice of the metonymic over the metaphoric pole of language (and vice versa) is a principal characteristic of aphasia, as Jakobson has shown.

The further analysis and comprehension of the opposition between these two poles is essential to the continued development of dialectical thought, to say nothing of the destiny of man. The building up of contiguous quantitative change in a system, for instance, may be explicable as an analog of the metonymic message processes (combinations) within it, and the dialectical "quantum jump" from quantity to quality seems to be an analog of a metaphoric change of code (substitution). Opposing messages and codes within the ecosystem may be related symmetrically or complementarily (or both); when the intensification of their deviations ("contradictions") can no longer be controlled by negative feedback, the system will remain viable – be maintained and transformed (aufheben) – only if the ensuing exponential schismogenesis leads to the emergence of a metasystem. But goal-seeking systems other than men in particular have no morality other than the morality of viability; depending on the order of complexity of the system and the semiotic freedom of its "characteristic response,"[18] it will manifest a synchronic tendency towards a steady state (conservation of entropy) and a diachronic tendency towards increasing orders of organization (negative entropy), levels separated from each other by code-switching quantum jumps. Dialectical reversal (aufhebung) is always the control of schismogenesis in the interests of the system – how do we equate those interests with our own? Or in other words, how do we control the controls of the system?

The answers to these questions will not come easily. But a materialist, naturalist, anti-reductionist theory of communication and exchange, such as that outlined in this paper, a radical cybernetic viewpoint which understands that the unit of survival and the unit of mind are isomorphic, at least points us in the right direction.

Cybernetic both-and thinking, combined with an understanding of the hierarchical orders of complexity of organization revealed in systems theory, enables us to deal rationally with the crucial problem of the distinction of logical types (the distinction of codes from messages and from metacodes), upon whose confusion the pathological communication of schizophrenic relationships depends. For any system which comments on itself and confuses metalanguage and object language generates paradox. And paradox, endemic to language and creativity, may be used, in the form of the PARADOXICAL INJUNCTION, on behalf of the control of some components of the system by others.[19] IF (and only if) I insist that A is either A or not-A, THEN you are either good or bad, normal or abnormal, sadistic or masochistic, American or un-American – and you will oscillate between these impossible contraries as long as you are prevented from metacommunicating about them by the control I exercise over your epistemology.[20]

Like the (so called) schizophrenic who cannot answer the double bind of identity, when the analytical epistemology is applied to open systems, it CONFUSES THE METAPHOR WITH WHAT IS MEANT, the Symbolic with the Real. It denies (verneinen) or rejects (verwerfen) the symbiotic relationship of sender and receiver by reifying difference, for this epistemology of identity and non-identity cannot properly account for either the synchronic transformation of difference (communication) or its diachronic transformation (change). Adequate principally in certain formal or static systems and for the energy-entity explanation of physical closed systems, it cannot account for non-lineal or non-causal theories of change, nor for the demonstrable teleology of the goal-seeking system, which is "moved" as much by what is not as by what is.[21]

The entities called genes in bioenergetics, for instance, become "questions about difference to which the answers are provided by their neighborhood" (Bateson) in cybernetic thinking.

The traditional epistemology will not be given up without a fight. All remonstrations to the contrary notwithstanding, its adherents manifest the self-alienating and anti-contextual "nostalgia for thingness, for the constancy of stone" which Sartre analyzed in his 'Anti-Semite and Jew': the same rock-like stability which Descartes tells us he sought. It is a manifestation of the short-circuited dialectic of narcissism, identity, and identification that Lacan calls the Imaginary.[22]

The problem is that we are all riddled with it. It is in our bones, in the corporeality of our own body images.

As Dostoevsky put it so succinctly in his 'Winter Notes on Summer Impressions' (1863) – if we ignore his Russian chauvinism:

> It seems that one cannot create brotherhood, because it creates itself, comes of itself, is found in nature. But . . . in Western nature generally, it does not readily appear. Instead you find there a principle of individuality, a principle of isolation, intense self-preservation, of personal gain, self definition in terms of one's own I, in placing this I in opposition to all nature and all other people, as an autonomous, independent principle completely equal and equally valuable to everything that exists outside it.

And whatever other changes we envisage as necessary to halt the one-way accumulation of biosocial surplus value at compound interest, so long as the prevailing epistemology remains a metaphor of "better dead than Red" or "better sick than Bolshevik", there's not much hope.

REFERENCES

1. This article is a direct result of conversations at the Oceanic Institute, Hawaii, with Gregory Bateson, made possible by a research grant from the University of California, San Diego. A summarized version of the ideas elaborated here was delivered at the Symposium on Language and Cultural Discontinuities at The State University of New York at Buffalo, April 15, 1970.
For the systems thinking underlying this essay, see my "Marcuse and the Freudian Model: Energy, Information, and 'Phantasie'," 'Salmagundi,' No. 10-11 (1969-70), 196-245.
A longer and somewhat different French version of the present article will appear in a forthcoming issue of Communications (Paris) on "La Crise."
2. Note that the subject of the philosophical cogito is always free, white, and twenty-one.
The significant passage for our purposes is the following:

> I knew [from the indubitability of "I think, therefore I am"] that I was a substance [i.e. "I" is a substance] whose whole essence or nature is only to think, and which, in order to be, has need of no locus and does not depend on any material thing, in such a way that this self or ego, that is to say, the soul by which I am what I am, is entirely distinct from the body . . . and even if the body ceased to exist, my soul would not cease to be all that it is.

3. Cf. Jacques Ellul, 'The Technological Society' (New York: Vintage Books, 1964). The topic of the book is not technology or machines, but la technique: the historical development of organization, classification, substitutability, taxonomy.

4. Cf. Newton, 'Principia' (1713):

    The motion of the whole is the same thing with the sum of the motions of the parts . . . and therefore the place of the whole is the same thing with the sum of the places of the parts, and for that reason, it is internal, and in the whole body.

5. The eighteenth century believed gravity to be a real, innate, physical force. Newton, however, says it is not. But his remarks about the vis insita of bodies (i.e. inertia) as not ESSENTIALLY including gravitation are contradicted by his (theological) assumption of a spiritus subtilissimus moving and affecting both the inorganic and the organic world. Clearly modelled on electricity, it is the bioenergetic "force" par excellence ('Scholium Generale,' 1713, last paragraph).

Cf. also William Gilbert on the magnet (De Magnete [1600], Ch. XII):

    The magnetic force is animate, or imitates the soul; in many respects it surpasses the human soul while that is united to an organic body.

6. In the 'Critique of Pure Reason' (1781) – "The Paralogisms of Pure Reason" – Kant adds a note to explain the possibility of multiple "identities" of the "I" over time (in spite of its logical "identity") by comparing the "substance" of the "I" in each state to an elastic ball impinging upon another ball representing the next state, and thus communicating its "representations together with the consciousness of them" (i.e. its energy) to the other ball.

7. As Lancelot Law Whyte points out in his remarkable analysis of the conception of non-conscious processes before the specifically Freudian discovery of the unconscious we speak of today, Descartes sought to "stabilize with final clarity an ancient dualism: the split of subject from object." Whyte regards this as possibly one of the most fundamental blunders made by the human mind (but as I have been at some pains to point out, it cannot be regarded as a blunder for the seventeenth century, whatever it may have subsequently proved to be for us). He goes on: "Until an attempt had been made (with apparent success) to choose AWARENESS as the defining characteristic of an independent mode of being called mind, there was no occasion to invent the idea of UNCONSCIOUS mind as a provisional correction of that choice." The adjective "unconscious," in the sense of not available to consciousness, appears in English in 1712, the noun in German in 1776, but the term "inconscient" does not appear in French until the 1820's. ('The Unconscious Before Freud' [New York: Anchor Books, 1962], pp. 23-25, 59). Note that the "primary process" discovered by Freud makes sense only in relation to the "secondary process," and it is, of course, a dialectical concept, not an analytical one. Nor is it in any proper sense "irrational."

8. The same error over substance and attributes and the same emphasis on entities – which are "moved" by the "internal contradictions" they "contain" as the primary "cause" of their development, their relationship with other "things" being secondary causes – obscures Mao's popular essay on contradiction (1937), which is an excellent example of the impossibility of dealing adequately with difference (which is what the essay is really about) from Lenin's mechanistic materialist conception of dialectics. It fails because, unlike Marxian epistemology, it is only another version of the atomistic epistemology it attacks. Contradictions, differences, oppositions, deviations, and paradoxes occur not in entities but in systems, that is to say, in patterns, in organization, in

relationships – which are nowhere. It is true that difference moves the cosmos, but the difference in energy levels which provide the motive power of closed systems tending towards entropy are not the same as the differences between information input and output which control the behavior of open systems tending to neutral or negative entropy (see below).

9. In a well known passage, Locke compares the understanding to a "closet wholly shut from light, with only some little openings left," which let in "ideas of things without" (1690). Perception, for Locke, is of course a one-way process: the subject contributes nothing to his perceptions; he is an entity which becomes "imprinted" with "ideas" (a tabula rasa).

10. Cf. Leibniz in 'The Principles of Nature' and the 'Monadology' (1714), writing AGAINST the Cartesians:

> Our thought of substance is perfectly satisfied in the conception of force and not in that of extension. . . .There is . . . no way of explaining how a monad can be . . . changed in its inner being by any other creature, for . . . the monads have no windows through which anything can enter or depart. The natural changes of the monads proceed from an internal principle. . . .Each monad is a living mirror. . . .

The interesting thing about Leibniz, of course, is that in spite of his monadology, his conception of entelechy sought to establish the principle of goal-seeking against Cartesian mechanism, and that he is one of the first modern thinkers to seriously pose the question of unconscious mental processes.

11. In the French version of this paper ('Communications,' Paris), I point out that Descartes is not in fact a solipsist, since his God is the mediator of all the communication between the individual cogito's. God is also the receiver of Descartes' message; He completes the circuit. However, the eighteenth century took over the Cartesian mechanistic materialism and ruled God out of the system, thus implying a more truly solipsistic, "individualistic" position, with no communicative mediator.

12. Cf. Darwin in 'The Descent of Man' (1871):

> At some future period, not very distant as measured by centuries, the civilized races of man will almost certainly exterminate, and replace, the savage races throughout the world.

13. All boundaries in this sense involve paradox, for they correspond to the digitalization of an analog continuum, and the DISTINCTIONS between the various components thus created can never be completely maintained. (It is this fact which accounts for phonological change in language, for the "distinctive features" making up the phonemes we invent to explain the system do not maintain their distinctions in relation to each other over time, because they are in constant interaction with the continuous substratum of acoustic frequencies.) Whyte points to the Godelian situation involved: "No rational system – logical, mathematical, scientific – can ever be used to define its own boundaries." (But the unconscious is not thereby the irrational, of course.) And further on:

> The temptation to treat static ideas as absolute, rather than as partial and provisional, proved irrestible to many Western thinkers; the apparent clarity of such ideas seduces the mind into dismissing change or transformation as a trivial secondary effect of interactions between the "real" entities. Static concepts proved to be very effective intellectual tranquilizers. . .(op. cit., pp. 40, 42).

I have tried to deal with the question of paradox and boundary, in relation to zero and "not," in my "Analog and Digital Communication: On Negation, Signification, and the Emergence of the Discrete Element."

14. It is information (a spermatozoa, a pin prick) which transforms the bi-polar radial symmetry of the frog's egg into the bi-lateral symmetry of the frog, by defining one meridian as different from all others (Brachet).

15. The concept of the body as the PROPERTY of the "self" – complete with fences and boundaries, and, presumably, amenable to real-estate speculation – is deeply imbedded in our culture. In the 'Second Treatise on Civil Government' (1690), Locke remarks that "though the earth and all inferior creatures be common to all men, yet every man has a property in his own person; this nobody has any right to but himself".

16. 'Being and Time' (1927). In an article whose title is a quotation from Locke – " 'But there is Nothing I Have is Essential to Me' " ('To Honor Roman Jakobson', [The Hague: Mouton, 1967], Vol. I, pp. 462-474) – Colin Cherry points out, as others have done, that the word individuum in the Middle Ages seems to have meant "indivisible from the community or unit", whereas the Ciceronian use is closer to present-day usage. The Trinity, for example, is "holy and individual". Cherry goes on to quote the 1888 edition of the New English Dictionary: ". . . Self-, as a living formative word first [appeared] about the middle of the 16th century, probably to a great extent by imitation or reminiscence of Greek compounds in . . . [whilst] the number of SELF-compound words was greatly augmented by the middle of the 17th century when many new words appeared in theological and philosophical writing, some of which had a restricted currency of about 50 years (e.g. 1645-90) whilst a large proportion . . . have a continued history down to the present time."

The noun "self" dates from 1595, according to the OED, with the philosophical sense of "ego" appearing in 1674. Similarly, the noun "moi" in French dates from the time of Montaigne (Robert).

17. Cf. R. Jakobson, "Two Aspects of Language and Two Types of Aphasic Disturbances" (Metaphor and Metonymy), in: R. Jakobson and M. Halle, 'Fundamentals of Language' (The Hague: Mouton, 1956).

18. Cf. M. C. Marney and N. M. Smith, "The Domain of Adaptive Systems: A Rudimentary Taxonomy", 'General Systems Yearbook' (1964), 9: 107-31.

19. In so-called "psychopathology" these are the paradoxical commands of the Other, the "double binds" of pathological communication, logically equivalent to a sign bearing the words "Disregard this sign" or to the command "Be spontaneous." Here the logical types (Russell) or what Carnap called the object language (statement about) and the metalanguage (statement about a statement or about a context) are confused. See: Bateson, Jackson, Haley, and Weakland, "Toward a Theory of Schizophrenia," 'Behavioral Science' (1956), 1: 252-64; and Bateson, "Toward a Theory of Alcoholism: The Cybernetics of 'Self,' " (forthcoming, 1971).

20. This pathological oscillation is very evident in Kierkegaard's "existential" opposition to the Hegelian philosophy of the mediation of oppositions, notably in his 'Either/Or' (1843). The text is full of double binds, especially as regards S. K.'s inability to follow through with his engagement to Regina. ("A man's duty is to marry: Marry, and you will regret it; do not marry, and you will regret it.") "Yea, either/or is the key to heaven," he cries, "both-and is the way to hell."

And in the 'Journals': "The idea of philosophy is mediation – Christianity's is the paradox" (1841).

21. R. G. Collingwood points to the essence of the problem at the end of his 'Idea of Nature' (1945):

> I conclude that natural science as a form of thought exists and has always existed in a context of history, and depends on historical thought for its existence. From this I venture to infer that no one can understand natural science unless he understands history: and that no one can answer what nature is unless he knows what history is.

Marx puts it another way in the 'Theses on Feuerbach' (1845):

> The question of whether objective truth can be attributed to human thinking is not a question of theory but is a practical question.

22. The Imaginary order is an interpretation of the mirror relationship between child (ontogenetically) or self (structurally) and other (cf. Marx on the mirror relation of men and commodities, 'Capital', I 52 and note). Although it begins in the complementarity of the necessarily dual relationship between mother and child (where he is a signifier in her discourse, a part of her whole), the Imaginary involves an attempted short circuit of the real difference between self and other, with a resulting paranoid symmetry of specular identification, projection, and opposition. This is true so long as the child is prevented (by the pathology of the parents) from entering the mediated, three-way relationship of the oedipal triangle. Here the Imaginary fixation on the duality of ego and alter ego is integrated into a relationship of displaced reciprocity, in accordance with the positive social injunction we call the prohibition of incest, where each locus in the triangle, including each parent, is protected from relating only symmetrically or only complementarily by the mediation of the others. (This is the domain of symbolic, unalienable exchange, the domain of the mediation of opposition, called by Lacan 'le symbolique' to distinguish it from "Symbolism" and from Freud's term, 'die Symbolik.') Without this constructive schismogenesis, the child will be trapped in the world of demand (always defined by others) and will never become the subject of his own desire. He will thus be caught in the double bind of EITHER self OR other, in which his own survival is (impossibly) at stake. See Jacques Lacan, 'Ecrits' (Paris: Editions du Seuil, 1966); the commentary on Lacan in my 'Language of the Self' (Baltimore: The Johns Hopkins Press, 1968); and Maud Mannoni, 'The Child, his "Illness," and the Others' (New York: Pantheon, 1970).

Man's Ecology, Ecology's Man

LEE THAYER

If there is an adult American alive today who is not aware that something is awry with his "environment," it is because he doesn't read newspapers or magazines, listen to the radio, watch television, talk to other people, or notice the graffiti on the walls of public toilets. Indeed, the condition of man's environment has become one of those things which one cannot NOT talk about.

A Social Issue is something that large numbers of people talk to each other about. I think it important to recognize straightaway that Social Issues serve an indispensable function in society. For example, they are useful for breaking the ice at cocktail parties. They are useful for establishing or enhancing one's social status as a knowledgeable, concerned observer of "what is going on." They are useful for identifying oneself with the "action." They give people something to talk about, something to rally 'round. With a Social Issue to talk about, even human in common, as if they belonged to the same exclusive club and these were the passwords.

Whatever else may be said about Social Issues, it must be said that they are remarkable engines for social interaction. If "hot" enough, they may even be used to launch a thousand conferences and symposia, where their pros and cons can be assayed and eulogized by "the experts."

Indeed, Social Issues are the commodities around which human societies live and grow.

The condition of our environment, our current "ecological crisis," will surely find its way into the hall of fame of Social Issues of all time. In saying this, I do not mean to belittle the problem, or its seriousness, or its urgency. As Social Issues rise and fall, the intensity with which they are talked about does occasionally lead to some residual action (for good or for ill). The realities of our present crisis are indisputable. I simply suggest that it may be therapeutic to begin with the recognition that Social Issues have, first and foremost, social utility. What is DONE about the conditions they purport to reveal is always, at best, secondary.

Having recognized this fundamental fact about Social Issues, however, I want to move on. For it seems to me altogether possible that all this talk about our "ecological crisis," that all of this talk about pollution and so forth, may be a smokescreen. Our ability to talk so easily and so readily about this particular Social Issue may be disabling us from seeing what's really at stake. I want to suggest, in fact, that we may be so polluting each other's minds by our endless talk about this Social Issue that the ultimate HUMAN issue may have eluded us.

SETTING SOME FACTS STRAIGHT

All evolutionary processes are irreversible. This means that, even though we were to succeed in levelling off the rate of increase in our production of pollutants, garbage, and people (which is altogether unlikely), we would have to continue to live with the same conditions we have now. No government policy is going to revive the withered trees, the dead and inedible fish, or the blight of urban sprawl.[1] To assume that things would somehow return to their original healthy state is simply wrong.

Second, if the deterioration of the physical environment continues to increase as it has (which it likely will), man as a species will either adapt to these changed conditions (and we have no definitive reasons to believe that, as a species, he could not do so),[2] or, as a species, he will cease to exist. The latter consequence – man's extinction – would certainly have the advantage of permitting recovery of the physical environment in the fastest, most expedient way possible, if that is our central concern. And who is to say that either of these eventualities is not man's destiny? Can there be any rational or moral reason why man SHOULD survive if the price of his survival is the extinction of all other living things on this planet?

Third, technologists will invent and construct artificial physical environments. This seems to me as inevitable as the invention and manufacture of synthetic food.[3] The thrust of modern man's technological history is toward the synthetic, the artificial, the vicarious. If man has come thus to nurture his mind and his spirit, why should it be different for the food he eats or the air he breathes?

Fourth, by some surely not so unbelievable coincidence, man's befouling of his own air, land, and water is greatest in those areas of highest human density (and consumption), even though these areas have the most stringent anti-pollution regulations and the most active recovery programs. What this means is simple enough: as the density of people in a given area increases, the over-all rate of befouling and blight will increase. Apparently, if we do not asphyxiate each other with our poisonous exhausts, we will manage to bury ourselves in our own garbage. Or, failing that method, we might succeed in reducing man to a quasi-vegetable state through the sheer ugliness of our man-made surrounds, the sheer common-placeness of our mass culture, the sheer kitsch we produce and the sheer pap we distribute, the sheer senselessness of much of what we do. That eventuality would, of course, have the ironic advantage of drastically decreasing human demand upon the limited life-support systems of "space-ship earth."

Our environmental resources, including our air, our land, and our water, are not limitless. We populate the earth as if they were. It is naive to believe that we face the ultimate crisis now. No matter what other steps we take, the QUALITY of the environment will vary with population density. The cry of every newborn baby today pulls the noose tighter around our necks. If more carbon monoxide than can be recycled is called pollution, shouldn't the production of humans beyond our natural life-support systems also be called pollution?[4] At a certain point, is the one form of intercourse any more dangerous than the other?

Fifth, in terms of his treatment of his physical environment, man is the earth's dirtiest animal. Will a hundred million dollars spent to revitalize Lake Erie change THAT? If we are looking for hard facts, here is one.

SCAPEGOATS AND SHORT-TERM SOLUTIONS

Man is the only living thing on the earth capable of placing the blame for his problems outside himself. Of all of his talents, this may be the most exploited, the most highly developed.

Recently, a Missouri family sued the American Tobacco Company for negligent complicity in the death of the husband and father by cancer. He was a heavy smoker. It no longer seems illogical to people to blame the producers of the goods we consume, for our inability to stop consuming them. When an incorrigible seductress cries "Rape!" it's difficult to determine what caused what.

People talk. When they do, they arrive at a perspective on Social Issues from which it seems perfectly reasonable that man's prerogative to ravish the earth is sacrosanct. Yet the same people may believe it equally reasonable that a man should not have the choice of doing something which might be harmful to himself. Incredible!

The search for the "Why?" of Social Issues usually becomes a witch-hunt. If something is wrong, there must be a villain somewhere.

If people get killed in cars, it could be because the manufacturer doesn't build cars that are "safe" enough. There are researchers who believe, on the basis of their research, that the most common "causes" of automobile "accidents" are drivers who are either careless, incapacitated, incompetent, or all three. But research findings seldom make good Social Issues. For these, a scapegoat is necessary. Why?

Pollution? The key villain, according to the great social grapevine at work on this one, is "big business." Aside from the fact that "big business" (and the "industrial-military complex") are simply the most popular villains behind all sorts of problems these days, one needn't be an expert to know that municipalities were long the worst offenders in all three areas of "pollution" (air, land, and water), and that the runoff from fertilized and herbicized fields and from feedlots presents a danger to life on this earth today altogether as great as industrial or human wastes. Who's the villain?[5]

Those amongst us who are opportunistically ambitious, and at the same time either naive or short-sighted, seem to be greatly attracted to the "causes" of Social Issues. The picture I have in mind is that of a group of otherwise intelligent and well-intentioned young people getting together for an "environmental sit-in." They protest the pollution of their environment. They leave behind several pounds of cigarette butts, candy wrappers, beer cans, pop bottles, and miscellaneous personal items no longer of use to them. They take with them the name of a scapegoat. Unwilling or unable to come to grips with the full complexity of the problem, they simply "solve" it. The shortest distance between an uncomprehended problem and an arbitrary solution is via a scapegoat.

People talk, and talk, and talk. Is there no danger of pollution therein?

It has often been noted that Americans are the untidiest of all the world's peoples. Is the fact that our pollution problems are worse even than those of Europe, where the industrial revolution began before the U.S. was settled, entirely a coincidence?

There are clearly some flaws in all our talk about our environmental problems.

First, "causes" are always energized by hindsight. And in any contest between hindsight and foresight, hindsight always wins. It's easy enough to see now that the manner in which surface coal was mined was likely not in our best long-range interests. It's easy enough now to see that nondecomposable plastics present a special disposal problem. It's easy enough to see NOW that something other than an internal combustion engine might have been better. But what else was there at the time "better" for powering automobiles and airplanes? Combustion driven power lawn mowers far outsell those powered in other ways (particularly those driven by man-power). Who's the villain?

How something gets designed or produced depends, among other things, upon the then-available technology. Milk and beer and soda bottles were originally returnable because they were expensive to produce, and re-use seemed justifiable ("economic"?) from both the producers' and the consumers' view. With a technology that can produce nonreturnable bottles cheaply, will consumers opt for the greater "inconvenience" of deposit-and-return?

It's easy enough now to see the long-range mistakes made in thousands of such judgments as these, all intended as solutions to some immediate problem or opportunity. But what of our present-day solutions to our present-day problems and opportunities? Will they have long-term effects or side-effects which, from the vantage point of hindsight, will be criticizable?

All technological solutions are short-term solutions. To be made "safe," organically contaminated water is treated with heavy doses of chlorine. What will be the long-term effects of that treatment on the living creatures of this world dependent upon natural waters? If highway driving is to be made "safe" with breakaway bridge abutments, what will the sleeping driver finally hit to stop him? Technologists say that the amount of radiation we receive today from our atomic testing and our atomic power plants and even from our household television sets is so slight as to present no hazard. Were these the same technologists who told us there was no hazard in the internal combustion engine – or in thalidomide? A synthetic environment would likely produce as unforeseeable and as drastic genetic changes in human life as would the pollution for which it is a "solution."

Second, as far as our ingenious and intemperate scapegoating is concerned, we should label it for what it is: potentially hazardous to our health and welfare.

Third, it should come as no great revelation to us, 300 generations after man first began to concern himself with the quality of his own thought, that simple minds are limited to the apprehension of simple problems and to the formulation of simple answers.

It is undeniable that if General Motors were indicted and shut down for producing automobiles in which people are killed, or from the exhaust of which they are poisoned, there would be a noticeable change in that one kind of social "problem" and that one kind of pollutant. But that solution might "solve" the problem in much the same way that Prohibition "solved" its problem. Those thousands disemployed by G.M. could get into the lucrative black market. And people would drive their old cars anyway, thereby increasing the highway hazard, and the pollution.

Any society worthy of the people in it is necessarily complex beyond their individual comprehension.[6] A complex system is an interdependent, dynamic, and evolutionary (and sometimes revolutionary) system. A simple or naive mind looking for "the source" of "a problem" in human systems must invent villains and scapegoats; it will never be able to see what is there – that, at any point in time, there is always an immensely complex, interdependence of all the elements and conditions of our physical ecosystem and of our total socio-ideological ecology. The parts and the whole exist in and through the other. Their dynamics and intricate relationships over time exceed our human comprehension. But that seems not to stop us from approaching our worlds as if they were cheap "Whodunnits."

Consider the water pollution problem in the midwest. Who is to blame?

– The technologists, who invented the fertilizers and herbicides? Weren't their interests generally to help increase the farmers' productivity and efficiency, which were clearly wanted and needed at that time?

– The farmers, who, like the rest of us, were only trying to do the "best" they could, through more "scientific" farming, to increase that much-needed food production and their own personal well-being?

– The government? Weren't the motives of those government policies which created "artificial" commodity markets in the interests of the then-struggling farmer and the then much abused consumer?
– The consumers? Is there any clear record that consumers have ever demonstrated for scarcity, for doing without, for high prices or unpredictable prices?

What were once successful technological solutions to immediate problems become ways of life. They become imbedded in the whole complex fabric of social, economic, political, and even aesthetic, life. They are no longer separable, no longer removable, no longer, in their cumulative and metamorphized effects, even the same.

If the individual farmers in the U.S. today were to cease fertilizing and applying herbicides, they would rapidly go out of business. With them would go a large portion of the world's food factory, as well as a sizable chunk of its contingent labor force. If the government were to establish laws against further fertilization and herbicization, not only would the nation's food production drop drastically, but all food producers, converters, and distributors would have to be substantially further subsidized. Technologists might be able to invent fertilizers and herbicides which would not contaminate our remaining fresh waters. But what would be the side effects of those NEW interventions? There would be side effects; yet they may be no more foreseeable than the side effects of the present fertilizers and herbicides. Consumers might be willing to do without, might be willing to pay high or wildly fluctuating prices for bread and butter and milk and potatoes. But has anyone recorded any strong public sentiment in that direction?

Everyone wants his. In the developed countries, getting one's own (regardless) is rapidly becoming a civil (if not a "human") RIGHT.[7] Can everyone have "his" without raping and befouling the environment, or without infringing on his neighbor's rights? Is there something which is unmistakeably "people" behind our environmental and social ills?

Littering? Where's the scapegoat? Where the short-term technological solution?

If people can justify in their own minds their part in the killing off of their living ecology, from what perspective in all of the universe could their own continuance be justified?

MAN'S ECOLOGY

All this is pertinent, but preliminary. It is the base upon which I would like to develop the main theme of this paper – which is that man's natural environment at its worst may be a far more suitable ecology for man than man is for himself.

It is, initially, useful to realize that man has two environments, two ecologies; his "natural" environment, and his "man-made" environment. Man's "man-made" environment has two quite distinct (but necessarily related) components. There are, first, the physical artifacts which he builds and produces (from his paintings and roads and buildings to his machines and bombs). There are, second, man's communicational realities – the beliefs, doctrines, ideas, and orientations which emerge, and prosper or perish, in and through communication.

Communicational realities are those realities which exist for us only because we can and do talk about them (such as what is "right" and what is "wrong", for example). All human relationships, from the simplest to the most complex, are built upon one or more communicational realities.. All social relations and all social acts, from a momentary act of courtesy to the largest and most diverse of human societies, are products of this most unique of man's inventions – his communicational realities. Their significance is this: that everything that is or could be done with them and through them add up to man's total man-made environment, his total socio-ideological ecology.

Western man's ideologies, which have provided the rich soil out of which science emerged and flourished, have also led him to conceive of himself as OVER AGAINST his physical environment. They have led him to think in terms of domination OVER, of control OVER, of achievement OVER. His mania for "progress," for "development," for that pie-in-the-sky known today as the "higher" standard of living, has evolved out of the Baconian ideal of power over nature, the cold logic of Spencerian determinism, the appeal of Deweyian pragmatism. Any resistance, any temporary failure of the natural environment to accomodate our most ravishing appetites and vilest wastes: this we look upon as a "crisis" – FOR MAN.

Our present ecological "crisis" has its origin not in what we have done to our physical environment, but in the faulty ways man has thus far invented for conceiving of himself-in-relation-to his environment.[8] Western man's basic ideologies have not led him to seek his destiny as an integral and evolving part of his physical environment, but in its mastery. It is THIS error, not his acts of pollution, which has brought about this confrontation between man and his environment.

Our technological hubris blinds us to the ultimate and inevitable interdependence of man and his ecology. Our ideologies betray us. Why is the problem always OUT THERE?

ECOLOGY'S MAN

The record, it seems to me, is clear. The fundamental ideologies upon which western man built his societies, and which are now pervasively imbedded in them, have served as remarkable lubricants for his industrial and technological "progress."[10] But they have simultaneously brought him to the very brink of disaster. Western man invented strikingly new definitions of himself and his environment which have enabled him to "master" that environment. But these same definitions have cumulatively and now critically increased his own unfitness for his ecology. Isn't it possibly insane to persist in the belief that we can now solve our ecological "crisis" dealing from the same ideologies that brought that crisis upon us?

We say, as I have just above, "man's ecology." How revealing this is of our ideological and technological hubris.

We use the term as if it were just another handy tool, and thereby misorient ourselves to the problem. The study of ecology came long before the popularity of "general systems" approaches. To those early investigators, ecology meant the RELATIONSHIPS between and among organic species and their physical surrounds.[11] If we are to persist in talking about man as OVER AGAINST other species as well as the physical environment which all must share, then we should at least look at it from both directions. The "crisis," in other words, is no more with "man's ecology" that it is with "ecology's man."

Are modern man and his societies no longer fit for the rest of nature?

How does man, individually and in his societies, measure up as a 'citizen" of the total ecology of this earth?

From what perspective – save the same ideological hubris which produced our ecological "crisis" – shall modern man hold to the illusion that he is justified in continuing on?

If it is the ecology we are concerned about, and not just man's "right" whimsically to indulge himself in it, there are additional questions which must be raised.

For example: How good as an "environment" are the minds of men for ideas?[12] Are men's minds even suitable for the evolution of those ideas, sentiments, feelings, and values which would make it possible for him to construct enriching and ennobling social environments, and at the same time to live in harmony with his natural environment? Of all that man COULD be, in spirit and mind and body, how good an environment are men for the feelings and the ideas and the disciplines that would redirect man's evolution down those paths? Science permits us to count much, but to "see" little. If the intelligence of the species is indeed decreasing,[13] are men's minds better or worse as the environment for ideas and ideals?

Second, western thought has increasingly led to the substitution of technology for man, in order to gain in productivity, efficiency, reliability, and precision. Assuming one day western man confronts a problem for which there is no technological solution – and our ecological "crisis" might well be such a problem – what will he be able to do about it? With each technological displacement, the range of individual man's competencies is decreased. At what point does he become incompetent to cope? Or at what point does coping become meaningless? From the view of the total ecology, at what point does man become, indeed, irrelevant?

All of the other species of this earth's ecology must adapt, or cease to be. If man ceases to be capable of adapting, having so peopled and polluted his environment that he can continue on only in a synthetic environment, is that to be counted as an achievement? If the birds and the fishes have no CHOICE with respect to limits on their numbers or with respect to the extent of their destruction of their environments, what possible justification can we invent for destroying them in the exercise of our free choice about those same matters?

Man's rationale, man's ideological position on all this, is paradoxical.

MAN'S FUTURE

In full recognition of the immediate and potentially lethal nature of our ecological "crises," I would like to risk the judgment that they are but symptoms of a much more fundamental human ill. We take the symptom for the disease partly because we approach all of our "problems" short-sightedly and myopically. We single out those "problems" which are "out there" (in the physical environment) and which therefore have some prospect of technological "solution."

Whether it be called "systems" thinking or just plain hard thinking, much more of it needs to be done. It might suggest this about man's future:

1. Although they will cure only the symptoms, the most likely solutions to our present environmental problems will be technological; our rush toward synthetic life with synthetic foods and a synthetic environment will simply be speeded up. Western man has always solved his problems thus. Is there any compelling evidence that he will not do so in this case?

Basic problems never reveal themselves. They can be read only through their symptoms. The long-range error is in treating the symptom. At this, man is prodigious and profligate.

So the present environmental crisis will be resolved in SOME way. The fundamental issue is not WHETHER but HOW. HOW it is resolved will have long-range cumulative effects which are not obvious in the solution – just as every solution to our past problems has had cumulative effects and side effects in our present. And these, of course, constitute our present crisis. The price we must always pay for short-term solutions is that of less soluble long-term problems.

IF our short-term solution is in the direction of further syntheticizing our physical environment, what will be the long-range problems created thereby?

There are certain inobvious risks. For example, we know from experience, from the record of history, and from laboratory evidence that the "harder" a species' environment, within limits, the hardier that species will be. The "softer" its environment, the less vigorous and adaptable the species.

The implications, it seems to me – if the solution to our present predicament is to build synthetic environments for humans – are that we need to take great care to build into them certain kinds of risks, uncertainties, threats and dangers which are necessary conditions for continued strengthening of the species. Even given that there might be no conditions of the physical environment which are NECESSARY to the actualization of man qua man, what long-range criteria for the actualization of man qua man are we following in the substitution of the "synthetic" for the "natural"?

If we were to gain the position of being able to be completely indifferent to our physical environment, would we then be superior to all of those creatures which CANNOT be aware of their environment? We are captives of our environments only to the extent that we are unquestioningly adapted to them. If we had such a "soft" synthetic environment adapted to US in such a way that we could ignore it, would we not then be in a position of being completely victimized by it?

2. One could hardly doubt or dismiss the significance, or the urgency, of our ecological crisis. It must therefore seem slightly gratuitous to argue that there exists a greater long-range danger to the quality human existence. Yet we have too long ignored it.

That greater threat to the quality of man's future exists in the relationship between men and their SOCIAL environments, their SOCIO-IDEOLOGICAL ecologies.

In our recurring skirmishes, love affairs, and now death-struggles with the byproducts of our technologies, it is perhaps neither so obvious nor so easy to recognize that, as man has gained his freedom from (control over) his physical environment through those technologies, he has inevitably substituted for that environment a social environment. Where once man was himself a product of the constraints, threats, and opportunities of his physical environment, he is more and more a product of the constraints, threats, and opportunities of his socio-ideological environment.

I propose that the same shortcomings of intellect, spirit, or sensibilities which have brought man close to physical extinction, but from which he will "escape," will ultimately snuff out that which is human about us. We will accomplish this not through wars or pestilence or even overpopulation, but through the subtle, irreversible evolution and development of certain conditions in our socio-ideological environments.

The environment within which modern, industrial man is in-formed, within which his unique HUMAN existence is created and nurtured, are his social and communicational environments. On balance, modern man has been unable to invent ideologies which would permit him to live synergistically with his physical environment. Consider how much less able man has been to invent ideologies on which he could build human societies as environments which would enrich and enhance his own humanity. That the latter failure is subtler and less tangible makes it all the more insidious.

Just as we have sought, through our technologies, ingenious ways to avoid the discipline of our physical environment, we have sought, through our social ideologies, ingenious ways to eliminate discipline from social existence.[14] Just as we have gained, through our technologies, ingenious freedom from concern about our physical environment, we have designed, through our social ideologies, ingenious means of socially-sanctioned self-indulgence and indolence.[15] Just as our technologies have provided us with "the knack of so arranging the world that we don't have to experience it,"[16] our social philosophies have increasingly provided us with human "cop-outs" of dazzling ingenuity. Is this what man is for?[17] – to figure out ways of avoiding the experience of living?

The cumulative effects of our social, economic, political, and humanitarian ideologies are as unforeseeable, perhaps, as the cumulative effects of our technologies have been. Yet, for the first time in his history, man has the wherewithal for designing himself and his ultimate destiny – through the design of his socio-ideological, his communicational, environments. Can be bring it off? My argument is simply that the same ideological postures which have led man to the mass destruction and defiling of his physical environment have led him to the creation and development of societies that will eventually define him out of existence as irrelevant.

The evidence is everywhere about us. Yet the scope of this ultimate "crisis" exceeds our grasp.

Our social policies move steadily in the direction of survival of the unfittest. We have confused freedom with self-indulgence. We have martyred the ill-formed, polarized mind. We have put mere existence before living to some end. We have enthroned tastelessness. We have mocked greatness and deified the put-on. We have made anything but short-sightedness appear "too philosophical." We have substituted the capacity to talk for the ability to say something worthwhile. We have made irresponsibility over into a state religion. We have outlawed "success," made failure illegal. God is a piece of technology, like a light switch. Love is an illicit or novel orgasm. Mediocrity has become a socially-sanctioned and government-subsidized life's work. Perhaps most insidiously, our social ideologies and policies increasingly penalize competence and reward incompetence. We take from the capable and diligent student to give to the incapable or indifferent student. Our popular movies lionize the inept, the ignorant, the incapable, the purposeless (e.g., 'The Graduate'). Competence and purposefulness are the butt of jokes; they are fit themes only for farce or satire. Excellence is rapidly becoming "unconstitutional."[18] "Security" is our social-policy answer to incompetence, incapacity, and indolence. We have made a tacit choice between being exploited by competence and being exploited by incompetence, and have chosen the latter.

The "critical mass" of a exponentially-increasing incompetence will someday completely paralyze the "developed" nations. A fantastic matrix of interlocked competencies must go into the production of a single loaf of bread, whether produced by one or by many. If there comes a day when the whole cycle that delivers bread so conveniently into our hands stops, we will exercise our incompetence, as indeed we do today, to redistribute our neighbor's bread. Having learned that what is his is ours, we will shoot him to get it. Competence will have become the last major obstacle in the achievement of the planetary "great society."

Indeed, humanity will not expire in a glorious fight against extinction by its natural environment. With but a whimper, humanity will simply grind to a halt, ignominiously, having ingeniously created and nurtured, in its man-made environment, the seeds of its ultimate irrelevance.

That is the ecological crisis we must eventually face. The diminishing of man by his own hand is the greater cause, just as it is the surer end.[19]

REFERENCES

1. Especially suggestive is Garrett Hardin, "The Tragedy of the Commons," 'Science,' Vol. 162, December 1968, 1243-1248.
2. In a speech adapted for 'Psychology Today' (March 1970), Rene Dubos said Man is not on his way to extinction. He can adapt to almost anything. I am sure that we can adapt to the dirt, pollution and noise of New York City or Chicago. That is the real tragedy – we can adapt to it. As we become adapted we accept worse and worse conditions without realizing that a child born and raised in this environment has no chance of developing his total physical and mental potential . . . if we do not understand what the environment is doing

to us, something perhaps worse than extinction will take place – a progressive degradation of the quality of human life.

Or, as other writers have suggested, as man is sucked into his hivelike cities, homo sapiens may evolve into homo convivens, "a new communal species that would control its reproduction and manage its ecosystem – the biosphere" ("The Stae of the Species," Special Supplement to 'Natural History,' January 1970).

3. It is as if whatever man can do, he must do. The case for the seeming inevitability of technologism has been well documented in recent years, and requires no further statement here. See, for example, Floyd W. Matson, 'The Broken Image: Man, Science and Society' (N.Y.: Braziller, 1964); Arthur Koestler, 'The Ghost in the Machine' (N.Y.: Macmillan, 1967); Ernest Becker, 'The Structure of Evil' (N.Y.: Braziller, 1968); and Jacques Ellul, 'The Technological Society' (N.Y.: Knopf, 1964).

4. It can hardly be without some interest that many animal and insect species, as well as some human societies, have (or had) population control mechanisms of one sort or another. Cf. Laura Thompson, 'A Self-Regulating System of Human Population Control,' Transactions of the New York Academy of Sciences, Series II, Vol. 32, No. 2, February 1970, 262-270; V. C. Wynne-Edwards, 'Animal Dispersion in Relation to Social Behavior' (Edinburgh: Oliver & Boyd, 1962); and D. H. Stott, "Cultural and Natural Checks on Population Growth," in A. P. Vayda (ed.), 'Environment and Cultural Behavior' (Garden City, N.Y. Natural History Press, 1969).

5. It would, however, be naive to suggest that all of our current environmental problems are causally related to population density. Unless we concern ourselves directly with the QUALITY of human LIFE, as Gordon Harrison says, ". . . there seems to be no reason why a society cannot be operated with a high quality environment at any population density" (in "The State of the Species," loc. cit., pp. 68-69).

5. For example, agricultural runoff, which eventually reaches the world's oceans, has already seriously inhibited the photosynthetic activity of large masses of green algae, on which the life of our waters depends. (See, e.g., P. A. Erickson and J. T. Reynolds, "The Ecology of a Reservoir," Natural History, November 1969). There are even subtler consequences of water pollution and diversion, such as potentially dangerous changes in underground water tables and in surface salinity. Such consequences of agricultural "progress" are not limited to the U.S. Similar consequences faced by the U.S.S.R. are described in the December 1969 issue of Natural History.

6. This condition, indispensable to complex social life, if not to science itself, is nicely developed in M. Polanyi, 'The Tacit Dimension' (N.Y.: Doubleday, 1966).

7. An engaging description of the pervasiveness of this issue is Walter Goodman's "On Doing One's Thing,"THE AMERICAN SCHOLAR, Vol. 38, No. 2, Spring 1969, pp. 240-247. Cf. "Freedom under Conditions,": Chapter 5 in Emmanuel Mounier, PERSONALISM (London: Routledge & Kegan Paul, 1952).

8. This argument is not, of course, new. Writers like Collingwood and Polak have made it central to their approaches.

9. This interdependence is dramatically described by John Bleibtreu (in 'The Parable of the Beast', N.Y.: Macmillan, 1968, p. 59):

Our technological control of our environment puts us into an ambiguous

> relationship with that environment; similar to that odd and destructive relationship that arises between a jailer and his prisoner. It seems as though the more control the jailer exercises, the more a strange kind of perverse love, a love that thrives on injury, grows between him and his charge. As the relationship develops, the prisoner often becomes the stronger of the two, inflicting by his very passivity the greater hurt.
>
> So we, as we control the environment, have gradually become the victims of our own control.
>
> Valery wrote, "On peut dire que tout ce que nous savons, c'est-a-dire tout ce que nous pouvons, a fini par s'opposer a ce que nous sommes."

It is interesting to speculate on the possibility that the concomitant of ease within the environment – perhaps social as well as physical – is loss of respect for it. As the social environment becomes "softer," the pollution and littering of men's minds and spirits may be inescapable. "In this world," said Oscar Wilde, "there are only two tragedies. One is not getting what one wants, and the other is getting it. The last is the real tragedy."

10. For a comprehensive review, see Charles Van Doren, 'The Idea of Progress' (N.Y.: Praeger, 1967); and, for a current hypothesis on the future of "progress," see Gunther S. Stent, 'The Coming of the Golden Age: A View of the End of Progress' (Garden City, N.Y.: Natural History Press, 1969).

In "Meru," W. B. Yeats describes the thrust of science and technology thus:

> Civilisation is hooped together, brought
> Under a rule, under the semblance of peace
> By manidold illusion; but man's life is thought
> And he, despite his terror, cannot cease
> Ravening through century after century,
> Ravening, raging, and uprooting that he may come
> Into the desolation of reality. . . .

11. There is a useful brief description of this background in Amos H. Hawley, 'Human Ecology' (N.Y.: Ronald, 1950), Chapter 1.

12. Man is both the producer and the product of his "noosphere" (Teilhard de Chardin, 'The Phenomenon of Man. N.Y.: Harper, 1959) – that is, of all of those "tacit dimensions" of ideas, feelings, etc., which frame and channel his humanness. Given that it is the few who produce the "noosphere," and that it is the many who are produced by it, the question might be: How adequate are the minds of the many to the greatest possibilities available to them? Cf. A. F. Bentley, 'Relativity in Man and Society' (New York: Putnam's, 1926). It was Pascal who wrote: "Thought makes the whole dignity of man; therefore endeavor to think well – that is the only morality."

13. See, e.g., George Gaylord Simpson, 'Biology and Man' (N.Y.: Harcourt, Brace & World, 1969), p. 125, pp. 145-146; R. C. Cook, 'Human Fertility' (N.Y.: Sloane, 1951); Harrison Brown, 'The Challenge of Man's Future' (N.Y.: Viking Press, 1954), p. 103; Wm. Vogt, 'People! Challenge to Survival' (N.Y.: Sloane, 1960), esp. pp. 86, 211.

14. In 1944, Philip Wylie wrote:

> Society is preoccupied with physical ease and only incidentally interested in certain minor aspects of inner discipline . .The human psyche. . .is intended to perceive the cosmos, adjust to it, struggle in it, and achieve there an individuation of itself. But our materialism has slain man's good opinion of

himself, stripped government of philosophy, taught the common people not to desire personal excellence but to barter in droves for the right to mediocrity (in 'Night unto Night', N.Y.: Harcourt, Brace, p. 148).

Yet how can the pursuit be justified? "Nothing," writes Ernest van den Haag, "can be more tiresome than the tireless, cheerless pursuit of pleasure. . ." ("Of Happiness and of Despair We Have No Measure," in 'Mass Media and Mass Man,' ed. Alan Casty. N.Y.: Holt, Rinehart & Winston, 1968).

An incisive variant is this from Abraham J. Heschel: "The failure of our culture is in demanding too little of the individual. . .in not realizing that there are inalienable obligations as well as inalienable rights. Our civilization offers comfort in abundance and asks for very little in return. Ours is essentially a Yes education; there is little training in the art of saying 'no' to oneself" (in 'Who Is Man?' Stanford University Press, 1965, p. 100).

15. "Do-nothingism: has become an accepted way of life in America, socially endorsed and parent-and government-supported. Having made failure essentially impossible in most human endeavors, this movement now stigmatizes success. Is what is left enough to evoke what is human about us? Perhaps few would agree with Jules Romains that we are faced with a "vast conspiracy" ("Open Letter Against a Vast Conspiracy," N.Y.: Heineman, 1968), but few could disagree with the observation that such changes are occuring, or that many give truth to Vico's prophecies.

16. Attributed to Max Frisch, on the title page of Daniel J. Boorstin, 'The Image: A Guide to Pseudo-Events in America' (N.Y.: Harper & Row, 1961).

17. A question dealt with more extensively in my "On Human Communication and Social Development," Proceedings of the 1st World Conference on Social Communication for Development, Mexico City, March 1970, reprinted in 'Economics et Societies' (Paris). Acutely or chronically, we substitute statism for "environment".

> It seems clear that the first major penalty man will have to pay for his rapid consumption of the earth's nonrenewable resources will be that of having to live in a world where his thoughts and actions are ever more strongly limited, where social organization has become all-pervasive, complex, and inflexible, and where the state completely dominates the actions of the individual.

Thus writes Harrison Brown of one of the social consequences of our present ecological "crisis," in 'The Challenge of Man's Future' (N.Y.: Viking Press, 1954), pp. 218-219. From quite a different perspective, Robert Frost wrote:

> The opposite of civilization is not barbarism but Utopia. Utopia can let no man be his own worst enemy, take the risk of going uninsured, gamble on the horses or on his own future, go to Hell in his own way. It has to concern itself more with the connection of the parts than with the separateness of the parts. It has to know where everyone is; it has to keep track of us. It can't protect us unless it directs us.

18. For a fictional speculation on this issue, see Kurt Vonnegut, Jr., "Harrison Bergeron," in 'Welcome to the Monkey House' (N.Y.: Delacorte Press, 1968). Ernest Hemingway described the present age as the millenium of the untalented. "We are deluged," he said, "by writers who can't write, actors who can't act and singers who can't sing. . ."

Nietzsche knew that, socially, merit itself is an affront. Cf. Goethe's prescription, "Confronted by outstanding merit in another, there is no way of saving one's ego except by love"; that of Nigel Calder, in 'The Environment Game' (London: Secker & Warburg, 1967); or that of Father Henri de Lubac:

> Nothing is more ingenious, more obstinate, nastier – indeed, in a sense, more clear-sighted, than mediocrity harrying every form of superiority that offends it.

Thus shall man go by his own hand, not as a tragic victim of his environment.

19. This is a slightly revised version of the paper originally prepared for the Joint Conference of the Society for General Systems Research and the Institute of General Semantics, Denver, May 1970. I wish to thank Sir Geoffrey Vickers and Mr. Bent Stidsen for their critical reading of the paper, and for their helpful suggestions for its improvement.

After the Communication Revolution — Then What?

E. S. SAFFORD

In August 1968 twelve outstanding scientists and academicians in the field of communications theory, psychology, sociology, and the neurosciences spent three days at a retreat in Colorado to discuss communication implications in today's world. Among these men was Dr. Warren McCulloch of MIT, considered one of the foremost scientists in the world in the field of brain research. In that session Dr. McCulloch pondered — "Can the world survive the communication revolution?" Then he added, thoughtfully, "My guess is, we have about twenty years in which to find out."

A year and a half has already passed in his twenty year allowance and, at this point, the developments in the technology of communication give us little comfort in the ultimate outcome.

When McCulloch talked of the communication revolution, he wasn't talking about the technology that has changed in the field of information transfer, information storage and retrieval, and data manipulation. He was really talking about the sociological and psychological results of a new environment which establishes completely new patterns of relationships among people, and a completely new relationship between knowledge and man. Man adapts rapidly to any new technology. For example, a Department of Commerce study indicated that television penetrated as many homes in ten years as it required radio twenty-five years, and as it required telephone eighty years. The impact of technological change is NOT that of technology on man, but is, rather, that of man upon man through technology. The technological advances and devices have merely extended man's capability of influencing man, and it is THIS phenomena that really composes the communication revolution.

Before we try to hypothesize on what will follow "the communication revolution," as we have defined it, we need to look at the elements of this revolution and the implications it is having upon mankind before we guess what next steps might be ahead.

While only seven life times stand between today and the day that Gutenberg activated his first movable type press (and we can say that the communication revolution started with that event), the principal effects of the technological changes in communication have been felt since World War II. The rapidly improving and rapidly spreading networks of telephone lines, trans-oceanic telephone cables, radio stations, television stations and satellites have, in a few short years, made it technically possible for a man on any point on the globe to contact another individual on the globe instantaneously. Within the next several years, the economics, the technology and the equipment will undoubtedly be available to encourage wide spread instantaneous, direct person-to-person contact regardless of geographic location. Following closely upon the heels of this capability will be the ability not only to have voice contact, but to have video contact on a person-to-person basis. This individual contact capability is in itself an extremely revolutionary occurrence to mankind's co-existence.

In addition to this capability of global, instantaneous person-to-person transfer of information, however, must be added the newly developed and rapidly advancing computer technology. This additional capability not only offers fantastic opportunities for storage, retrieval and random access of vast volumes of information, it also introduces the opportunity for massive mathematical computations, creative search resources and increasingly complex decision-making capabilities far exceeding man's abilities. The "revolution" then becomes one of instantaneous world-wide access to all kinds of information and knowledge, and the manipulation of that information and knowledge in incredible volumes at incredible speeds for instantaneous application. In a few short years, the advantages and the power which comes from owning knowledge has passed from individuals to machines. At this moment, the WISE use of knowledge still remains a human prerogative; however, with the rapid advances of computer technology, this advantage to humans will also be short lived. At some reasonably near point in time, we can expect that computers will be better able to assess the given alternatives of any given situation, from any body of knowledge, and produce the "wisest" solution to apply to that situation. The future opportunity for human intuition, or the human ability to assess the unknowns of the future, and to reasonably forecast the impact of such unknowns upon present actions will be considered as unreliable as witchcraft.

Many technological developments are already out of the laboratory, yet not in general use. Such items would include the picture telephone, the electronic video recording for home television sets, the direct satellite to personal television set transmission, the home computer, the newspaper printed in the home by the home television set, the use of lasers for computer storage, computer memories of pictures and pictures in color, the laser beam transmission of mail, the holographic picture transmission for business or entertainment, the computer networks linking libraries and other information sources, the micro image reduction of visual material for hard copy storage and use, and the microminiaturization of highly complicated electronic communication devices with resultant high mobility and individual use capability.

If we go beyond things which are already out of the laboratory, but yet not completely economically attractive, we can look at some of the things IN the laboratory with great anticipation and some concern. For example, the work going on in the exploration of the chemical transfer of knowledge, the electrical control of emotions, the developing theories of genetic engineering and the neuro-scientific discoveries of how the brain functions – what is memory and what causes learning. How far away are we from the laboratory to application in today's rapidly changing scientific world? What sudden breakthrough may occur in some remote lab tomorrow morning which may make practical extrasensory perception as a standard communicating device between humans? What break-through may occur in creating a symbiosis that would connect computer databanks directly with human brains so information, introduced electrically, may be humanly assimilated and utilized? How far away are we from the ability to produce the right frequency of radio waves that will energize the basic emotion centers of the brain so we can control human behaviour by buttons?

The communication revolution then is not a revolution of technology – it is the revolution of KNOWLEDGE and the AVAILABILITY AND UTILIZATION OF THAT KNOWLEDGE – through new technology. It is this revolution, I believe, that worried Dr. McCulloch. This knowledge, or information revolution, is creating a whole new set of problems and frustrations never before experienced by mankind. Never before in the history of civilization has man been so closely in touch with all of his fellow human beings, and so conscious of other's problems, disasters and accomplishments as today's individual. For the first time in man's existence, through the medium of TV, he now PARTICIPATES in history as it is being made throughout this globe, and even off the planet upon which he exists. No longer does he receive this information at some point in time after the event occurred, and no longer does he have the luxury of having these events described and explained by qualified observers.

Modern man is confronted with the new problem of personally interpreting world events as to their meaning in his own life. Add to this the rapidly expanding networks of information which are feeding him constantly greater volumes of data on subjects, places and events about which he previously never thought, and about which, previously, he had no concern. As the economics of the technologies evolve, it will permit wider and wider introduction of information networks and increasing numbers of information input devices such as, radio, TV broadcasting stations, mobile radio broadcasting, radio telephones, and personal mobile phone capabilities. The amount of information procured for the worldwide networks will increase at geometric rates.

Within a very few years, the combined communication network capabilities of the world will provide this globe with one, grand, global nervous system, and all the world's population may react to inputs much as an individual body presently reacts to an input. Just as a human reacts instantaneously when he steps upon a tack barefooted, so will the world react when a disaster, catastrophe or substantial aberration in social behavior occurs at any one point on the surface of the globe. The rest of the globe will know of that event immediately and may very well react impulsively.

We already are seeing the first evidences of this kind of "global nervous system" phenomena through the limited, although extensive networks that currently exist in the United States. When the racial disturbances first started, a few hot summers ago, we found that revolution in the streets occurred in Watts and Detroit, New Jersey, and Georgia within a few short days of each other. When the tragedy of Kent University occurred, 340 universities had strikes, shutdowns or demonstrations within a week after that event. The nervous systems of these particular cultures had reacted as a system – as one body – and they did so through the capabilities of instantaneous communication. The knowledge of these events, the information about these events, which were mass distributed, were seized upon by individuals having similar thoughts, similar concerns, similar worries and similar ambitions – and these individuals created similar actions in their own geographic area. While the general information about these events was widely dispersed and was available to all ages, cultures, races, colors and all levels of intelligence, the REACTIONS to the distribution of this information were highly definitive. In other words, cultural groups, groups of individuals with deep cultural ties acted in concert, and this phenomena is descriptive of what is evolving out of this knowledge revolution.

Marshall McLuhan has made the statement, "Electricity has eliminated geography," and the events described above confirm and illustrate what McLuhan was saying.

There is a temptation to extrapolate current reactions into broad generalities in attempting to anticipate what the future will be. Yet, in today's complex technology, such an exercise can, indeed, be fruitless and foolish. Short term extrapolations may have some validity, however, and if we are to project the concept of cultural identification through communication technology into a worldwide pattern, we may very well see mankind changing its loyalties, its philosophies, its concerns, its ambitions, its power structures, and its interpersonal relationships very drastically from historical precedent.

Up to this point in time, man has lived pretty much in islands, whether they be geographic islands or political islands. Groups of people with similar political motivations and philosophies, similar economic systems, similar languages, similar colors or similar needs have joined together for practical reasons to defend themselves or to enlarge their sphere of influence through joint efforts. With the introduction of a worldwide communication network capability, (a worldwide nervous system), however, we may very well see mankind becoming polarized into CULTURES, rather than into economic, or race, or language, or color, or political groups. Individuals of similar cultures, regardless of position on the globe, regardless of color and regardless of language, will find a kinship among others on the globe of similar makeup, and they will find that they can intimately share this kinship on a very personal basis through satellite TV, through picture-phone satellite projection, and through all the other instantaneous communication media which will exist in the years ahead.

If this occurs, then a whole new set of social problems, inter-relationship problems and general coexistence problems develop. For example, if the philosophers of the world and the top scientists of the world find a very strong cultural unity (and there is certainly evidence that this is occuring now in the scientific fields), then this group, worldwide, may develop its own community of conversation, of shared experience, of personal fulfilment through association with like kind in the medium of global instantaneous communication. As a matter of fact, such a worldwide cultural relationship has firm roots already extending prior to World War II among certain groups of specialists. The so-called hidden colleges of top scientists and philosophers in specialized fields have existed for quite some time, but have done so in an awkward, inefficient fashion, with a great deal of individual isolation, and almost no visibility. As it becomes more practical, more economical and more convenient for individuals with like interests to obtain personal relationships on an increased frequency basis, regardless of location on the globe, the appetite for personal discussions particularly exciting to that specialty, will force as a natural consequence increased interpersonal involvement among that group. If we assume this is true, then we must recognize that there will be a considerable erosion of national loyalties, community loyalties, and company loyalties as we have known them in the past. Countries and nations are expedients by which man has made it possible for groups of like-minded people to exist and progress together. As civilization evolved, the original immobility of humans, and their inability to communicate with other humans geographically separated from them, forced the invention of countries or nations upon a geographic basis. With a worldwide communication network, or nervous system, interlocking individuals of similar cultures, the whole conceptual need for nations, countries, states, counties and cities begins to change rapidly and radically. No longer are human beings grouped together by geographical constraints. Almost in a generation, humans have become more intimately connnected with the KINDS of humans they find most comfortable, most exciting, most personally fulfilling with whom to live.

If we find, then, a deterioration of political structures which have served humans up to this point, such a change should not be surprising. If traditional political structures are no longer sufficient for regulating relationships, we will find a new kind of conflict arising. Cultural groups, bound together worldwide, but existing individually in different kinds of political environments may find such political constraints too confining. As political structures were designed by humans, they will be redesigned by humans. For practical and personal reasons cultural groups will not tend to cluster together physically, as there will be no need for this through the capabilities of worldwide communication networks. Instead, individuals in cultural stratas will insist that their political environment accommodate a practical living relationship for them.

Pursuing the idea of the philosopher-scientist combination as a cultural group or strata, let us say that a principal part of the wisdom of civilization, as well as the awareness of knowledge, currently lies within this group of people. If this group develops a strong and intimate global kinship, and become, in effect, the gatekeepers of the superior knowledge of the earth, at some point, they may elect to NOT place their knowledge and their wisdom in the service of politicians or military organizations. They may realize that geographic national units and traditional economic systems no longer have any valid reason for destructive power struggles or exploitations. This group may recognize that the conflict hazards ahead are inter-cultural conflicts, not international, or interracial, or inter-economical, and therefore, such traditional struggles are now archaic and meaningless. Should this occur, the development of weaponry would be removed from political manipulation, and ultimately the power and influence of politics would deteriorate. A replacement structure for governing inter-personal relationships would naturally evolve. Such a structure would be designed around cultures, not around geography, or economics. Whole different perspectives of physical needs, of human recognition by levels of cultures, of the interdependency of cultural stratas would form the criteria of design of new "governments."

This example merely illustrates that while the technology has developed a nerve network for the globe, THE IMPORTANCE OF THIS NETWORK IS IN THE KNOWLEDGE THAT WILL BE CARRIED BY THE NETWORK. Knowledge will become the power, and those who can control the knowledge and can apply the knowledge are the ones who will have the power. This knowledge is not just knowledge of things, knowledge of electronics, or nucleonics, or medicine, but it is knowledge of psychology and sociology and philosophy. In fact, the emerging strength of new knowledge may be in the field of sociology rather than in any of the hard sciences, although the hard sciences will be the conveyor, the universal carrier, for the superior knowledge.

As with all technological developments, we want to translate them into "What will it mean to INDIVIDUAL human beings – not just human civilization?" Because this revolution of knowledge and information is a revolution of human inter-relationships achieved through new capabilities, then we need to consider what it can do to individuals. It can do several rather dramatic things. For example, in interpersonal relationships a natural law seems to exist similar to a certain physical law; namely, that for every force there is an equal and opposite force. As we generate a global communication network capability, a global nervous system, and we overlay this on a world population which is growing in numbers by leaps and bounds, we are compounding the concept of a "universal public." AS A "PUBLIC" GROWS, AN EQUAL AND OPPOSITE NEED FOR "PRIVACY" WILL DEVELOP. The human beings, the individuals on this earth, will HAVE to escape periodically, this worldwide public and will have to have a privacy more severe than exists today.

This privacy may be achieved by several means. Eventually it will mean the development of a fairly formal sense of individual selective curiosity, and a means by which the individual can tap into the worldwide communications network on a selective basis. In other words, an individual will be able to purchase a communications filter, if you will, designed to his own profile of interests, concerns and ambitions, and this filter will screen from him all the other extraneous information about which this individual doesn't want to know. The individual, in this fashion, can selectively design his world of interest, his level of culture and his sense of personal being by NOT exposing himself to the overwhelming flood of data, of knowledge, or miscellaneous information constantly washing across the world in greater and greater densities.

Secondly, I think that the one-to-one relationship among people will become more and more important. When mass communication first came into real being after the turn of the century, heroes were made in all kinds of specialties. Names like Madam Schuman-Hienke, Babe Ruth and Lindbergh were not just momentary heroes; thay were a part of that age. As mass media increased in its volume and capability, heroes expired more quickly. How many can remember the name of the first man into space, although that flight was a much greater achievement than Lindbergh's? Each World Series season produces its current stars, yet a year later, new ones take the spotlight. It is increasingly difficult for any individual anywhere on the globe to obtain prolonged public recognition currently, unless he is the leader of a major political state. The frustrations of the millions of individuals who seek some kind of acceptance, some kind of individual recognition, must find its satisfaction, not in the eyes of the public, but in the eyes of other individuals. Eventually, this gratification may come from only one individual, or a very small group of individuals, rather than from a public. The never-changing need of a human being for love, for appreciation, for recognition, for acceptance, may find increasing realization in the concept of marriage because the "universal public" will become more and more overwhelming, competitive, unsympathetic and impersonal.

Third, individuals will probably find increasing satisfaction in the contemplation of spiritual matters. The human spirit must find its companionship just as human emotions must find their satisfactions. The increasing awareness on the part of individuals of the awesome knowledge which has been discovered already, and the increasing awareness that man has merely scratched the surface of what there is to be known will drive more and more individuals to seek a reason for personal existence, through a search for something greater than man's own works.

What will all of this mean to us who are in the business of technical publishing? Several things will come into focus. The information needs of the engineers or the professional man will change, and his systems for getting such information will also change. For one thing, he will become identified as a part of a cultural stratum because professional people have similar ambitions, similar interests, and similar motivations which gather them into a comprehensive group. I emphasize that this is a CULTURAL stratum, not a scientific or technological stratum. This is the human being, as a human in his cultural environment, and a part of such cultural strata is his professional involvement. Editorially, we will have to consider this professional individual in his cultural environment, as well as in his professional area, and we will have to direct information to him in that context. Certainly we will have to realize that as concepts of loyalty change in relation to cultural strata versus political strata, this will have an implication for loyalties to company involvements also. While there will be proprietary information sacred to the individual enterprises and important to the existence of the enterprise (which, incidentally, will be honored by the professional); nevertheless, his feeling of involvement will be beyond that of an individual company. Because of this, he will seek information from individuals outside his immediate employer.

Studies available today indicate that the major input for new ideas and new information into an enterprise come, basically, through "gatekeepers" who are individuals primarily attuned to picking up information outside their organization, evaluating this information and adapting it to their organization's needs. At the present time these gatekeepers obtain their information from seminars, from personal acquaintances with whom they are in regular contact, from technical literature and from additional schooling, either formally or informally. With the expanded communication capabilities, however, I expect that these people will feel no constraint about communicating directly and frequently with other acquaintances throughout the country or throughout the world who are in their cultural and professional level, and whom they identify as also being "gatekeepers". In other words, flow of information should increase dramatically on an interpersonal basis, and the initiative for the information will come, more and more, from the user rather than the producer of information.

Some publications will be produced to be read only by computers, either on a daily or weekly basis, and the computer reading the information will have in its memory, profiles of interests of the individuals it serves. It will abstract or flag such information, for the individuals and then have in its own storage bank all of the backup detail to the depth required by the human interested in that subject at that point in time. The processing, therefore, will become more and more a machine activity with the human exercising his selective curiosity to instruct the machine to pick out the material of interest for human use. This will require a different handling of material on the part of the information producers and the information brokers. In turn, this is going to require the producers and the brokers of information to identify parameters of information so that the material can be most useful to the professional audience when called out of storage or when assembled with other data. This categorizing and processing of information cannot be left in the hands of computer technicians, and this will impose a completely new level of responsibility upon the technical communicators. The importance of this difference is difficult to emphasize properly.

I posed a question in the title of these comments, so I am obligated to suggest a possible answer. What will follow the communication revolution? For lack of a better description, I suggest that civilization will be caught up in a "Cultural Revolution". It will be the child of the communication revolution because it will see mankind polarized in totally different dimensions with resultant totally different problems. If Dr. McCulloch were concerned about the ability of civilization to survive the Communication Revolution, I wonder, if he were alive today, what chances he would give mankind for surviving the "Cultural Revolution"?

The Worm and the Wampanoag

ALTON BARBOUR

Let's begin with some speculation about origins. Deep in man's past, sometime after ape-like creatures came out of trees and began to hunt on the ground, they began to walk upright, possibly to see prey. Since they no longer used their hands in walking, this upright stance left their hands free. They began to grasp and hold sticks and rocks and to use them for tools. When these two events occurred (upright walking and grasping tools) the creatures' brains began to grow in size and to develop.

From the dawn of man's pre-history, brain use has been linked to tool use, and the two have been associated with humanness.

In a simple sense, technology is tools. This would include the implements of a primitive culture as well as the machines of an advanced society. It would also include intellectual tools, computer languages, and analytic and mathematical techniques. The pervasive influence of technology is generally what distinguishes primitive cultures from contemporary society. A culture which has not yet learned to make a bowl is thought of as being "backward", and a society which is capable of sophisticated biological warfare is thought of as being "advanced".

As a species, man is a relative newcomer on this earth. But in the span of two million years, he has transformed an inhospitable planet into his own private domain. Much of this transformation is the result of his unwillingness to accept the world without questioning it or to leave the world as he found it. His questions and his search for answers has enabled him to diminish the mysteries, riddles, and myths about the world and to enable him to achieve some certainty and control over his environment. He developed both science and technology to systematize his accumulation of knowledge. This knowledge has grown so imposing that man believes he can do virtually anything with it.

Our scientific view of the world for the past 300 years can be described as being "Cartesian". The modern age has taken important cues from René Descartes, an early seventeenth-century philosopher. Descartes made a twofold contribution to systematizing knowledge.

First, he provided a basic axïom about the intelligibility of the universe. Grossly oversimplified and inelegantly stated, it says, "the whole is the sum of its parts". Secondly, Descartes contributed a method for making his axiom effective in organizing knowledge. In his "Analytic Geometry", he unified the world in simple quantitative relations that efficiently described such processes as motion, change, and the flow of time. He united "knowing" and "measuring", and influenced the widespread use of mathematics as a universal language.

The idea that the whole is the sum of its parts implies that the existence of the whole is brought about by the completeness of the parts and that there is no "wholeness" without the sum of the parts. It further suggests that one can know about the whole by measuring the parts and their relations. Although this sounds simple enough today, it was a radical innovation when Descartes first propounded it and it had considerable impact on man s view of himself in a rational, plausible universe.

Many people are likely to think of science as a body of knowledge. But it is not so much a body of knowledge as it is a method. Moreover, that which was thought to be "scientific" in the early seventeenth century is not usually thought to be scientific today. Although the Cartesian world view was extremely useful at one time in advancing science, it is questionable that it is still that "scientific" nowadays.

EVERYTHING IS CONNECTED TO EVERYTHING

We are coming to believe that a more useful way of reviewing the world is not as a "sum of parts", but as a "system". Peter Drucker notes that biologists now talk about "immunity", "metabolism", "ecology", "syndrome", and "homeostasis". Psychologists talk about "self", "personality", "behavior", and "gestalt". Business talks about "automation", "cybernation", and "management". Economists talk about "national income", "economic growth", and "productivity". The physical scientists talk about "systems", and "quanta". Theologians talk about "existence".

These terms and many more like them are creeping into the language. They represent a new world view. None of these words had any standing among scientists or scholars fifty years ago. What is particularly noteworthy about them is that they suggest QUALITATIVENESS. They don't refer to "how many", but they do concern themselves with pattern, configuration, and order. This new world view emphasizes PROCESS. It stresses not the sum of the parts, but an interrelatedness in which something that happens in one part of the system is felt elsewhere in the system. In other words, everything is connected to everything, and everything is changing. Technology has provided man with the power to transform the world, but in spite of his potential for being the master of his fate, man is not yet master and has, in fact, enslaved himself. He seems unable to free himself from Cartesian blinkers. He still sees the world in terms of its "sum of parts". He seems unable to visualize any order except causality. He is not yet process oriented. He does not understand the system. He misses the unifying order. When he attempts to be "scientific", he reverts to the static and mechanistic methods of 300 years ago. When he "knows" something, he organizes it into tiny bundles, the more bundles, the more knowledge (sum of the parts).

It was over 2500 years ago that Heraclitus explained that one cannot step in the same river twice because by the time one had stepped a second time, the river had changed. His comment illustrated the inevitability of change. Nowadays, our reactions to change are mixed. Some would like to speed the river up. Others would like to slow it down or stop it altogether. Some praise the flow and others condemn it. Not a few worship it. Some are terrified at the flood they perceive moving downstream. Others say that it is not nearly enough to do anyone any good.

WHO WILL SAVE MICKEY MOUSE?

Although change is the fundamental process of all existence, a growing number of scientists now believe that history will record the United States as a technological society which was struck down by a lack of ecological understanding. That would be a lack of understanding of how something which happened in one part of the system had consequences throughout the system.

Thinking about the future is a mighty lever for progress. There seems to have been too little of this thinking about the future as a consequence of the actions of the present. One major area which has not been adequately investigated has been that of technology.

It is possible to build a very good case to support the assertion that science and technology are the principal sources of change in our society, even more so than economic or political events, because of the PHYSICAL impact of science and technology. They are at the very roots of change.

Five fundamental machines, the screw, the inclined plane and wedge, the wheel and axle, the pulley, and the lever have enabled man to do work, to make war, and to create diversion. They've provided a "hand hold" for progress. This has been particularly so in the Western world, and notably so in the United States. Steam, petroleum, electricity, and atomic power have merely accelerated the process.

If the "American Dream" can accurately be described as the potential for improving one's condition, then the fabric of that dream is machine-woven. Machines have become a part of the American heritage. Americans seem to believe that there is nothing that a nation of doers cannot do, and moreover that machines can be built to do every job and answer every problem. This fascination with machines runs from pride of identification in the successes of the space program to the approximate 100 million buses, trucks, and automobiles which Americans use. Americans have grown up with machines, have become used to them, have come to depend on them, have come to love them.

American society's love affair with technology might be characterized by the mythical worm Ouroboros which ate its own tail and became the symbol of a world which survived by devouring itself. Primitive man survived by accumulating things which were a measure of security, and now modern man is virtually possessed by those things which he once possessed for his own survival. Although on the surface, that sounds much like Mickey Mouse's dilemma in the "Sorcerer's Apprentice" sequence in Walt Disney's 'Fantasia', we have no reason to believe that science as the sorcerer will arrive in time to save us from ourselves. It was in 1948 that William Vogt, in his book 'Road to Survival', said that we were plundering the earth that we have to live on. How long can the worm Ouroboros eat its own tail before there is nothing left?

SOLUTIONS CREATE PROBLEMS

Emmanuel G. Mesthene, the Executive Director of the Program on Technology and Society at Harvard University, has said that technology induces or "motors" social change in a couple of ways. It creates new opportunities for men and societies. It also generates new problems for them. Thus it has both positive and negative effects at the same time. In a sense it provides us with solutions which create problems.

The critics of technology are fond of saying that automation deprives people of work and will reduce the population to idleness and unemployment. There is virtually no evidence to support such an assertion. Each new invention or tool creates a new opportunity to do something better or differently, or something that wasn't possible before. In that way it provides new opportunities for societies. It is true that a new machine, product or process can make some jobs obsolete. But workers can be displaced from one job to another. This displacement is eased if the individuals involved are willing and able to change or if the society has means for assisting in retraining.

Societies can fail to respond to the opportunities created by technology. This would mean that potential lies unused. Problems are unsolved and goals are unreached because technological opportunities are wasted or used inefficiently.

NO THINKING ALLOWED

In terms of waste and inefficiency, one of the most curious things about science and technology is its own apparent resistance to science and technology. The history of discovery is often a record of closed minds. When Marco Polo returned from China to Venice, no one would believe him when he said that there were such things as paper money, printed books, nuts as big as a man's head, and black rocks that burned.

In 1875, the Director of the U.S. Patent Office sent in his resignation. He said that there was nothing left to invent. Marcellin Berthelot, the great chemist, said in 1887, "From now on there is no mystery about the universe." Also in the late 1800's, Professor

Samuel Langley was expelled from the renowned Smithsonian Institute for imagining flying machines propelled by the then recently invented internal combustion engine.

A splendid example of this resistance to scientific or technological change was uncovered by Rep. Melvin Price (D) of Illinois. He dug the story out of forgotten U.S. Navy files and told it to his colleagues in Congress. It is the sad story of the Wampanoag.

During the Civil War, it appeared that the British might side with the Confederacy, so the Navy set about designing a new and speedy vessel that might be used against British shipping if the need should arise.

The need never did develop so the construction of the Wampanoag continued at a somewhat leisurely pace and it was not finished until 1868. When it went on its trial run, the Navy discovered that it averaged an amazing speed of seventeen knots when the best ships on the seas were capable of no more than fourteen knots. This was clearly a significant achievement in ship design so the experts in the Navy put their heads together to decide what to do about it.

This was a time when the admirals were a good deal more comfortable with sailing vessels than with steam ships. The Wampanoag was a steam ship. The four smoke stacks of the ship outraged the aesthetic tastes of the admirals, so they took out two of its boilers and the reduction gears. Then they reduced the propeller from four blades to two blades to reduce "drag".

When they were all through, the speed of the Wampanoag was no more than that of any of the other ships on the seas, so it was scrapped. As a result, the U.S. Navy did not get a ship capable of the "unbelievable" speed of seventeen knots until some twenty years later. The Wampanoag was doomed because it was "unconventional".

There is every reason to believe that this same sort of blindness continues today. Theodore Gordon's article "Bucking the Scientific Establishment", and Dr. Ian Stevenson's article, "Scientists with Half-closed Minds", concur on this. Mausner and Mausner's research on "A Study of the Anti-scientific Attitude", supports this analysis.

However, just as it is important to remember that the world is changing, it is important to remember that science and technology are changing also. "Science and technology" is a very abstract label representing a multiplicity of events and experiences. Moreover, "science and technology" is not what it used to be, nor what it is likely to be. "Science and technology" is not a huge purposeful monolith, moving relentlessly toward our salvation or destruction, but a number of individuals and organizations attempting to make some sense out of the world, using the best means they have at their disposal.

HOW FREE IS FREE

Technological opportunity is needed now to solve problems which confront us such as pollution and over population. Technological innovation is so sought after that the Federal government puts some $16 billion a year into research and development. The United States usually invests its money in crises, however, not in long range planning. Evidence the military technology in World War II and the sums spent in the space race to beat the Russians to the moon. Truly, "haste makes waste", and long term problems continue unsolved. The fault is not only with the government however.

In the United States, freedom of individual decision making is a value that has long been cherished. It has been built into the institutional scheme of our society. Thus the negative effects of technology are often a result of individual and irresponsible decision making which does not give attention to matters which affect all of us. With each person given technological capability and some mystical autonomy, it is possible to "freedom ourselves to death".

In a sense, if a man is free, he ought to be free to make mistakes. In another sense, if his mistakes affect the lives of all of us, it may be necessary to restrict his mistakes to those things which do not make the rest of us his victims. This is particularly so when those mistakes are based on ignorance and stupidity. Voltaire once said that people who believe in absurdities are likely to commit atrocities. A glance at man's history would appear to bear this statement out. An isolated atrocity in a historical setting may not command much attention now, but one which is likely to have impact on future generations is one to watch and probably one to do something about.

This sets up a tension between our desperate need to control a technology which may destroy us, and our desires to preserve our values which may destroy us. Either alternative left unchecked precludes any easy solution coming from the other. Technology, for all of its benefits, is a threat to some of the human values that we hold dear. However, those same values are just as harmful if one fails to consider the results of the actions which go with them. One way of describing the pollution problem is that everyone has been allowed too much freedom to smother one another.

One of the things that misleads us is that we often believe that values are eternal and unchanging. That is, in fact, a way of defining values. However this is terribly misleading. Consider some of the values that societies have held unchanging such as the notion that punishment is a deterrent to crime or that some races are superior to others. One might easily question now whether these are still values, although they were respectably held by a large number of people for a considerable length of time. These once immutable values have fallen into disfavor because of recent information. It might be said that as knowledge increases, old values are challenged and new values arise to fit new situations.

According to Milton Rokeach, values are "bureaus of standards" for making decisions and choices. The choices are limited to the options one has available to choose from. An example would be the food available in a restaurant. These options are changed and expanded by technology. Technology, says Emmanuel Mesthene, has the effect of calling society's bluff, of testing society's values.

For example, communications technology makes "instant feed-back" possible in corporations and institutions. It can cover large distances quickly making instant voting possible. In this way, each citizen could have a direct voice in political decisions as in the idea of the "town meeting" form of government.

However, this same communications technology can lead to a feeling of impotence in the face of a machine society, by a lack of personal privacy and an increasingly manipulative use of the media. Each use of communications technology is possible. But popular values dictate the acceptance of one alternative over the other. George Orwell's '1984' depicts such a technological society of the future which is appalling and dehumanizing. Thus individual privacy and dignity can decline in a mass society, and even more so in a technological one. It appears that many people are perfectly willing voluntarily to trade their privacy for the benefits of some value they believe to be more important, for instance, being listed in the phone book. Of course, there are a number of ways in which individual privacy declines involuntarily, such as abortion laws, the selective service system, credit checks, income tax returns, the "no-knock" law, and wiretapping. Arthur R. Miller, professor of law at the University of Michigan Law School, says that today's laser technology makes it possible to store a twenty-page dossier on every American on a piece of tape that is less than five thousand feet long. If it is possible, will it be done? If our values allow for it, almost certainly.

In Victorian times a citizen could be said to have had more "privacy" than a citizen enjoys today, principally because no one much cared what happened. A citizen was free to remain ignorant, starve, fall ill, and die in complete privacy. In those times there was less value placed on individual worth. However, even though such a value exists now, the depersonalization of mass society generates a feeling of anonymity and worthlessness. The technology can be used to preserve democratic values in a multi-billion person society. The computer, mass data processing, and communications can make us free or enslave us. Much depends upon how we use our technology.

UNPLUGGING THE COMPUTER

It is curious that the word "change" comes from the Celtic and suggests both "to bend" and "to make crooked." It has long held both positive and negative connotations. Change is a mixed blessing. It gives and it takes away. This has been particularly so with technological change. It is often the worst offender of the future in terms of creating waste and consuming resources. Technology combined with "production" becomes a voracious filth-maker.

It has been said that the pro-technologist, upon finding out that there were more feet than shoes would try to find a way to make more shoes. The anti-technologist would encourage people to become reconciled to going barefoot. If those who had supported "the simple life" had always been listened to, the average age of death might still be eighteen. Now it is close to seventy. By 1600, the beginning of the scientific revolution it had risen only to 33.5.

Buckminister Fuller tells us that if all of our politicians were sent on a space ship to Alpha Centauri, leaving the technology intact, there would be no appreciable change in human welfare. If, on the other hand, the politicians were retained and the technology were sent into space, that at least two billion people would starve to death within six months. Fuller pointed out the relative importance of our political institutions and our industrial-technological complex. That while one is useful and beneficial, the other is critical and crucial to survival. Science and technology need no apologies.

Whereas the abject poverty of non-industrial Calcutta is a prime example of a lack of technology, Hiroshima is a clear example of technology gone wrong, evidence that although it is necessary to our survival, technology is not a savior, and not all technological change is progress.

Since, as Waldemar Kaempffert has said, much of our scientific and technological progress has been devoted to war, it is no wonder that such progress is suspect now. Many people assume a natural relationship between the arms industry and technology.

Kenneth Boulding has said that if the human race is to survive, it will have to change its ways of thinking more in the next twenty-five years than it has done in the past 25,000 years. But it appears now that there is a greater strain between the advancement of science and man's adjustment to it than ever before. We appreciate the benefits, but are concerned about the threats. We are uneasy about the power which science gives us without the control necessary to restrain the power. We are suspicious of the challenges it makes to established values and its inclinations to dehumanize us.

It is no wonder that scientific change results in what Toffler has called "future shock" or what Cronin has called the "sting of change." It is no wonder also that this has resulted in a distrust of technology and what Roszak terms "technocrats" and a desire to resist or control the change so as better to evaluate it. Just as it appears foolish now for "experts" to have resisted change represented by the speed of the Wampanoag, it now seems just as foolish to accept every change that comes along. In the past, scientists could only do what the societies allowed, and no doubt societies restrained unreasonably a great many advancements which might have been of tremendous benefit. It is also true that scientific "solutions create problems," and a "science of man" is sorely needed for a critical assessment of technology in this process universe. It may be a time to "unplug the computer" and start thinking.

Anthony Standen, in his book 'Science is a Sacred Cow,' said that the world is divided into (1) scientists who practice the art of infallibility and (2) non-scientists who are taken in by it. However, the child-like faith that Americans have long had in technology may very well be over. One night in November of 1965 an 80,000 square mile area of Northeastern United States experienced a power failure that blacked out several major American cities. It was a surprise to everyone. No one could explain how it happened or why, or could promise that it wouldn't happen again. And since that time, it has happened again.

Although the products of technology are often awe inspiring, people are not necessarily "uplifted" by them. They often view such developments as gimmicks, toys, or diversions. Willy Loman, in Arthur Miller's 'Death of a Salesman,' discovered that the subject matter preserved on his boss's new tape recorder was trivia. It sometimes appears that the whole of science is aimed at producing either weapons or Barbee dolls.

CHANGING CHANGE

It ought to be clear now that the universe is in process, that change is inevitable, and that not to change in such a world is to be out of phase with the system. One who stands still will be passed by. The concern here is not with change PER SE, but for the nature and rate of change. Society, including science and technology which induces such change, needs to be concerned with what changes or who changes, how the change takes place, at what speed, to what end, for what reasons. What is the guidance of direction of the change and what is the resistance or inhibition? According to what value standard does the change take place, and what are the consequences on our environment and our humanity? Society has within its power the potential for changing change and for changing the way in which change is regarded.

Abraham Sachar has told us, "The future is not what it used to be." Man needs to consider what the future will be. Man is both agent in and product of change on this planet. Everything he does influences everything else that he does. He is in the system and cannot escape the results of his decisions about how he survives or how he perishes. G.K. Chesterton has said, "We are all in the same boat in a stormy sea, and we owe each other a terrible loyalty." Man is again at the point of grasping tools, standing on his own two feet, and using his brain.

REFERENCES

Commoner, Barry. 'Science and Survival,' New York: Ballentine, 1966.

Cronin, Constance. 'The Sting of Change,' Chicago: University of Chicago Press, 1970.

Drucker, Peter F. "The New Philosophy Comes to Life," 'Harper's Magazine,' August, 1957.

Glorfeld, Louis E. "The Rhetoric of Science" in Akin ET. AL.(Eds.), 'Language Behavior,' The Hague, Holland: Monton, 1970.

Gordon, Theodore. "Bucking the Scientific Establishment," 'Playboy Magazine,' April, 1968.

Keampffert, Waldemar. "Science, Technology and War," 'The Yale Review,' Spring, 1942.

Mausner, Bernard and Judith. "A Study of the Anti-Scientific Attitude," 'Scientific American,' February, 1955.

Mesthene, Emmanuel. 'Technological Change; Its Impact on Man and Society.' New York: Mentor, 1970.

Mesthene, Emmanuel. 'Technological and Social Change.' Indianapolis: Bobbs-Merrill, 1967.

Miller, Arthur R. 'The Assault on Privacy.' Ann Arbor: University of Michigan Press, 1971.

Rokeuch, Milton. 'Beliefs, Attitudes and Values,' San Francisco: Jossey-Bass, 1968.

Roszak, Theodore. 'The Making of a Counterculture,' Garden City, New York: Doubleday, 1968.

Smith, M. Brewster. 'Social Psychology and Human Values,' Chicago: Aldine Publishing Co. 1969.

Stevenson, Ian. "Scientists With Half-Closed Minds," 'Harper's Magazine,' November, 1958.

Taylor, Gordon. 'The Biological Time Bomb,' New York: New American Library (Signet), 1968.

Toffler, Alvin. 'Future Shock,' New York: Random House, 1970.

Vogt, William, 'Road to Survival,' New York: Sloane, 1948.

Zajonc, Robert. "Brainwash: Familiarity Breeds Comfort," 'Psychology Today, February, 1970.

Zimbardo, Phillip and Ebbe Ebbesen. 'Influencing Attitudes and Changing Behavior,' Reading, Mass.: Addison-Wesley, 1969.

On Producing the Epicosm Model Reiterantly — To Mirror the Cosmos to Men

STUART C. DODD

PLAN OF THIS PAPER

This paper reports the methodology or behavior of the scientist in developing a testable theory of the cosmos called the "Epicosm Model." It reports three versions or forms of that methodology in three parts as follows:

Part I presents a CONSPECTUS SERIES. Each conspectus tries to analyze and resynthesize (such as to restore and improve on) one section of the modeling. The e Conspectus pages of Part I are listed and characterized below.

1. Abstract summarizing the set-product or semantic system called "the Reiteratings Matrix" that produced the model
2. The Reiteratings Matrix generating the 4 Epicosm submodels in 4 tenses
3. Row 1 of the matrix generating MAN'S SYMBOLS
4. Row 2 of the matrix generating ALL SYNTAX
5. Row 3 of the matrix generating 4 KEY UNIT PROCESSES
6. Row 4 of the matrix generating the 4 SUBMODELS by tense
7. The Reiteratings Matrix for Cosmists-Expanded Version
8. The Epicosm Model in Graphic and Algebraic Versions
9. The Epicosm Theory Restated in prose
10. Dodd's Publications Relevant to Project Epicosm

Part II presents a PROSE ANALYSIS AND RESYNTHESIS of the Reiteratings Matrix that models our methodology in the Epicosm Modeling. This Reiteratings Matrix aims to so describe the cosmists' verbal behavior that replicating it will reproduce the Epicosm Model of the cosmos.

Part III presents an EVALUATIVE REVIEW, in statement outline form, of how far the modeling fulfils the aims of scientists if taken as: "To so describe whatever is studied as better to explain, predict, and even control it."

PART I

To produce a model that will mirror the cosmos to man, A SEMANTIC SYSTEM [1] is here constructed in four phases; namely, Inputs, Thruputs, Outputs, and "Reputs" (testing and using the previous phases). This paper explains the Thruput phase.

I. INPUT

As input for this modeling, look at the cosmos extensionally as the universal set (=U0) of all things-namable (called "actants" here). Exhibit B graphs this cosmos.

II. THRUPUT

As thruputs for this cosmic modeling, take the intersect, or set-product, of the set of all actants (i.e., the cosmos) and the set of four reitering operations, specified in the columns of the Reiterating Matrix, Exhibit A. [2] These operations compound the SET of cosmic actants into a SYSTEM of cosmic activity. This build-up can be stated most briefly; thus, ACTANTS' INTERACTS ORGANIZE COSMOS.

This four-word summary is extended to forty words in the next paragraph (underlined), and to nearly 400 in Exhibit A, while its article describes this transition from cosmic set-to-system in some 4,000 words – which are further extended to 40,000+ in the 250 mimeographed working documents that explore the Epicosm modeling preparatory to publishing it fully.

SUCCESSIVE "REITERING" OPERATIONS ON ACTANTS ORGANIZE THE EPICOSM MODELS THUS (see matrix in Exhibit A):

ROW 1 CAN GENERATE ALL MAN'S SYMBOLS;
ROW 2 CAN GENERATE ALL SYNTAX INTEGRATING SYMBOLS;
ROW 3 GENERATES 4 KEY UNIT-PROCESSES;
ROW 4 GENERATES 4 STOCHASTIC LAWS, DEFINING 4 EPICOSM SUBMODELS.

EXHIBIT A THE REITERATINGS MATRIX to help mirror the cosmos to man. Showing how REITERATING builds the EPICOSM

MODELS that try to tell how "Interacting actants organize the cosmos." Where "cosmos" is a name for the universal set of all things namable, U°=1 "actant" is a name for any thing-namable, however primal or compounded "interacting" is a name for any repeating, combining, or permuting "organizing" is a name for compounding of interactings "reiterating" is a name for any compounding of reiterings (defined in Row 1)

4 Degrees of Reiterating ↓	↓	4 Degrees of Reitering ↓	↓	↓	The 4 row acts produce ↓
4 Factors in a → Reiterating	*Recurring* or REPEATING of things-namable ="re-" → ↓	*Naming* or COMBINING with a name → "reit-" ↓	*Standardizing* or non-PERMUTING → among people "reiter-" ↓	*Symbolizing* or INTERACTING in 3 forms "reitering" ↓	⇒ produce all SYMBOLS S
4 rounds of a Reitering, i.e., →	*Listing* into SETS → ,,,	*Adding* into SUMS → +	*Multiplying* into PRODUCTS → x	*Self-Multiplying* into POWERS 1^x ↓	⇒ produce all SYNTAX sSs s s ↓
4 Self-Reiterings i.e., →	↙ *Self-sums* or PAIRS 2n ↓	↙ *Self-products* or SQUARES n^2 ↓	↖ *Pair-powers* or NORMS 2^n ↓	*Self-powers* or FULFILLMENTS n^n ↓	⇒ produce 4 key PROCESSES K^s ↓
4 Log Reiterings i.e., →	Explaining as HELIX, Past cycles $(e^{\pi i})^{2n}$	Describing as NORMAL, Present probabilities $(e^{-.5})^{n^2}$	Predicting as ENTROPIC, Future growths $(e)^{2^n}$	Controlling as COMBINATORIC, Anytime self-governings $(e)^{n^n}$	⇒ produce 4 key LAWS S§ ↓
Column acts model the tenses, i.e. →	Past →	Present →	Future →	Anytime →	of the Cosmos
	TIME AXIS, $t = \frac{\log A}{c} = a/c$ in $A_t = C^t = 10^{ct}$			of the COSMOS	Cells record the ACTION AXIS, A

ORGANIZATION AXIS, $\log C_n = c = \lg^2 10 = 11$

The Epicosm modeling of the material and mental universes can be stated:
Algebraically, in the Epicosm equation, a = ct, where, in logs, a = actants, c = creation rate, and t = time; and in the 4 submodels by tense as in Row 4 above.
Geometrically in the Mass-Time triangular graph of actants organized in 8 levels; in Prose, as the hypothesis: "If all n actants interact in their n^n possible ways, then they perpetually form and unform the cosmos and all its parts – as Epicosm models try to spell out."

The REITERATINGS MATRIX, R_{4x4}: ROW 1 producing Epicosm submodels that graph the cosmos by tenses

ROW 1 tells how the FACTORS of a reitering form every symbol man uses

ROW 1 of R_{4x4}		and	and	defines a unit	called a
	REPEATING an it, a *thing namable,* an actant,	COMBINING name, and "it,"	PERMUTING it not among people	INTERACTING in a man·symbol·thing system,	SYMBOL, S the element of all language
	= re–	= re·it–	= reit·er–	= a reitering	

Read across

and also down.

The REITERATINGS MATRIX, R_{4x4}: ROW 2. ROW 2 tells how 4 ROUNDS of successive reitering form all syntax, s_g^o, or the elements of the set called:

	As Round *0*	As Round I	As Round II	As Round III
Thruputs→	Repeating namings or *Listing,*	or Combining namings or *Adding,*	or Permuting namings or *Multiplying,*	= Interacting with symbols or *Self-multiplying,*
Inputs ↓	i.e., naming like items, forms ↓	i.e., counting like elements, forms ↓	i.e., adding like addends, forms ↓	i.e., multiplying like factors, forms ↓
Bases				**a POWER, S^{III}** of a base
Factors			**a PRODUCT, S^{II}** of factors	
Addends		**a SUM, S^{I}** of addends		
Elements	**a SET, S^{0}** of elements			
Inverses→	Not listing forms a Complement	Subtracting forms a Difference	Dividing forms a Ratio	'Rooting' forms a Root

Row 2 develops *Syntax* which structures speech and language, helping to evolve man's mental life.
Note the System here: Row stubs = Inputs; Column stubs = Thruputs; Diagonal array = Outputs (boxed).

The REITERATINGS MATRIX, R_{4x4}: ROW 3. Row 3 tells how SELF-REITERING each of the 4 rounds of reitering form the 4 Key operations, K^s(= 2n, n^2, 2^n, n^n), of combinatoric algebra, as below:

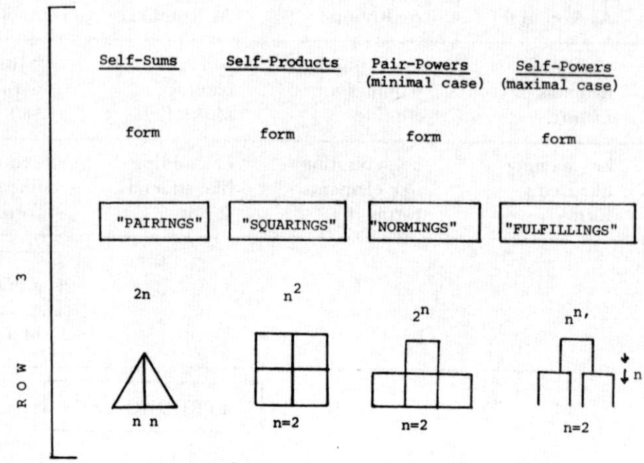

The 4 Key constants, ^sK, and 4 Key operations, K^s, form a set-product called "The Key Periodic Table Generating Cosmic Constants," (of which the Mendelyev Periodic Table is one array).

The REITERATINGS MATRIX, R_{4x4}: ROW 4. Row 4 tells how LOG-REITERINGS, e^{Ks}, by exponentiating the self-reitered Key operations, $Ks = \{2n, n^2, 2^n, n^n\}$, help to generate the 4 Epicosm submodels by tense to

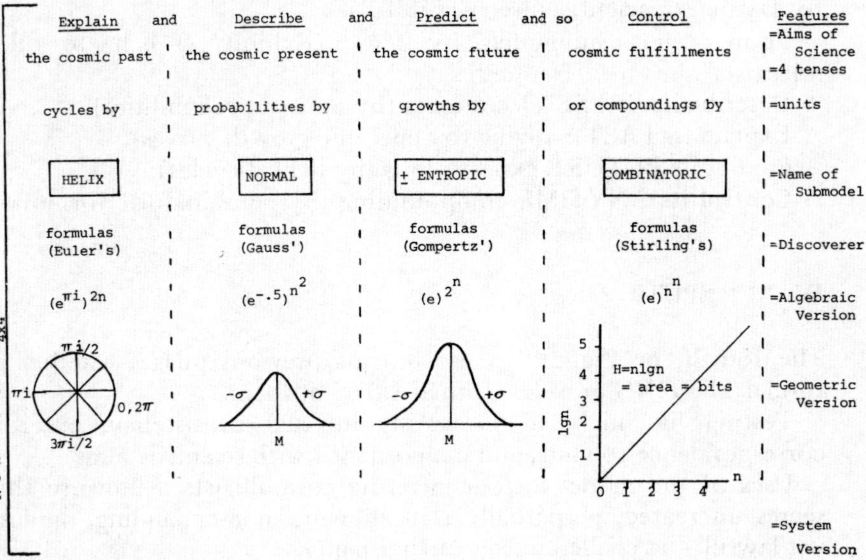

These self-reiterings as logs form the 4 cosmic submodels by tense.
(Inputs) (Thruputs) (Outputs)

III. OUTPUT

The formulated Epicosm Model, or output phase of the present modeling, can be semantically restated in four versions in algebraic formulas (like: $U = 1 = a/ct, = e^{2\pi ixat}$), geometric graphs, and prose for laymen or scientists (see Exhibit B).

Four cosmic submodels by tense (Exhibit A) help to fulfil scientists' aims to:

Describe its PRESENT contents (by normal probabilities);
Explain its PAST evolving (by entropic growth curves);
Predict its FUTURE expectations (by helical cycles);
Control its ANYTIME compounding (by combinatoric formulas).

IV. "REPUT"

The fourth, or "reput," phase repeats previous phases whether in formal TESTING or in less formal USES of them.

Testing has included measuring internal consistency, external correspondence to facts, and concordance with scientific aims.

Uses of the model include offering cosmologists a universe that seems uncreated, perpetually self-renewing, non-expanding, random yet lawful – as spelled out in further papers.

THE REITERATING MATRIX, R_{rc} FOR COSMISTS, EXPANDED VERSION

This 4 x 4 matrix offers, for criticism and testing, a methodological model, R_{rc}, of the cosmologist's speech acts when building a substantive model, a/ct=1, for the cosmos. It describes the actions of the cosmist when he describes the actions of the cosmos. Both of these actions in this cosmic modeling transaction are hypothesized to be well explained and predicted by REITERANTS. Evidence on how reiterants help cosmists to model the cosmos is ordered in this Reiteratings Matrix as follows.

This Reiteratings Matrix in 4 rows, R_{rc}, states 4 operations, called 4 "reiteratings," which tell the cosmist how to produce:

SYMBOLS of all sorts from the 4 factors of 'reitering' (in Row A below)

SYNTAX and language from the 4 rounds of reitering (in Row B below)

AMEY RULES or formulas from the 4 self-reiterings (in Row C below)

COSMIC LAWS or submodels from the 4 log-self-reiterings (in Row D below)

This reiteratings matrix in 4 columns, R_{rc}, state 4 operations, called 4 'reiterings,' for cosmists who build models for:

CAUSES of the cosmos in the past, in terms of repetitive reiterings mostly (in Col. 1 below)

CONTENTS of the cosmos at present in terms of combinative reitering mostly (in Col. 2 below)

CONSEQUENCES of the cosmos in future as permutative reitering mostly (in Col. 3 below)

COMPOUNDING in the cosmos at any time, as combinatoric reitering wholly (in Col. 4 below)

The 16 cells (called "reiterants" = R_{rc}) as a set-product of rows and columns, hypothesize that:

"REITERANTS CAN BEST MIRROR THE COSMOS TO MAN"

This hypothesis will be confirmed insofar as reiterants produce models of the cosmos that are more (1) comprehensive, (2) exact, (3) reliable, (4) explanative, (5) predictive, (6) controllative, etc., than rival methodologies.

NOTE ON FITTING THIS MATRIX TO THE MIT (= MASS-TIME △)

This Matrix above expands Level Ai for symbolic systems in the Mass-Time Triangle into 4 quadrants, if the 4 matrix Rows translate into 4 powers of i in the helical submodel, $(e^{\pi i/2})^t = 1^t = \pm 1, \pm i$; t = 1,2,3,4

Row A forms SYMBOLS = $1^1 = \sqrt{-1} = (e^{\pi i/2})$

as imaginary names of real things-named.

Row B forms SYNTAX = $1^2 = i.i = -1$ (= $e^{\pi i}$)

as products of symbols, i.i, of Operatings.

Row C forms KEY FORMULAS = $1^3 = i.i^2 = -i$

as names (= i) for Key Processes of Syntax.

Row D forms SUBMODELS = $i^4 = +1 = e^{2\pi i}$, each viewing the cosmos as a unit life-cycle of

(10^{32} years among 10^{176} primal actants??).

THE REITERATINGS MATRIX, R_{rc}, FOR COSMISTS

(Expanded Version)

Semiotic
HYPOTHESIS A: If the *4 sub-reitering acts* (underlined below) recur jointly, then any SYMBOL can recur— as most *liked* by the users.

	Col. 1	Col. 2	Col. 3	Col. 4	Col. 5
ROW A How Subreiterings produce SYMBOLS	The act of Recurring REPEATS a thing-namable, i.e., an item or "It" called an "Actant" (in Time), making it likely. This rule Serializes It ="re-"	The act of Naming it COMBINES a name with a thing-named among People, making it liked. This rule Socializes It ="re-it-"	The act of Ordering it PERMUTES the repetitions of that naming in diverse Contexts, making them alike. This rule Standardizes It ="re-it-er"	The act of Reitering it INTERACTS this trio as a product," or item, of speech Behavior, i.e., any symbol, making it "enliked." This rule Symbolizes It =a "re-it-er-ing"	Note 7 versions: ←in Lay terms ←in Combinatoric " ←in Operational " ←in Dimensional " [B_s=APTC] ←in Likability Theory terms ←in Scientific terms ←in "Reitering" terms

Semiotic
HYPOTHESIS B: If the 4 reitering acts (underlined below) recur cumulatively, then any KEY OPERATING can recur - as most alike to the facts.

	Col. 1	Col. 2	Col. 3	Col. 4	Col. 5
ROW B How cumulative reiterings produce SYNTAX	Reitering Namings as in Listing items forms A SET of elements called an "entity" e.g. (1,2,3,4) $o=n^o=1$ This round $\bar{o}$ of reitering builds Qualitative speech, $S^{\bar{o}}$ as in Language	Reitering Sets as in Adding counted elements forms A SUM of units called a "natural no." e.g. $1+2+3+4=10=n^I$ This Round I of reitering builds Quantitative speech, S^I as in Mathematics	Reitering Sums as in Multiplying forms A PRODUCT of factors called an "intersect" e.g. $1x2x3x4=24=n^{II}$ This Round II of reitering builds Relational speech, S^{II} as in daily Life	Reitering Products as in Self-Multiplying a number of forms A POWER of a base called a "log" e.g. $2{\cdot}2{\cdot}2{\cdot}2=2^4=16=n^{III}$ This Round III of reitering builds Systemative speech, S^{III} as in Scientific Laws	←In terms of: ←Math. operations ←Algebra ←Units ←Technical names ←Examples ←4 Rounds of ←Reiterings build ←Dimensional ←Speech, S^X, or ←Analysis

Semiotic
HYPOTHESIS C: If the 4 Self-reitering acts (underlined below) recur pairingly, then any KEY FORMULA can recur - - as most *likely* for prediction.

ROW C How Self-reiterings produce Key Formulas, Ks	Self-Sums form minimal	Self-Products form maximal pairings, or a	Minimal Self-Powers form a	Maximal Self-Powers form a	←Self-Reitering terms
	PAIRINGS, 2n,	SQUARING, n^2,	NORMING, 2^n,	FULFILLING n^n,	←Connotative terms ←Key terms
	as simplest Repeating	as simplest Permuting	as simplest Combining	as simplest Full-Interacting	←Combinatoric terms
	of n things, if taken 2 at a time.	of n things, if taken 2 at a time.	of n things, if put into 2 states.	of n things, if in any of the 3 ways.	←Operational terms
	General formula=mn Minimal formula=2x2=4 Maximal formula=n x n=n^2	n^m $2^2 = 4$ n^n	m^n $2^2 = 4$ n^n	n^n $2^2 = 4$ n^n	←General formulas ←Minimal cases ←Maximal cases

Semiotic
HYPOTHESIS D: If the 4 log-self-reitering acts (underlined below) recur empirically, then any SUBMODEL can recur- - as best fits the cosmos.

ROW D How Log-self-reiterings produce SUBMODELS of cosmos, A = Bn, or A/Bn = 1	Exponentiating 2n to the half-cycle base, $(e^{\pi i})$, forms Euler's Helical coil with n loops:	Exponentiating n^2 to the natural root-base, (e⁻ .5) forms Gauss's Normal probability curve:	Exponentiating 2^n to q or the natural log base, (e), forms Gompertz' Entropic growth and decay curve	Exponentiating n primal, real actants to n as log base forms Stirling's Combinatoric formula for interacts	Exponentiating time, t, to a Creation Rate, C, as a cosmic log base, forms Dodd's Epicosmic model, graphed in MIT,
	$e^{2\pi in} = cis 2\pi n = 1^n = 1$ ($cis\,\theta = \cos\theta + i\sin\theta$)	$y = 1/u_1(e^{\cdot 5})^{n^2} = 1$ (n=x) ($u_1 = \sqrt{2\pi \cdot 1} = 2.50662$)	$y = e^{2n}$ or $q_t = q_0{}^{2n}$ so p=1 (n = t = no. of time units)	$n^n = e^n n!/u_n = e^{n \ln n}$ ($u_n = \sqrt{2\pi n} = 2.50662 \cdot \sqrt{n}$)	as $A_t = C_n^t = 10^{ct} = 2^{256}$ t=7 or a = ct = 11 t-- in logs.
	This Epicosm Submodel₁ helps in EXPLAINING the past ceaseless life-cycling of all actants as they seem to form and unform at all 8 levels of an eternal universe.	This Epicosm Submodel₂ helps in DESCRIBING the present random distribution of all actants as kept up by finite self-renewal at all 8 levels within an infinite (?) domain	This Epicosm Submodel₃ helps in PREDICTING the future reiterant evolution of all actants as successive pair-permutings at all 8 levels of a self-governing whole	This Epicosm Submodel₄ helps in CONTROLLING at any and all times, a probabilistic design for actants, overall constant and unitary, at all 8 levels of a self-fulfilling cosmos	This Epicosm Master Model may help men analyze and synthesize all cosmic interacts. Its hypothesis says: A C T A N T S I N T E R A C T S O R G A N I Z E C O S M O S

NOTE ON FITTING THIS MATRIX, R_n, TO DATA

The 16 cells (called here "reiterants") in the matrix above need further tests of:

1. their RELIABILITY correlations with reobservings, requiring $r > .9$.
2. their VALIDITY correlations with empirical facts, requiring $r > .8+$.
3. their PREDICTIVITY correlations with outcomes, $r > .7+$.
4. their PROBABILITY correlations with math. theorems, $r = 1.00$
5. their AGREEMENT, above 99.6% when fitting cosmic constants.

THE EPICOSM THEORY RESTATED: A MODEL FOR THE COSMOS IN TERMS OF SETS, BITS, AND CYCLES

Let cosmos = $U^\circ = 1$ = the UNIVERSAL SET OF ALL NAMABLE ELEMENTS, n in number, and let them be called "actants".
Hypothesis 1: If n actants interact, reiterantly, randomly and ceaselessly, then they will always be forming and unforming compound actants, each in its own life-cycle, by trying out all their n possible combinations, or permutations, or repetitions, and thus operating the cosmos and all its parts largely by the "key" self-reiterant processes (i.e., 2n, n^2, 2n, n^n) as uncreated creators of all that was, or self-made makers of all that is, forever working towards greater fulfilment (= n^n) of what may be.

More exactly: if all n actants thus interact, they form the cosmos overall at a fixed Creation Rate, C_n, in 8 levels of organization, each a normal distribution, with its average life-cycle in one eon or log century of time – all as more fully hypothesized below;

1. Cosmos seems formed at a (LOG) CREATION RATE, c(= log C_n = 11.+ = $\lg^2 10$) of actants organized per eon or log century as graphed by the 2 slopes of the Mass-Time triangle (= MTT) Exhibit B:

Exhibit B

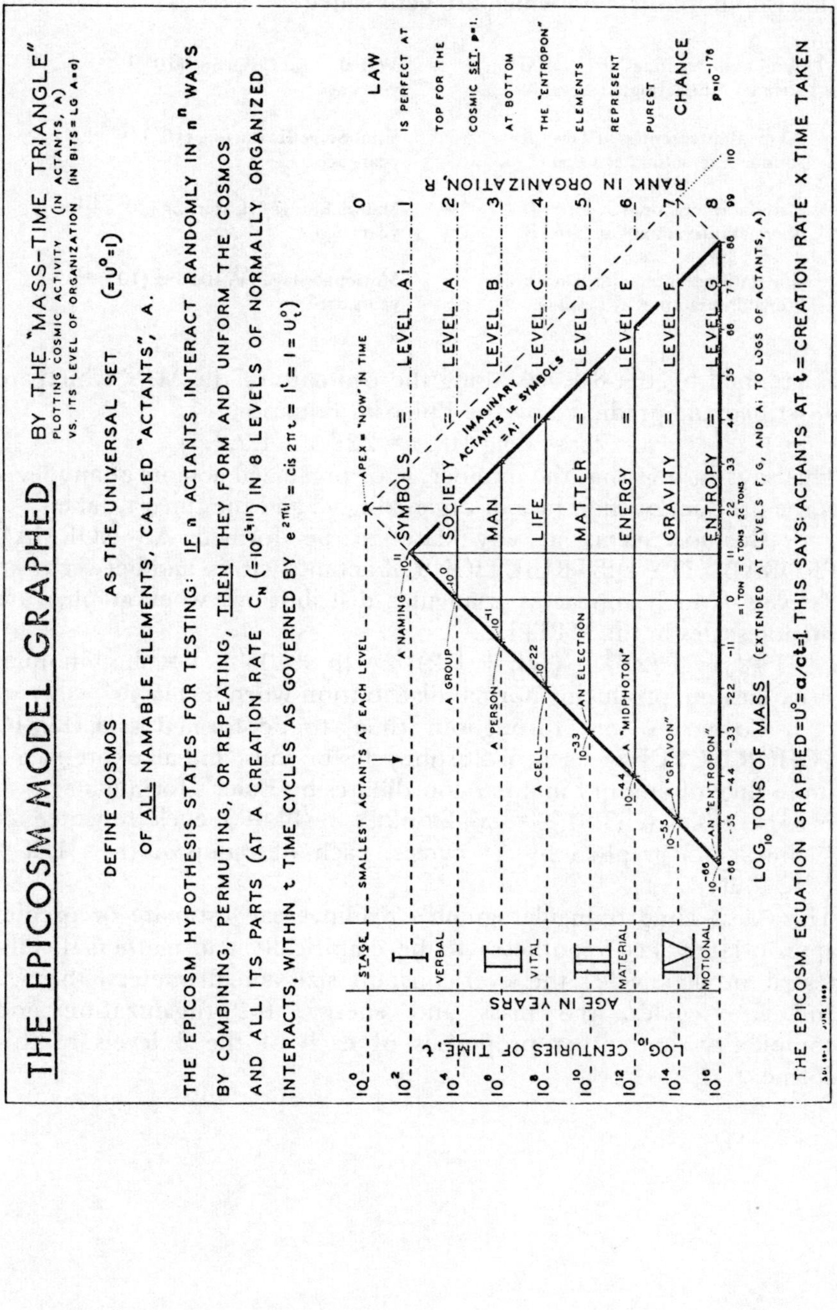
THE EPICOSM MODEL GRAPHED BY THE "MASS-TIME TRIANGLE"
PLOTTING COSMIC ACTIVITY (IN ACTANTS, A) VS. ITS LEVEL OF ORGANIZATION (IN BITS = LG A=a)
DEFINE COSMOS AS THE UNIVERSAL SET (=U⁰=I)
OF ALL NAMABLE ELEMENTS, CALLED "ACTANTS", A.
THE EPICOSM HYPOTHESIS STATES FOR TESTING: IF n ACTANTS INTERACT RANDOMLY IN nⁿ WAYS
BY COMBINING, PERMUTING, OR REPEATING, THEN THEY FORM AND UNFORM THE COSMOS
AND ALL ITS PARTS (AT A CREATION RATE C_N (=10^{c≈11}) IN 8 LEVELS OF NORMALLY ORGANIZED
INTERACTS, WITHIN t TIME CYCLES AS GOVERNED BY e^{2πit} = cis 2πt = I^t = I = U⁰_t)
LOG₁₀ CENTURIES OF TIME, t
AGE IN YEARS
STAGE
I VERBAL
II VITAL
III MATERIAL
IV MOTIONAL
SMALLEST ACTANTS AT EACH LEVEL
APEX = "NOW" IN TIME
A NAMING
A GROUP
A PERSON
A CELL
AN ELECTRON
A "MIDPHOTON"
A "GRAVON"
AN "ENTROPON"
IMAGINARY ACTANTS i.e. SYMBOLS =Ai
SYMBOLS = LEVEL Ai
SOCIETY = LEVEL A
MAN = LEVEL B
LIFE = LEVEL C
MATTER = LEVEL D
ENERGY = LEVEL E
GRAVITY = LEVEL F
ENTROPY = LEVEL G
RANK IN ORGANIZATION, R
LAW IS PERFECT AT TOP FOR THE COSMIC SET, p=1.
AT BOTTOM THE "ENTROPON" ELEMENTS REPRESENT PUREST CHANCE
p=10⁻¹⁷⁸
LOG₁₀ TONS OF MASS (EXTENDED AT LEVELS F, G, AND Ai TO LOGS OF ACTANTS, A)
=1 TON =C TONS 2C TONS
THE EPICOSM EQUATION GRAPHED = U⁰ = a/ct = 1 THIS SAYS: ACTANTS AT t = CREATION RATE X TIME TAKEN

2. Cosmos seems formed IN 8 EONS (= LEVELS) or 4 stages (= t) of organization, one c apart, which cumulate in complexity (measured in bits) and which are here called:

Symbol interactings at Level Ai Societal interactings at Level A	Verbal Stage I, starting $(10^4)^{I}$ years ago
Personal interactings at Level B Animal interactings at Level C	Vital Stage II, starting $(10^4)^{II}$ years ago
Material interactings at Level D Energial interactings at Level E	Material Stage III, starting $(10^4)^{III}$ years ago
Gravitational interactings at Level F Primal interactings at Level G	Motional Stage IV, starting $(10^4)^{IV}$ years ago

as graphed by the 8 levels along the ordinate of the MTT which are described and predicted by the Epicosm Equation:

$$a = 2ct = 4lglg\ 10 \,.\, t = 22t \quad (t = 1,2,3,4)$$

This says in logs that the number, a, of organized actants at any level is the product of their rate of organizing, c, and the time, t, taken.

3. Cosmos seems at any date to be formed AS NORMAL PROBABILITY DISTRIBUTIONS of actants within and between the 8 levels (which appear as triangular distributions when graphed on the log scales in the MTT) i.e.,

$(U^\circ/2 + U^\circ/2)^U = (1/2 + 1/2)^U = (p + q)^U = 1$ = the binomial expansion, producing normal distribution when n is large

4. Cosmos seems, throughout time, to be formed as COSMIC COHORT CYCLES each of 2t subcycles or eons, and all expressing a finite slice of eternity in polar coordinates by Euler's formula:

$a^T = e^{\theta i} = (e^{2\pi ni})^{2t} = \text{cis}\ 2\pi' n2t = 1^{2nt} = 1$ cycle repeated 2t times and graphed by 2t cycles, each of radius n (n = 4, t = 1,2,3,4)

This Mass-Time triangular graph's readings can estimate or predict appropriately (as hypotheses to be empirically and mathematically tested progressively) the ever-constant size and diameters, the age and life cycles, the mass and energy, the organization and complexity, and other properties of each of the 8 levels of this cosmos.

DODD'S PUBLICATIONS RELEVANT TO PROJECT EPICOSM

Of these 150 research publications, those most relevant to Project Epicosm are:

1. "A Mass-Time Triangle", 'Philosophy of Science', Vol. II, No. 4, October, 1944.
2. "An Isosceles Distribution of Material Entities", 'Main Currents in Modern Thought', Vol. 7, No. 4, Winter, 1950.
3. "Something Out of Nothing", 'Main Currents in Modern Thought', Vol. 8, No. 4, February, 1952. (U:61-52)

4 "Human Dimensions — a research for concepts to integrate thinking, 'Main Currents', Vol. IX, No. 4, February, 1953.

5. "A Dimensional system of Human Values", with C. Turabian, Transactions, Second World Congress of Sociology, International Sociological Association, 1954, pp. 100-105.

These present the discovery of the isoscles triangular distribution of all material entities in the cosmos and its hierarchic development in levels.

6. "The Reiteration Rule — a cyclic system for syntax, neurograms, and all laws", 'Syntheses', Vol. XI, No. 1, March, 1959. (U:56-143)

This paper presents a universal rule for generating every symbol and for expressing of the regularities of logic, mathematics and all science. Compoundings of this rule therefore syntactically develop all Epicosm formulas for empirical testing by fits to observed data and principles.

7. "The Transact Model — a predictive and testable theory of social action, interaction and role-action", 'Sociometry', Vol. XVIII, No. 4, December, 1955. (U:55-71)

This paper develops the 8 factors of the transact model for describing and predicting human behavior. These 8 factors were later generalized to represent the 8 levels of cosmic organization.

8. "How Random Interacting Organizes a Population — exploring a simple chance model to relate diffusion theory to information theory", 'Syntheses', Vol. VII, No. 1, March, 1960.
9. "The Logistic Law in Communication", with Marilyn McCurtain, 'The Symposia Studies Series No. 8 of the National Institute of Social and Behavioral Science, Series Research in Social Psychology', Washington, D.C., Sept. 1961, pp. 1-9 (U:59-182 Rev.)
10. "How Momental Laws Can Be Developed in Sociology — by deducing testables and predictive 'Actance' Models from Transacts", 'Syntheses', Vol. XIV, No. 4, December, 1962 (SD-67-16).

11. "Products Predict Interaction Where Sums Do Not", 'Sociological Inquiry', Vol. 36, No. 1, Winter, 1966. (U:58-101)

These four papers develop principles of behavioral interaction or statistical joint probability from empirical experiments on message diffusion in logistic growth curves, etc. Their generalization from human actants, to any actants led to the central hypothesis explaining the cosmos as forever self-generating by interaction of its n actants in their possible n^n ways.

12. "Scient-Scales for Measuring Methodology", Supplement Issue, June, 1966, 'American Behavioral Scientist', 1965.

This paper explores and develops the highest level Ai dealing with scientific symbolic systems which organize all else (and themselves) in the cosmos.

PART II
THE REITERATING MATRIX: A METHODOLOGY FOR COSMISTS

A. ROW 1 OF THE REITERATINGS MATRIX (SEE CONSPECTUS)'

The Reitering Rule says, in the imperative mood:

1. REPEAT a thing-namable
2. COMBINE it with a name, and
3. PERMUTE 1 and 2 in either order
4. To make a "naming" or symbol.

This Reitering Rule or four-fold speech act, can be restated, in the conditional mood, as a semiotic[3] hypothesis, thus:

1. If Repeating a thing-namable	=	"re-"
2. and if Combining it with a name	=	"reit-"
3. and if Permuting 1 and 2 in any way	=	"reiter-"
4. are jointly done	=	"reitering"

(i.e., if the intersect of 1, 2, and 3 are taken), then a reitering, a naming, or a symbol is thereby formed. Among these three synonyms or equivalent names, let the word "symbol" stand for its ordinary name; let the word "a naming" stand for its operationally specified name; and let the word "a reitering" stand for its analytic-operational name.

Every symbol man makes or uses must have these four necessary pre-conditions or factors, whatever further sufficient preconditions or factors may operate in selecting the particular thing-named and the particular name for it in any specified language.

In larger context, a reitering thus denotes a symbol-occurring, a word-used, a naming asserted, a recorded symbol recognized, a SPEECH ACT or element of the set called "language", "communicated human thought", or symbol-mediated "mental activity".

Every symbol can then be analysed and resynthesized as an ordered pair = (a,b) or as a complex number (= a +bi) consisting of the real thing-named (= a) and the imaginary name (= bi) for it. Then all compounding (or "reiterating") of symbols in sentences and equations or other sets, in book, files and libraries – whether stored externally in print or internally in the minds of men – all symbolizing, in short, can become, with semiotic research, increasingly mathematized by the algebra of ordered pairs or complex numbers. These semiotic technics may open up vast vistas for more scientifically guided and humanely pursued social and mental evolution of man.

B. ROW 2 OF THE REITERATINGS MATRIX (SEE CONSPECTUS 4)

Consider a first step in orderly compounding of symbols. Let us see what reitering a reitering leads to. Let us exhibit this reitering of any symbol through four cumulating cycles (called "rounds" here). Let these four rounds of reitering be spelled out in the tabulation below called "the Reiterings, Matrix".

ROUND 0 OF ROW 2 OF REITERING, forming S E T S (and all words). Given any "thing-namable as input or undefined primitive term",
the Reitering Factors (or steps), which specify the thruput are:

1. REPEAT any thing-namable n-1 times, making n occurrences of it.
2. (a) and COMBINE each occurrence with the name, "an element";

(b) and COMBINE all occurrences with the name "a set".

3. and PERMUTE each naming either as "a thing-named and its name" or as "a name and its thing-named"
4. and TAKE FACTORS 1 TO 3 JOINTLY as thruput operations on the input thing-namable, thus producing as unit output of this basic semiotic system:

A SET OF N ELEMENTS, each element being an instance of "a name-and-a-thing-named." Thus every set as well as every symbol man uses (which are usually assumed as given primitive terms in set theory) are both here derived by the reitering rule as a four-fold speech act.

ROUND 1 OF ROW 2 OF REITERING, forming S U M S (and all numbers). Given the output of Round 0 (namely "a set of symbols") as input for Round 1, then the Reitering Factors or steps, which specify the thruput are:

1. Repeat any elements of a set in some order as in a sentence or equation.
2. and Combine each ordered element with a general name for its position such as "one, two, three" etc.
3. and Permute each name with its thing-named in either of their two possible orderings.
4. The joint or mutually conditioning occurring of factors #1 to #3 in this round are called "adding of units" or "counting" while going on. When completed this output is called A SUM OF UNITS whether a counted union of sets of one, or any sum of the natural numbers.

Thus the natural numbers which are usually assumed as primitive terms in the foundations of mathematics are here derived by the reitering rule or four-fold speech act.

ROUND 2 OF REITERING, forming P R O D U C T S (and all relations). Given the output of Round 1 (namely sums and natural numbers) as inputs for Round 2, then

The Reitering Factors or Steps which specify the thruput of Round 2 are:

1. Repeat any sum or number, n, of units, m times.
2. Combine this identical repeating of a sum, or addend, m times with the name, "multiplying", when viewed in process as a thruput, and with the name, 'a product,' when viewed as completed or output.
3. Permute each name and its thing-named both ways at will, as above.

4. Take Factors 1 to 3 jointly, thus producing as output, A P R O DU C T of sums whether an intersect of sets, or any arithmetic or algebraic or other form of product.

Thus all forms of mathematical products and all forms of manufactured or otherwise compounded products seem derivable by the reitering rule from their corresponding form of sums whether seen as unions of subsets, or additions of quantities, or systems of alternatively occurring subsystems.

ROUND 3 OF REITERING, forming P O W E R S (and all laws). Given the generalized output of Round 2 (namely a Product or Outcome of many sorts) as input for Round 3, then

The Reitering Factors or Steps, which specify the thruput of Round 3, are:

1. Repeat any product identically in time through t periods.
2. Combine this identical repeating of a multiplying with the name 'self-multiplying,' or 'empowering,' if viewed as in process, or with the name "a power" (or "an exponent") if viewed as completed.[4]
3. Permute Steps 1 and 2 as above.
4. Produce in Steps 1, 2, and 3 together as their joint product or output: A POWER of products including a repeated (i.e., continuing) self-intersect of sets.

This becomes a joint occurrence which repeats identically under identically recurring conditions. Insofar as the recurrences approach identity and the n repetitions are many, this empowering may be called 'a law.' All scientific laws seem to have this implicit empowering of sets or symbols as a necessary factor whatever further sufficient factors may produce the particular law as phrased.

ROUND 4 OF REITERING, forming T I E R S (and cosmic order). Given, on the vast averaging, an exact output of Round 3 (namely a Power or integer exponent) as input for Round 4, then

The Reitering Factors or Steps, which specify the thruput of Round 4, are:

1. Repeat such empowering on the highest exponent.
2. Combine this empowered round of reitering upon a power with the name 'entiering' if viewed as in process in the present tense, or the name "a second tier" (of exponent, i.e., e^{n^n}) if viewed as completed or in the past tense.
3. and Permute the name and the things-named in either order, as before,
4. and thus Produce, as joint product or output of Steps 1, 2, and 3 of Round 4:

A TIER of powers or powered power – as in $2^{2^2} = 16$ ($= 2^4 = 4^2$) for a minimal case. The most important case of tiered powers is the S-shaped Gompertz growth and decay curve, $p_t = 1-q^{2^t}$, which expresses the Entropic Law of cosmic evolving and devolving.

Successive tiers can define a hierarchy of cosmic governors. Thus consider a minimal case where each governor, or successive tier of exponent is 2.

This forms the four-tier hierarchy of bits, $2^{2^{2^2}}$. This seems to measure the real activity of the cosmos as either read off from the bottom Level G of Primal Actants in the Mass-Time Triangle or as predicted by the Epicosm Equation, in antilog form, $A = C^t = 10^{ct}$, or $a = ct$ in log form, where $a = \log A_t$ and $c = \log C$. If t here is maximal at 7 (for the seven levels of the real cosmos), the c^7 converts, via the equation series below into 256 bits of rising primal actants and a balancing number of falling or disorganizing actants ($A = C^7 = C^t = 10^{ct} = 10^{11t} = 10^{77} = 2^{77 \lg 10} = 2^{77h} = 2^{77 \times 3.32193} = 2^{256}$). The interaction of these 256 bits, or the squaring of 256, gives $2^{256^2} = 2^{2^{8.2}} = 2^{2^{2^4}} = 2^{2^{2^{2^2}}}$, the four-tier hierarchy of bits of primal actants that seems to randomly govern the real cosmos, according to this case of the Epicosm modeling. This bits-hierarchy (supported by other lines of converging evidence detailed elsewhere) here illustrated how the reitering rule can build up in successive rounds to describe the randomly determined law and order of vast aggregates that seem to govern every cosmic cohort cycle.

Now let these four factors and four rounds of reiterings (Row 1 and 2 in the Reiteratings Matrix) be summarized in systems version. This reitering theory of human speech behavior and consequent mental life of man can be described in systems theory by the language of dynamic subsystems, co-systems and super-systems, self-steering by feedbacks and "feed-forwards", etc. Let any one reitering, or single symbol-in-use-by-a-human-brain, be called 'the system-at-issue' or 'nodal system'. Then the subsystems of this nodal reitering would be anything namable within its boundaries whether additive alternative parts or multiplicative joint aspects (such as the three combinatoric factors of a reitering, namely, the repeating, combining, and permuting of the name and the thing-named by a namer). This product of the three reitering factors may also be called a man·symbol·thing system, or unit of semiotics.

Then the super-system of this nodal reitering would be the product (i.e., joint, mutually modifying occurrence) of the nodal reitering AND other factor reiterings (as in the joint occurrence of the nodal words along with other words in a sentence). The super-system of a reitering thus includes its successive rounds of reitering or replicating the cumulative reiterings to date. The four rounds of reitering sketched above thus specify four enlarging super-systems.

Up to this point in the Reiteratings Matrix we have operationally described and explained Row 1, the reitering or element of all symbolism and Row 2, the operator or element of all syntax that relates symbols together in language.

C. ROW 3 OF THE REITERATING MATRIX (SEE CONSPECTUS FOR ROWS 3-4).

Now let us continue describing the Reiteratings Matrix by seeing in Row 3 how "Self-reitering" helps in one case to develop general qualitative language into the more exact quantitative terms of what is here called the four "Key processes, K^s."

If the four reitering Key operations of Row 2 are each reitered on itself, the output is the four "Key self-reiterated processes." This product or intersect from Row 2 of each cell entry's Key operation done upon itself produces the four cell entries of Row 3 as follows:

3:0) n^o, n^o, the set of n elements is reasserted or repeated.

3:1) $n^1 + n^1 = 2n$, the n counted elements are added to themselves yielding a minimal sum, called a "pairing."

3:2) $n^1 \cdot n^1 = n^2$, the number n, is multiplied by itself, yielding a minimal product, called a "squaring."

3:3) $2^1 \cdot 2^1 \cdot 2^1 \cdots \frac{1}{n} = 2^n$, the bit, 2^1, is self-multiplied n times yielding an exponent (or log to the base 2) of n units. This is called "n bits." It simultaneously measures:

a) the number of combinations of n elements if taken zero up to n at a time;

b) the number of combinations of n things if put into 2 cells or 2 states (such as "present or absent").

c) the area under the binomial curve and under the normal curve (as n enlarges).

d) let us call this process "norming" ($=n \rightarrow 2^n$) whereby n units become n bits or the nth power of 2 and thus GOVERN the normal probability distribution.

3:4) $n^1 \cdot n^1 \cdot n^1 \ldots n^1_n = n^{1+1+1+\ldots+1_n} = n^n$. This "self-power of n," is defined by n becoming both the exponent and the base.

This self-power, or "n-to-the-n^{th}" n^n, seems a very important concept in the Reiterating Matrix and its resulting Epicosm Modeling for the cosmos.

Note some of its meanings.

a) n^n is the maximum number of COMBINATIONS of n things put into m cells by the formula m^n, at its upper limit when m = n.

b) n^n is the maximum number of PERMUTATIONS of n things taken m at a time by the formula n^m, at its upper limit when m = n.

c) n^n is the maximum number of REPETITIONS of m things if self-multiplied m times by the formula m^m, at its upper limit when m = n.

d) n^n seems thus the upper limit for each of the THREE COMBINATORIC PATTERNS, of forms of interaction, among n things. It thus measures the FULFILMENT of combinatoric possibilities in any set of n things.

e) n^n is the number (for n large) of joint combinatoric unit patterns by Stirling's formula (rearranged) namely: $n^n = e^n\Gamma(n)\sqrt{n}/\sqrt{2\pi \cdot 1} = e^n \; n!/\sqrt{2\pi n}$ where e^n represents the COMBINATIVE process or factor:

$\Gamma(n)$, (the gamma function), represents the permutative process;

$\sqrt{n}$ represents the root of the repetitive process;

$\pm\sqrt{2\pi \cdot 1}$ represents the root unit-cycle or elementary unit of the three COMBINATORIC processes jointly.

An important variant of the self-power is the self-power-with-unit-growth, n + 1. If re-expressed in relative terms, $\frac{n+1}{n} \overset{df}{=} n$, and then raised to the nth power this is called a "generator," $((n + 1)/n)^n$. When n is large this is the formula for e = 2.71828, the natural base of logs. So n^n = e. When n is minimal at ±.5 this generator variant of the selfpower yields two further combinatoric constants, c_y and i. $c^{-\frac{1}{y}} = ((.5 + 1)/.5)^{.5} = \sqrt{3} = 1.73205$ so $c_y = .5772$; and $i^{-1} = ((-.5 + 1)/-.5)^{-.5} = (-1)^{-.5}$. This Euler constant, c_y, usefully converts between domains of continuous and discrete variables [5] while the imaginary, i, converts between the positive and negative domains.[6]

When n is 2^1, (i.e., one bit or the binary unit) its successive self-powers produce a remarkable series of critical limits, thus:

2^2 = 4, the bit bit or common unit for the four reitering rounds of sets, sums, products, or powers since 4 = 2,2 = 2+2 = 2x2 = 2^2

$4^4 = 256$ = the mass of Mendelevium, marking the upper natural limit of the atoms

= 256 = the number of bits of real entropons marking the upper limit of the smallest elements of energy in the cosmos according to the reading of the Mass-Time Triangle's baseline where $C^7 = 10^{7c} = 10^{77.24654} = 2^{7c}\,1g10 \doteqdot 2^{256}$

$256^{256} = 2^{8.2^8} = 2^{2^{3+8}} \doteqdot 2^{2c} = 2^{2^{2r}}$, the hypothesized upper limit of potential combinatoric compounds of bit-entropons in the cosmos (where c is the "Creatant Rate" ($c = 1g^2 10 = 11.03522 = 2^r$) r is the "Crux" constant or mode of all cosmic combinatoric patterns since it is the unique point where the number of combinations equals the number of permutations i.e. $2^r = r! = r\Gamma(r) = c$

Many further examples can be cited of how the self-power, n^n, sets critical constants both combinatoric and empirical throughout the Epicosm modeling of the cosmos. But such detail would unduly overburden this sketch of how the reiterating processes, summarized in the 4x4 Reiterating Matrix (see Exhibits), build up the cosmos.

In summary of Row 3 of the Reiteratings Matrix, the self-reitering operations of this row build up and define the four "Key" processes. These processes are symbolized in algebraic form as $K^s = (2n, n^2, 2^n, n^n)$ and are here named respectively the self-sum or "pairing" (=2n); the self-product or "squaring" (=n); the pair-power, or "norming" ($=2^n$); and the self-power or "fulfilling" ($=n^n$). These four Key processes exhibit, as their name implies, key functions in organizing the cosmos.

One of these functions creates "The Key Periodic Table Generating Cosmic Constants."[7] This periodic table for the cosmos is a matrix in which the Mendelyev Periodic Table for the atomic elements becomes just one array or submatrix. This Cosmic Periodic Table, or 4x4 matrix, was generated and discovered by forming the set-product or intersect of (a) the set of four key processes, K^s, with (b) the set of four key cosmic constants, ^sK. These cosmic constants, mark the four stages in the evolving of every cosmic cohort of actants (see the Exhibit of the Mass-Time Triangle). They are symbolized by: $^sK = (c^{-2}, \pi$, h 2r) where:

$c_y = .5772$ = The Euler-Mascheroni constant $\doteqdot 1/\sqrt{3}$;

$\pi = 3.14159$= the semi-circle to radius ratio;

h, =3.32193= $1g10 = \log C_n = \sqrt{c}$ = the "root-Creatant" constant;

2r =3.46 = lgc = the "Crux" constant $\doteqdot 2 \tan 60^\circ = \sqrt{12} \doteqdot \lg r\Gamma(r)$, etc, etc. The sixteen cells of this 4x4 matrix (that is the set product of the four Key processes and the four Key constants) specify the major constants and ordering relations among them for the cosmos as hypothesized by the Epicosm model.

The Self-Reitering or Key processes of Row 3 in the Reiteratings Matrix have many more uses, interpretations, etc., too numerous to develop here.[8]

In conclusion, note that self-reitering provides a standardizing symbolic and empiric operation in common that "helps to so describe as to explain and predict and even control much about the cosmos – as Row 4 demonstrates further."

D. ROW 4 OF THE REITERATINGS MATRIX.

The four cell entries in Row 4 are formed by exponentiating the four cell entries of Row 3. The self-reiterants, or indices, of the Key processes, K^s become exponents or logs. These Key indices become "empowered" as e^{Ks}. Their base is the base of natural logs, e_j (modified as flagged by the j subscript). They form the four formulas called "submodels" for the cosmos when looked at in the four tenses. Each model, or testable formula, looks at cosmic events as in the past tense, or in the present, or the future, or as at any time.

These four self-reiterant processes when taken as logs can neatly state formulas for all cosmic activity in the past, present, future, or any time as spelled out below. Together they can express special subforms of the general exponential model, $A = e^{\theta}$, in natural log units for any and all organization ($=\theta$) of any and all cosmic action ($=A$) or phenomena whatsoever.

But before describing how the four submodels by tense in Row 4 coalesce into one simple and general model for the cosmos (as $A = C^t$), let us describe how each submodel, in turn, in Row 4 is synthesized by exponentiating the four self-reiterants, 2n, n^2, 2n, and n^n into e_1^{2n}, e^{n^2}, e^{2n}, e^{n^n}.

(1) The Helix Submodel, or Col. 1 in Row 4 of the Reiterating Matrix.

When the self-sum, 2n, is exponentiated to e^{yi} (=–1) as base it becomes, $e^{2\pi in} = +1$, the formula (if n is an integer) for either the helix of one cycle or coil with radius n, or a helix of n cycles each of unit radius. (The i factor in the exponent converts rectangular to polar coordinates.) The helix also expresses the single random process and since it equals unity, a constant, it also represents a stochastic process wherein each total state, or set of subsets, probabilistically determines the next total state however much shifting may go on among the subsets. If one assumes the cosmos to be compounded from a random set of all elements of energy, as the Epicosm model does, then the helix predicts perpetual recycling of that total random cycle.

Readings of the Mass-Time Triangle tentatively assign a cosmic cohort cycle of $C^7 = 10^{77+} = 2^{256}$ entropon units of real activity and a total cohort life cycle of 10^{32} (=10^{25} = 10er $\doteq$ 10^{4u}; $u_1 = \sqrt{2\pi} = 2.50$) earth years, i.e., a half life of ten quadrillion years (=10^{16}). Thus the helix applied to a cosmic cohort cycle predicts the perpetual recurrence of such cycles as its future tense or long run prospect for the cosmos as a whole. This pictures a perpetually self-renewing constant cosmos with no sudden creation ever. (This non-expanding cosmos is harmonized with the red shift by the semantic hypothesis of antilog time scaling.)

(2) The Normal Submodel of Col. 2 in Row 4 of the Reiterating Matrix.

When the self-product, n^2, is exponentiated to the base $e^{-.5}$, or the root reciprocal of the Naperian base, it becomes the formula for the standardized normal curve (with unit midordinate). On double log coordinates this bell-shaped curve is transformed, or rectified, to reappear as the Mass-Time Triangle. (See Conspectus #8.) Thus taking natural logs of both sides of the standardized normal curve, $y = e^{-.5^2}$ straightens out the lips of the bell and transforms the normal curve into a parabola, $\ln y = -.5x^2$. Then to straighten out the shoulder curvature one may take logs again, or more interpretably, one may take the first derivative. This "rectifies" the bell-shaped normal curve to the two symmetrically sloping lines that form the sides of the Mass-Time Triangle. Their slope measures the rate of change or here, for the cosmic distribution of all actants, their slopes measure the rate of organizing the cosmos by negentropy and the rate of disorganizing everything eventually for each cohort by the physicist's heat death or positive entropy. The Mass-Time Triangle thus is hypothesized to express the entropic law of cosmic evolving and devolving within each successive stochastic cohort of ever-streaming actants. It thus seems to solve the problem of sudden creation by replacing it with perpetual cycling of random interacts of actants, forming and unforming in due life cycles every possible compounding. This maximum is the self-power, the fulfilment of all combinatoric possibilities in any n-set of actants.

By the Central Limit Theorem the normal curve is expected to be the upper limit for the distribution of all means however diverse. So it should describe the cosmos at any moment since it includes everything by definition. That the Mass-Time Triangle's left side fits the cosmic growth rate so well (in jumps of C, i.e., one level or eon) evidences empirical support for this normal hypothesis for the distribution of all cosmic actants.

(3) The Entropic Submodel or Col. 3 in Row 4 of the Reitering Matrix.
When the pair-power, 2^n, is exponentiated to the natural log base as e^{2^t}, it becomes the formula for the assymmetric S-shaped Gompertz curve of growth and decay. This seems to describe and predict within each cohort cycle, the course of evolution and devolution of cosmic interaction. This entropic curve is built up by successive self-products, or squarings, as $(((q^2)^2)^2)\ldots)^{2t} = q^{2t}$ where t = time. Note that in random interacting of any set of independent entities the cross-products will cancel out and only the self-products or squarings will tend to persist. This seems to explain how the law of ± entropy arises from random interacting. Note also that the first derivative of the Gompertz curve gives H, the complexity, i.e., $\frac{d2^{2}t}{dt} =$ $t\ lgt = lgt^t$ = bits of fulfilment in a t-set. This complexity, t lgt, is graphed as the area of the Mass-Time Triangle, computed as the abscissa (=Activity, A) times half the ordinate (=Organization=θ=1gA), so H = .5 A 1gA = .5 $1gA^A$ = bits of fulfilment of the actants, A.

(4) The Combinatoric Submodel, of Col. 4 in Row 4 of the Reiteratings Matrix.
When the self-power, n^n is exponentiated to some base, it becomes a combinatoric governing formula, such as 2^n in bits, or e^{n^n} in nits, or $2^{4^4} = 256$ in bits of entropons. These entropons are the ultimate real elements of cosmic activity or energy. (Each entropon is defined by MTT as 10^{-60} ergs, as read off from the baseline of the Mass-Time Triangle (Conspectus #8). This self-power in its upper reaches when n is large, factors into the three combinatoric processes of combining, permuting, or repeating by Stirling's formula (re-arranged), namely; $n^n = e^n \cdot \Gamma(n) \cdot \sqrt{n}/\sqrt{2\pi}.\overline{1}$. When n is small (or any integer), the self-power is the maximum for each process alone.

This self-power figures in many empirical cosmic constants. Thus the minimal self-power, $22 = 4$, a purely syntactic constant, gives the number of unit cycles as empirically observed in MTT. These govern the Creation Rate, C_e, of the cosmos as estimated by the cyclic version formula, $C_e = e2\pi.4 = e8\pi$.

The most important interpretation of the self-power seems to predict the trend or destiny of the cosmos and all its parts to continue random interacting ceaselessly and thereby tend to form in life cycles any compound of actants. This pictures a vast perpetual cycling of concurrent and successive and mixed phases of stochastic processes. These overlapping cycles of organizing and disorganizing mean that every cohort of actants interacts in varying degree (inverse to their adjacency) with all more mature and less mature cohorts. This seems to explain all feedBACK as interaction of the more mature on the less mature as in human education of the young. Given the organizing of memory in human speech and nervous systems, this vast process of feedback then can account for the cumulating culture and civilization of man and his learning how to dominate over all animal species in biological and ecological and psychological evolution.

This cycling trend to randomly fulfil combinatoric possibilities in any set of actants seems able broadly and on the vast averaging not only to explain past cosmic on-going but also to predict future end-states (or "goals," if stated in terms of human aspirations). Such ergodic processes lend support to the humanist philosophy: Insofar as man learns (by scientific methods of testing outputs applied to the social and value sciences increasingly) how to achieve his ends progressively, in just so far mankind can tend to fulfil that which most men like most deeply and durably. This highly tautological inference from Epicosm Modeling can be useful insofar as its antecedent and consequent terms can be observed in partial independence, each used to stimulate, guide, and motivate the other with augmenting feedback.

E. THE 4 ROWS OF THE REITERATING MATRIX.

With the above glimpses of the implications of each of the four rows of the Reiteratings Matrix in mind (Conspectuses #1–9), let us go on to sketch some implications of the four rows jointly, i.e., their intersect or system of interrelations.

Note, first that the 4x4 Reiteratings Matrix is a verbal tool for scientific methodology. It standardizes 16 steps in the procedure of the cosmist studying the cosmos. It can be itself studied in detailed focus Row by Row in Conspectus #3–6; or as a summary in Conspectus #1 & #2; or by formal and expanded analysis in the semiotic hypotheses of Conspectus #7.

Note next, the ordering produced among the key submodels or self-reiterant stochastic laws for cosmic processes. The six stochastic processes studied here are jointly ordered as exhibited in the table below:

1) by their combinatoric factors involved;
2) by their successive Rounds of Self-Reitering;
3) by their successive statistical moments;
4) by their successive tenses of time implied;
5) by their time spans.

TABLE 1 Ordering of Stochastic Models for the Cosmos

COMBINATIVE SET OF MOMENT LAWS

	Expontential Form	Key Factor	Self-Reiterant Rounds	Mo-ment	Time-span	Tense
Expontential	e^n	n	Listing $p^t_, q^t$	1st	Any period	Past or Future
Normal	$(e^{-.5})n^2$	$-n^2$	Adding $(p+q)^t$	0th	A Date	Present
Logistic	$1/(1+e^{-kn})$	$-n$	Multiplying $(p \cdot q)^t$	2nd	A Half-cycle	Past or Future
Entropic	e^{2^n}	2^n	Powers $(p^q)^t$	--	A Half-cycle	*Past* or Future
	COMBINATIVE AND PERMUTIVE SET OF CYCLIC LAWS, i.e., with π included					
Helix	$(e^{2\pi i})^{2n}$	2n		--	All cycles	Past or *Future*
Self-power	$n^n = e^n n!/\sqrt{2\pi n}$	n^n		--	One cycle	Timeless

Note thirdly, the cumulative achievement of the four rows.

Row 1 operationally defines the general factors common to ALL SYMBOLS man uses (leaving the particularizing factors to be further defined as needed).

Row 2 defines the four basic operations called "reiterings" which with their inverses and compounds, build ALL SYNTAX, or structures of symbols, from the most regular in mathematics to irregular mixtures in daily speech.

Row 3 operationally defines the four-Key self-reiterant or "Key" processes by successively compounding pairings (=2n), squarings (=n2), normings (=2n), and fulfillings (=nn).

Row 4 operationally defines four sub-models of the Epicosm Model, or testable theory, of the cosmos, each formed by exponentiating a key process.

Then Row 4, Col. 5 in Conspectus #7 synthesizes the four submodels in the Epicosm Power Equation as:

$A_t = C_n^t$ where A_t = no. of actants at Level t (in MTT)

$C_n = 10c$ = the Creation Rate

and c = lg210 = 11.03522 = the "Creatant"

or log Creation Rate in bits

More simply in logs, the Epicosm Stochastic or Unitary Equation is:

a/ct = 1 where a = log A_t and c = logC_n

This says: "The number of log actants, a, at each of the eight (=t) levels is a product of the cosmic creatant rate, c, and the time taken."

It estimates, as nomograph readings from the Mass-Time Triangle (amplified), the number of actants, the diameter and age, the mass and energy, the number of organisms, humanoids, and planetary societies at each of the eight levels that possess such properties, and the degree of organization (=θ=1gA) and complexity (=H=A1gA) of every level.

Is there any other model of the cosmos that THRUOUT ITS MATERIAL TO MENTAL RANGE, even claims to estimate so much information, so precisely and so testably as does this Epicosm Model,1971?

Note, lastly, how the Reiteratings Matrix is itself formed as a sort of squared set-product of the four Key reitering operations cross-classified against themselves.

re-	it-	er-	ing	Row 1 repeats reitering elements in a set
Sets	Sums	Products	Powers	Row 2 cumulates reitering rounds into sums
Self-sums $n+n = 2n$	Self-products $nxn = n^2$	Pair-powers $2\cdot2\cdot\cdot2_n = 2^n$	Self-powers $n\cdot n\cdot\cdot n_n = n^n$	Row 3 produces self-reiterings into products
$(e\pi i)^{2n}$	$(e^{-}.5)n^2$	e^{2^n}	e^{n^n}	Row 4 empowers self-reiterings into powers
Helix	Normal	Entropic	Combic	= submodels

Though not a "smooth" conceptual intersect, this matrix shows how simply and swiftly the four reitering acts of the scientist build up all symbols and syntax in Rows 1 and 2 and the Key processes and cosmic submodels in Rows 3 and 4. Can any alternative methodology model so much content with so few steps?

PART III HOW EPICOSM MODELING STUDIES THE COSMOS AS THE WHOLE SET OF ACTANT-ELEMENTS INTERACTING IN TIME

ABSTRACT

Epicosm models of cosmos offer, for testing, an integrated man–symbol–thing system of hypotheses, summarized in the Epicosm Equation, a = ct[9] and graphed in the "Mass-Time Triangle" (attached). This is aimed to help men synthesize their partial knowledge, their diverse World-Views and Value-systems (or preferring behaviors) into one whole picture.

I 3 TENSES

Epicosm modeling studies the cosmos in three TENSES.

It tries to describe cosmic activity at present so as to explain better its creation in the past and to predict better its destiny in the future.

II 3 PARTS: ACTS, BITS, CYCLES

Epicosm modeling studies the cosmos as seen in three PARTS that are called its "three axes".

It looks at all cosmic phenomena or ACTIVITY, a, as the set (= a0) of all elements or "things-namable" called "actants" here.

It looks at cosmic ORGANIZATION as the set of all interacts of the actants, observing the interacts as products, measured in bit-logs like $lgC_n^1 = 2h3 \div rr$. which expresses in bits the amount of cosmic activity at any one of the four stages, of cosmic evolution as the self-power or fulfilment of the "Crux" constant, r(= 3.46410).

It looks at cosmic ordering or sequencing as the set of TIME CYCLES (such as the daily and annual cycles of the Earth, or the life-cycles of persons, or particles, or stars, etc.).

It measures cosmic activity by trying to count its actants (=a1) whether primal actants or "gravons", photons or particles, mice or men, nations or names of any actants (to cite the eight kinds or subsets of actants organized at the eight levels).

It measures cosmic organization by trying to count its interacts in a period, calling this "a Creation rate" ($=c^1$).

It measures cosmic sequencing by trying to count its number (=t), of appropriate time cycles.

Epicosm modeling, then proposes (as a semiotic hypothesis to be tested) to measure the cosmos, or any major piece of it, by computing the interacting of its actants in time as specified in the Epicosm Equation: a = ct. This says that any major sample of cosmic activity, a, is a product of its rate of creation, c, and the time, t, taken to create it. This creation formula sums up the Epicosm modeling. It is graphed in the Mass-Time Triangle (= MTT) above. It is spelled out by all of the Epicosm modeling that tries to tell what it can imply and mean.

III ROUTES

Epicosm modeling studies the cosmos by three logical ROUTES:

A. It INDUCES the Epicosm's generalizing formulas (like a = ct) from myriad particular observations of actants interacting in time.

B. It DEDUCES the Epicosm's particular implications (like "Cosmos creates itself continually") from (a) a set of primitive terms (e.g., "actants", "interact", and "time-cycles") and (b) assumptions as to their nature (e.g., "The actants interact randomly, ceaselessly, and reiterantly"), and (c) postulates or working procedures (e.g., observing extensionally in seeing every word or name as standing for A SET OF elements – whether instances, ingredients, uses, etc., of whatever is the thing-named).

C. It PRODUCES the Epicosm's applications (like resolving the dilemma of an exploding universe by using antilog time scales) chiefly by forming and testing useful set-products among its subsets.

How far the three routes diverge or converge tests the Epicosm modeling as to its: empirical correspondence or fit to external facts, logical consistency or equations among internal parts, practical concordance or application of the scientists' means to their ends. Testing to date shows high converging of these three routes.

IV 3 TIME SPANS: A: A DATE, B: A PERIOD, C: ETERNITY, D: ANY SPAN

Epicosm modeling studies the cosmos over three TIME SPANS: (1) for one date, (2) for a life-cycle, (3) for eternity.

A. A "One-Date" submodel describes the ongoing cosmos of any current instant as a vast NORMAL PROBABILITY DISTRIBUTION of all actants. This can be induced by plotting all weighable actants by their mass against their time sequence in evolving as in the Mass-Time Triangle and observing their resulting normal distribution (that appears as an isosceles triangle on double log coordinates). This normality can be deduced convergingly by the central limit theorem. The fit of the normal curve to the data now available seems impressive.

B. A Life-Cycle submodel can explain the creating of the cosmos over its first Half-Life cycle as a vast SIMPLEX GROWTH curve. (This is defined by $a = u^{2t}$ where $u = \sqrt{2\pi} = 2.50662$ and t = 1,2,3,,,8 eons of time.) This reflects how actants of each level form self-products or squares (= u^2) in the eight successive eons of time.

C. An All-cycle Sub-model predicts the destiny of the cosmos as perpetual cyclings of all its parts within an eternally existant and constant whole. This submodel studies the cosmos as a vast yet stable stochastic process that is continually recreating itself. It seems forever self-renewing by the ceaseless and cyclic interacting of all its namable actants.

This all-cycle submodel can be measured by Euler's formula, $e\theta$ i. When written explicitly for any number, n, of cycles from zero to infinity this is: $e^{2\pi ni} = \cos 2\pi n + i \sin 2\pi n = 1^n = 1$.

A special case of it is the One Life-cycle model when $\theta = 2\pi$.

This all-cycle submodel can be graphed:

in polar coordinates as a helix of n coils;

in rectangular coordinates and antilog units as a sine-cosine wave;

in rectangular coordinates and log units as a series of two-way Mass-Time Triangles (MTT).

D. Epicosm modeling finally studies the cosmos as synthesized by the antilog-log formula, or submodel $A_t = C_n^t$ This exponential equation relates the number of actants, A_t, to their least organized actants (called actons) at each of the eight levels. This equation helps to predict from MTT readings, the total cosmic activity that is organized in, or up to, any eon of time.

In simplest log terms, this "Log Epicosm Equation" is:

$$a = ct \text{ where } a \doteqdot 11; (t = 1,2,3,4,5,6,7,8 \text{ in turn})$$

This says that the total amount of cosmic activity (in logs i.e., $a = \log A_t$) in any level or eon of time, t, is a product of the Creatant Rate, c, ($c = \log C_n = \lg^2 10$) and the time taken to create it.

In simplest terms of sets, this "Set Epicosm Equation" is:

$$a_u^o = c^o\ t^o = 1$$

This says that the unitary cosmos (= 1) is here analyzed into the set (= a^0) of all its actant-elements which are further specified as the intersect or joint occurrence of its t^0 subsets (i.e., the eight substages) and the c^0 Creatant elements of each subset.

In terms of the physicist's dimensional formulas, the number of actant-elements measuring cosmic activity represents the physicist's Action, A, or product of Energy (E) and Time (T) in erg-second units. $[A = ET = ML^2T^1]$

V TESTS

Epicosm modeling studies how to test itself as research proceeds, as follows:

A. Inductive Fits

1. By Measuring the closeness of fit between the cosmic normal curve or Mass-Time Triangle, and such points as may be observable among the eight minimal and eight maximal actants at the eight levels, the mean and sigma, the variance and the range, the midordinate and area, the inflection point (at $\pm\sigma$); the first five statistical moments (see rm + 2rm Tables), the first and second derivatives, its bits logs, etc.
2. By Measuring the closeness of fit between the simplex growth and decay curve, $A^{II} = n2^t$, and its 16 empirically observed or inferred increments in actants when organized in each of the eight eons of cosmic evolving.
3. By Measuring the closeness of fit between Euler's cyclic curve ($e^{\theta i} = \text{cis}\ \theta$) and the corresponding empirically observable or estimatable points.

B. Deductive Equations

4. By Calculating the implications, as Arnot did in 1941, from the transformation, $a^t = e^{t_s}$, of sidereal clock time, t_s, into their antilogs, at, called "atomic clock time",
5. By Calculating the fit of the relevant known constants involved in cosmic modeling to these deduced from, or predicted by, the "Key Periodic Table Generating Cosmic Constants".
6. By Calculating the fit of the deduced cell entries to the actual constants in the 4 x 4 "Key" matrix and its 8 x 8 Extension which exhibit the chief/real cosmological constants interconverted in integral terms, each of every other. E.G., $2, \pi, h_1, r_e; e, 10, c_y u_1$.

C. Productive Findings

7. By Exhibiting the fruits of Epicosm modeling
   ——— as the next section below starts to do under twenty headings or items of fruitage.

VI FRUITS

Epicosm modeling by studying the cosmos as a vast man-symbol-thing system has produced the following output to date:

A. Epicosm studies have helped to DESCRIBE the present cosmos;

1. As a UNIFIED FIELD of interaction for gravitational and electro-magnetic forces as also for all eight levels of cosmic interactings.
2. In simple INTERCONVERTIBLE EQUATIONS which translate the chief 16 constants of the Epicosm models into each other in integral terms, thus providing a test of their logical consistency.
3. As wholly GOVERNED BY NATURAL LAW throughout, without any supernatural forces, or acts of creation from nothing, at any time.
4. As an ORGANIZATION OF ACTIONS that can be measured by logs(t) and their antilogs, A, in the Epicosm Equation, $A_t = C^t$, in THREE TIME SPANS by three submodels in "three tiers of logs", namely:
   a. at one date, as a vast normal probability distribution of actants; $[A^I = 2^t = \log A^I]$
   b. in one cohort's life-cycle going through eight stages of forming self-products or successive squarings:$[A^{II} = u^{2t}$ or $t = \text{loglog}\ ^{II}A]$

B. Epicosm studies have helped to EXPLAIN the cosmos:
1. As a NONEXPANDING UNIVERSE which is made consistent with the "red shift" by using anti-log time scales;
2. As an active curved space, or gravitational field, in which every point (called a "spatial actant") moves in curved paths governed by π like the centers of two marbles rolling around each other when swapping places in the randomly permuting subset of all cosmic actants at Levels G and F in the MTT; as formally stated in $\Gamma\ {}^2_1/2) = \pi$
3. As CONTINUALLY EVOLVING and equally devolving in all its parts by actants interacting ceaselessly and randomly, pairingly and reiterantly;
4. By a CONSTANT CREATION RATE, C_n (where $\log C_n = c = \lg^2 10 \doteqdot 11$ and $\lg^2 c = 12$) of actants organized per eon by forming successive self-products ($u^2 = 2\pi$ and u^{2t}) and self-powers, etc. ($u^u = 10$ and $10^{10} = C_{10}) = C_n/10$)
5. Including MAN'S KNOWLEDGE at its apex by spelling out how self-reiterant stochastic processes seem to build up every actant as an apex of its isosceles triangle and normal curve.

C. Epicosm studies can help to PREDICT, in the cosmos;
1. The exact FISSION AND FUSION of particle-pairs and

photon-pairs by simple readings of the resulting wave lengths and frequencies or equivalent masses from the Mass-Time grid for all interlevel transformations throughout the organizing or disorganizing halves of the cosmos.

2. The eight MINIMAL ACTANTS (or least masses, called "actons" and defined by $A_t = C_n^{-1}$) that mark the lower boundary of each of the eight levels of cosmic organization as pictured in the Mass-Time triangular grid (= "MTT")
3. the eight MAXIMUM ACTANTS (or largest masses, called "maxtons" and defined by $A_t = C_n^{+t}$) that mark the upper boundaries of the eight levels (see MTT)
4. The seven TRANSITION EQUATIONS each expressing a self-product or squaring, $A_i^2 = A_{i+1}$, that forecast from readings of the Mass-Time grid the organizing at any ith level that produces the level above it (see MTT). A special case, transforming energy elements called photons into material elements called electrons is Einstein's formula $C^2 = E/M$.
5. The FUTURE DESTINY of the cosmos as perpetual interacting of actants that forever maintain the Mass-Time Triangle's picture of n actants, normally distributed at eight levels of organization, each formed as a self-product or squaring of the level below it, and unformed eventually as it completes its life cycle, within the ever-constant streaming of the on-going cosmos.

D. Epicosm studies help to CONTROL in the cosmos:

1. The GOVERNING OF THE COSMIC CONSTANTS, or most invariant combinatoric patterns, (such as: $e, \pi, i, c, 10$; r, t, α, C, G, F, N; $2n, n^2, 2n, n^n$, etc.) as tabulated in the "Self-reiterant Key Periodic Table for Generating Cosmic Constants."
2. The SELF-FULFILMENT of any n actants by forming all their n^n possible compounds;
3. The FULLER SELF-GOVERNMENT implied by actants rising to higher levels of internal organization and maintenance in the Mass-Time Triangle.
4. The trend to VERBAL AND INNER-DIRECTED cosmic evolution implied by the MTT wherein the interacting of actants rises from the Inorganic to the Living levels, from animal Man to civilized Society and on towards synthesis in symbolic systems at the level (Ai) of Science (as in

Research and Planning)

5. The CRITERIA OF HUMAN FULFILMENT as optimal organizing of those actants which prove:
   a. Most desirable to humanity in respect to their kinds, amounts, balances and timings;
   b. Most sharable among humanity by assuring equal opportunity for all (as in Bills of Rights, etc.);
   c. Most durable into posterity by extending their life--cycles (as for longevities since 1900)

REFERENCES

1. A system is here defined as at least the intersect (i.e., product, not sum) of its parts including all relations and sequences. A semantic system is here defined as the intersect of namers, names, and things-named, i.e. (a man x symbol x thing), system.
2. By-products of this modeling are in zero exponent notation for sets ($X^{o} = 1$) and the potential application of pair algebra, (a,b) = a + bi, to all symbolizing and consequent mental life – since every symbol comprises the pairing of a (real) thing-named (= a) and its (imaginary) name (= bi).
3. Semiotics is here defined as the science of man-symbol-thing systems. Semiotics thus studies intersects or products of sets of symbolizers, symbolizings, and things-symbolized.
4. Or with the name "a governor" if viewed explicitly in every tense of time.
5. By the formula: $\in n^{-1} = \ln(n + .5) + .5772 = \ln(n + .5) \times e^{.5772}$).
6. By the formulas: $e^{\pi i} = -1$ and its square $e^{2\pi i} = +1$.
7. This as yet unpublished Matrix is undergoing expansion refining and testing. As EpiDoc 144 a copy may be obtained from the author for criticism, but not for publication.
8. A "shearing" in the matrix should be noted whereby the columns of Row 3 are moved to the left one column compared with the column cells of Row 2. To simplify the useful ordering in the 4 x 4 Reiteratings Matrix, the unimportant self-set at the left is thereby ignored and the self-power cell at the right end of Row 3 is split into 2 important cells – one for the minimal power, 2^n, and the other for the maximal power, n^n.
9. This equation states the expected amount of cosmic activity, a, in logs of erg-seconds, at each of the eight levels in which the cosmos organizes itself. These levels of increasing complexity, measured in bit-tons, are called "Entropy (completed), Gravity, Energy, Matter, Life, Man, Society, and Science". The equation predicts the number of actants at Level or Stage t in a cohort's lifetime as the product of the Creation Rate c($\doteqdot$ 11), of actants organized per eon and the number t(= 1,2,3,4,5,6,7,8) of time periods required.

Eco-Cybernetics: The Nucleus of Unified Knowledge and Curriculum Patterns for the Space Age

JERE W. CLARK

THE MISSING COMPONENT

A great deal of imagination would be required to picture the task of assembling the component parts of an atom if the nucleus either were missing or were considered to be merely an incidental part. The task would be reminiscent of the efforts of "all the king's horses and all the king's men" to put Humpty-Dumpty together again.

Unfortunately, little if any imagination is required to picture the task of assembling a general curriculum without a nucleus. All we have to do is open our eyes and look around us today. There are hundreds of intelligent, experienced, creative, well-educated, and conscientious individual educators each working ever harder with ever more money to perfect his own component of an obsolete curriculum. Except for making an occasional gripe about other specialists not putting the other pieces in the proper place relative to his own piece, the rank-and-file educator tends to ignore most other specialists.

This is a serious indictment to make against the highly sophisticated educational system of 1970. Unless there is something we can put in place of what we have criticized, this criticism will merely add to the junk heap of negative criticism already polluting the environment of the educational system.

Against this background, we can now:

1. Identify the nature and role of a possible nucleus around which a curriculum might be designed to meet the needs of the complex, dynamic, and interdependent world of the technotronic age;

2. Identify some common operational features of the most common types of organized activity (in the physical, biological, and social worlds) which can be combined with the nucleus to form a

conceptual model of knowledge and action that is simple, flexible, and operational; and

3. Provide some idea of the need for a general systems curriculum and some perspective regarding the institutional arrangements which will be required to globalize the program.

THE NUCLEUS OF KNOWLEDGE

As a springboard into a discussion of the nucleus of functionally-related curriculum patterns, we might define the term "nucleus" as the term is used here. A nucleus is the central, basic core of an organization which not only provides a basis for organizing the various components but also provides the rationale (or force) that "holds the components together."

With that definition in mind, we might ask ourselves next: What is the most basic condition in nature which leads to – in fact, requires – organization? A leading answer seems to be that it is the condition of SCARCITY OF RESOURCES WITH WHICH TO ACHIEVE WHATEVER ENDS MAY BE INVOLVED. Whether we are talking about a physical, a biological, or a social organization (or system) need not matter. All of these can be viewed as having ends or goals or objectives such as survival, security, growth (regardless of whether they are conscious of them or not). In order to achieve these ends in the face of scarcity (of time, tools, raw materials, energy, information, etc.) the entity must organize – or be organized – in accordance with the principles of economizing – i.e., the effective utilization of scarce resources in achieving goals. Only in this way can the main problems posed by scarcity be solved.

In its efforts to survive or grow, for example, the organization must somehow utilize such ECONOMIZING PRINCIPLES AS PRIORITY SETTING, COMPARATIVE ADVANTAGE, OPPORTUNITY COSTS, SUBSTITUTION, DIMINISHING RETURNS, DIMINISHING UTILITY, AND "SPREADING THE OVERHEAD." By applying the principles of priority setting, for example, a mouse may decide to run when a cat approaches (thereby implying a priority on survival). By applying the principle of comparative advantage a worker ant is able to specialize in some of the particular activities (or "niche") for which it is best suited.

A simple analogy between the structure and functioning of a

human economic system and that of an ant society will help to clarify the idea of applying economizing principles to biological organizations. For this purpose we quote from one of six short radio lectures on the economics of ant societies given by the author on the New Haven radio station, WAVZ, in December, 1964.

> You no doubt have thought of ants as being "social insects," but have you ever stopped to think of an ant hill as being an economic system?
>
> In this system we find production and consumption of food, shelter, and protection – just as is true of any human economic system. Just as the human economy has natural objectives such as economic growth, freedom, and security, the ant economy can be conceived to have goals such as survival, growth, and security from the elements or its enemies. As the economy of men is faced with a scarcity of resources with which to achieve its goals, so is the ant colony faced with the same set of problems. Because this scarcity of means or resources affects both the economy of men and ants, we find both men and ants organizing themselves into patterns of specialization or division of labor. Economic principles are indispensable for success in each type of society.

In a similar vein we might think of a machine or mechanical system such as a home heating system as being an "economizing" organization. The producers and installers of the system organize the many components (sub-systems) such as the furnace, air conditioner, thermostat, pipes, radiators, etc., for much the same reason a business executive organizes people into effective economizing teams. Within this context, we can even think of the "mind" or controller of the system – the thermostat – being conditioned or set (by the installers) "to prefer" a relatively stable temperature to a widely fluctuating one. Is this not, in principle, what an advertiser tries to achieve with men – i.e., does he not try to condition people's minds to prefer the advertised product?

Enough has been said about the economizing process to indicate the indispensable need for the priority-setting and decision-making framework of economics in systematic efforts to overcome scarcity. Two dimensions or aspects of general ecology are equally vital. One is the BROAD SCOPE of the total environment (entire universe); the other is the BALANCE-OF-NATURE FUNCTION. Instead of emphasizing the viewpoint of the observer (ecologist), however, we emphasize the decision-making viewpoint of an individual "decision-making unit" such as an individual ant or an individual ant society. We are especially concerned with the processes whereby the individual person, animal, plant, or perhaps atom "decides" to perform the function for which it is made. That is, what kind of physical, biological, and social pressures or forces lead it to "choose" one course of action instead of another?

In this behavioristic study of organized activity, we are concerned primarily with the effects of action rather than with the concrete nature of the actions themselves. Consider, for example, that a metallic ball moves from point A to point B six times on a table. The first time it moves because an object strikes it from behind; the second, because an invisible magnet pulls it to point B; the third time, it moves because of air blown on it from behind; the fourth, because of the suction of a hose; the fifth, because of the operation of a small motor hidden inside the ball; and the sixth, because it simply rolls down hill. In all six cases, the effect is that the ball has moved from A to B. In no case did the ball successfully resist the pressures which led it from A to B. The effect is the same as if it had "preferred" to move to point B instead of remaining still or instead of moving in some other direction. The ball behaved as if it had "chosen" this movement.

The third and final indispensable ingredient of the nucleus of organized activity (which enables the system to solve the problems posed by scarcity) is the INFORMATION-HANDLING CAPABILITY. In the final analysis, it is the information-handling capabilities which enable a business manager, a queen ant, a cell, or governmental agency to make the decisions necessary for survival, growth, etc. This is where the relatively new discipline of cybernetics comes in. Here we use a broader, more heuristic, and less rigid notion of cybernetics – which we prefer to call PARA-CYBERNETICS – than do professional cyberneticians. In this context, we might say that para-cybernetics is the systematic, behavioristic, metaphoric study of the process of communications in biological, physical, and social organizations with emphasis on regulation and control through feedback and related processes. [1]

By combining the indicated basic elements or dimensions of ECONOMICS, ECOLOGY, and CYBERNETICS, we have what we call ECO-CYBERNETICS. These basic elements are, by way of summary: (1) the decision-making, priority-setting framework of economics; (2) the scope and species-balance function of general ecology; and (3) the information-handling capabilities highlighted in

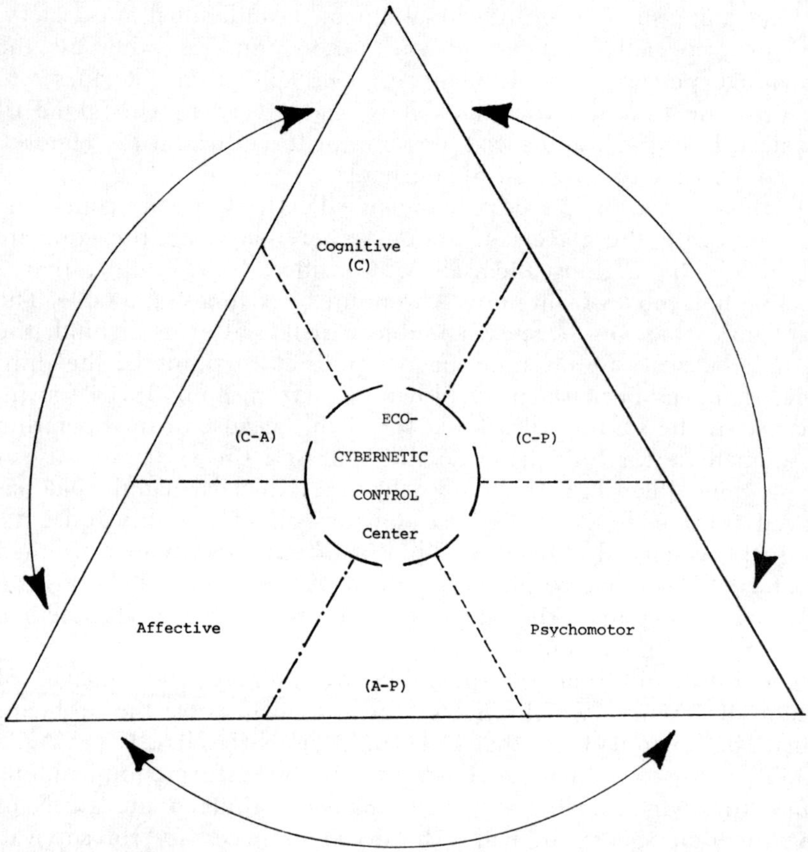

Figure 1 The Eco-cybernetic nucleus of knowledge. The figure symbolises the overlapping nature and interdependence of the affective (feeling), cognitive (thinking), and psychomotor (doing) domains.

cybernetics and related disciplines. Put these ingredients together, and we have the basis for organizing so as to overcome scarcity.

One convenient means of portraying the general functioning of the suggested nucleus of knowledge – eco-cybernetics – is to put it into the context of the three domains which the psychologist uses to describe human behavior. These are the psychomotor, cognitive, and affective domains as shown in Figure 1. In an oversimplified and generalized way, we might say that the organization (person, animal, tree, cell, business firm, governmental agency, family, etc.) must be equipped with some kind of COGNITIVE DOMAIN (intellect or its counterpart) for the purpose of thinking, making decisions, adapting, etc. It must also have some form of – or substitute for – the AFFECTIVE DOMAIN where its preferences or values and feelings are thought to reside. It is easy to picture these affective agents as they operate within the human realm. It is a bit more difficult, however, to picture a thermostat being programmed by a person or machine to "prefer" a stable to an unstable temperature. If the entity survives, it must somehow perform AS IF IT HAS CERTAIN VALUES which can be translated into a set of priorities. Equally important is the PSYCHOMOTOR DOMAIN (muscles, machines, tools, etc.) for effectuating or translating the decisions into programs of action.[2]

The crucial point for this purpose is that the effective operation and coordination of all three of these domains are essential to solving problems posed by scarcity. Hence, they are all essential to survival. That is why eco-cybernetics is portrayed in Figure 1 as the nucleus or control center of interaction among the three domains. If we may adapt some words from Robert Eberle, Assistant Superintendent of the Edwardsville, Illinois, School System, we can say that the eco-cybernetic control center is designed to "blend, mix, organize, orchestrate, and direct intellectual and physical energies into disciplined purposeful patterns of actions, ideas, and ideals."

APPLYING ECO-CYBERNETICS TO EDUCATION

Against this background we can now picture the embryo of an integrated body of knowledge, curriculums, and administration. To do so, we turn to the context of action which we view as a problem solving process. Problem solving is viewed in the broad context of performing a task or striving to achieve one or more objectives, as well as to "correct" something that is "wrong." Within this context, all of the following activities are viewed as examples of problem-solving behavior:

- a businessman modernizing his plant;
- a teacher preparing a lesson;
- a child learning a lesson;
- a school superintendent preparing his budget;
- an amoeba acquiring its food;
- a plant transforming sunlight into energy;
- a monkey fighting another monkey;
- a mayor fighting urban decay;

— a curriculum supervisor designing a curriculum;
— an artist painting a canvas;
— a minister preparing a sermon;
— a programmer programming a computer; and
— a philosopher evaluating the computer.

In each of these situations, the decision-making (controlling) unit of the organization in effect goes through a series of "steps" or phases of the scientific problem-solving procedure. Although there are many ways of classifying and listing this sequence of "steps," it will be convenient for our present purpose to organize them into the following sequence:

1. SENSING (observing) the relevant features of the environment;
2. INTERPRETING observations of the environment provided by the sensory agents;
3. HYPOTHESIZING — developing a set of alternative courses of action;
4. DECIDING which of several possible alternative courses of action are to be tried;
5. EFFECTUATING the decisions — i.e., translating the decisions into action; and
6. ADAPTING and improving as a result of experience and the evaluative feedback following each "trial."

Although all of these steps may not be used on any given problem, and although the sequence of steps may vary from time to time or from species to species, the eco-cybernetic process seems to be at work at each and every step that is taken successfully (apart from mere chance). In sensing the relevant features of the environment, for instance, the affective, cognitive, and psychomotor (or their physical counterparts) are all basic "actors" in the drama. Not only must certain aspects of each of the domains be utilized, but the right aspects have to be selected and blended into a coordinated whole. The selection, emphasis, and blending are believed to vary considerably from step to step, but the general result is believed to be the same (in all successful efforts).

It will be convenient to imagine an eco-cybernetic triangle (as pictured in Figure 1) for each of these problem-solving steps, all interwoven into one larger network such as shown in Figure 2.

In that figure, the small circles represent the eco-cybernetic control center in each of the six triangles, while the large circle in the center of the hexagon represents a master coordinator for the hexagon. This picture might be supplemented with an imaginary

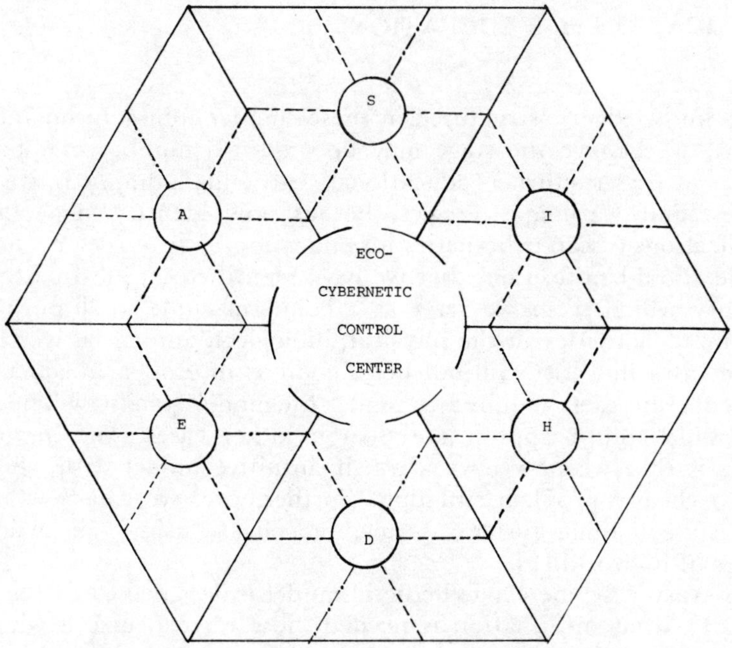

Figure 2 The problem solving process viewed symbolically as a sequence of six interdependent, eco-cybernetic steps.

The eco-cybernetic interplay of the affective, cognitive and psychomotor domains not only controls the process at each step but also coordinates the steps.

Code:

S = Sensing	D = Deciding
I = Interpreting	E = Effectuating
H = Hypothesizing	A = Adapting

X-ray picture of the figurative network of "communication wires" lying underneath the hexagon which make the eco-cybernetic coordination possible. It may be helpful at this time to turn back to the suggestive list of illustrative problems on pages 7-8 to which this model can be applied.

We should keep in mind that the over-all function of all of these problem-solving activities is, in effect, to minimize the undesirable effects of scarcity. Stated positively, the purpose is to solve the problem (achieve the objective) with a minimum of resources.

IMPLICATION FOR EDUCATION

With knowledge re-structured in these simple, common, operational terms, we believe the stage may be set for a number of quantum leaps in re-structuring educational curricula, administrative processes, and learning strategies. Let us now identify some of the implications of eco-cybernetics for education.

We should note, first, that we have identified an ideal pattern of activity which seems to serve as a common guide to all purposive, organized activities in the physical, biological, and social worlds. In some cases the "fit" will not be as good as in others. In some cases the concrete facts required to render the model specific will be more accessible and/or appropriate than in other cases. The important thing is that we now have a rough, intuitive model with which to approach any problem situation. Furthermore, we can view an ant hill, an economic system, a tree, or cell as "using" an analogous image of its world.[3]

In every case the eco-cybernetic model is designed to enable us to identify what information is needed, how to recognize it when we encounter it, and how to organize it when we get it. Likewise, some form of the general approach is assumed to enable each purposive organization in the natural and social worlds to "navigate" in its environment.

Figures 3, 4, and 5 have been developed to emphasize the near universality of the eco-cybernetic problem-solving procedures. In interpreting Figure 3, we want to keep in mind two ideas:

1. All organized entities – especially the purposive ones – in the physical, biological, and social domains are assumed to observe and adjust to their environment in ways that are analogous, although they are quite different on the surface.

2. Since the process of sensing, for example, is essentially the same function in all areas, it is advantageous for the student to learn to observe the field of each subject in a way that will help him learn how to observe any other field.

Hence, what stands out in Figure 3 is the common denominator of the eco-cybernetic process: Sensing; Interpreting, Hypothesizing, Deciding; Effectuating; and Adapting.

The same common denominator is at work in Figure 4. Emphasized is the fact that the learner, instructor, and administrator are all using the same process (while doing their respective jobs) as

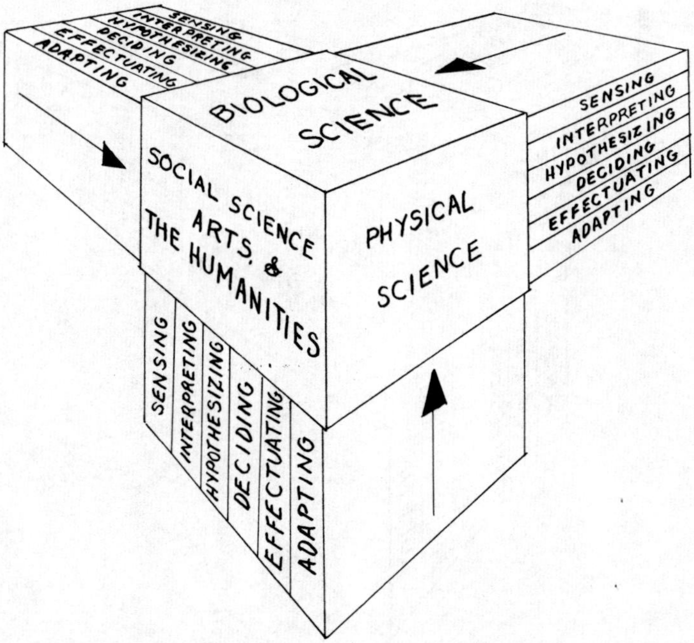

Figure 3 Eco-cybernetic patterns in the behaviour of entities studied in content areas of the unified curriculum, pre-K, to 19+.

the amoeba, tree, cell, atom, and human society are assumed to be using in Figure 3.

Figure 5 is designed to portray two fundamental ideas:

1. All educational growth is improvement in the eco-cybernetic operations (broadly conceived).

2. Although educational growth of the learner is the main justification for schools, basic, long-range growth of the learner is not very likely to take place unless the instructional and administrative staff are also growing in the process.

THE POINT OF NO RETURN

In an unchanging, fairly well-known, and isolated community, survival might be possible if only a few leaders, teachers, and supervisors should know how to perform all of the six steps in the

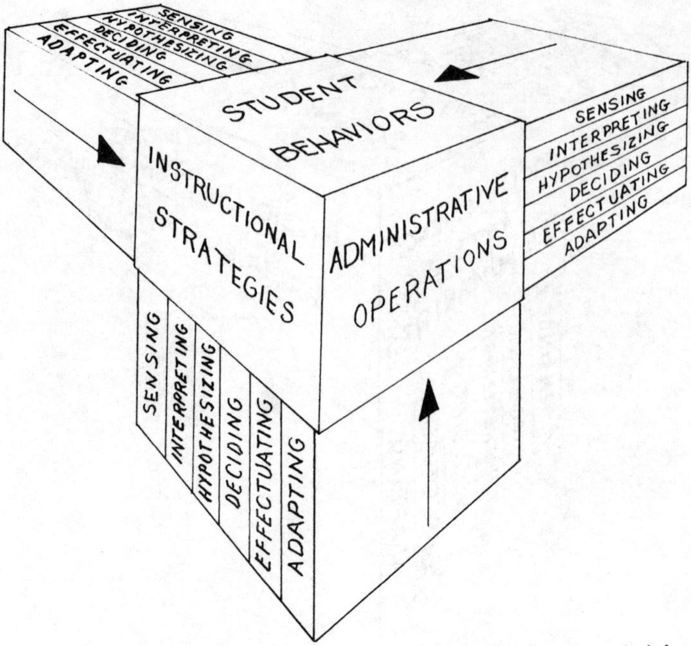

Figure 4 Eco-cybernetic patterns of students, teachers and education administrators.

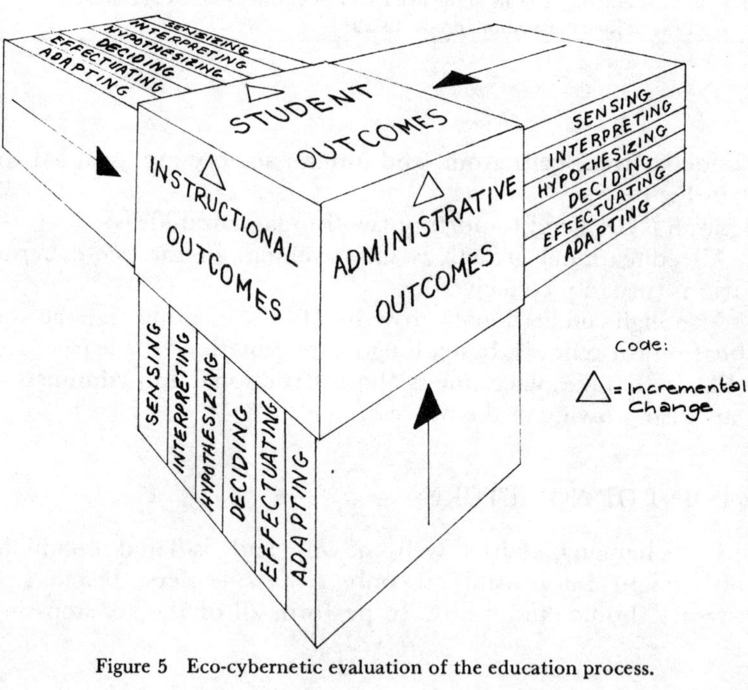

Figure 5 Eco-cybernetic evaluation of the education process.

eco-cybernetic learning process depicted in Figures 3, 4, and 5. The teachers could then pass on to the masses of people knowledge at the "E" (effectuating) stage of the game. Alternatively, in such a setting, learners might be able to get by with textbook-lecture versions of relevant knowledge which had been sensed, interpreted, hypothesized, decided, effectuated, and adapted by educators and scientists in the past.

In a relatively unknown, interdependent, and dynamic world, the strength or value of our education might be likened figuratively to that of a chain. We know that a chain is only as strong as its weakest link. Thinking of each of the six eco-cybernetic steps referred to above as "links," is it not true that there is a real tendency for the strength of the educational chain to be limited to the strength of its weakest link?

Today, however, we are living in an entirely different world. In the first place it is not isolated. In fact, the world-renowned engineer and architect, Buckminster Fuller, recently stated that in his opinion we have only a few years in which to fathom at least the embryo of a global society – if our civilization is to survive. In a similar vein, Glen Seaborg recently stated in an address that if man is to survive it will be because he has successfully moved "from man to mankind" in fairly short order.

Nor is our society unchanging. Consider the case of the information explosion. Assuming that the volume of accumulated information doubles each decade, when the present first-grade child enters his last decade of professional work (prior to retirement), the volume of information will probably have multiplied thirty-two-fold. Hence, present efforts to stretch the volume of the mind or to fill it fuller are not only inadequate but also are almost irrelevant to the requirements of the future. No matter how well the student absorbs tested information about some specialty that may be obsolete before he begins practice, he may be shackled by his education rather than liberated by it. We must help him LEARN HOW TO NAVIGATE IN THE RELATIVELY STRANGE, DYNAMIC, INTERDEPENDENT, AND COMPLEX WORLD OF TOMORROW. Or, to put the matter differently, we want to use eco-cybernetics to HELP THE STUDENT PROGRAM HIS MIND (with the use of eco-cybernetic procedures and concepts) SO HE WILL BE ABLE TO RE-PROGRAM IT THROUGHOUT LIFE whenever and wherever the need may arise.

Nor are we living in a known world. Buckminster Fuller often points out in his addresses and publications that to date man is able to detect less than one-millionth of the total spectrum of phenomena waiting to be discovered and explored.

This line of reasoning becomes especially important when we consider the chances for the survival of democracy – or even the survival of society in any viable form – with or without the kind of means provided through eco-cybernetics. Figure 6 portrays an idea of the fact that we may be approaching a decision point of no return in the freedom versus slavery game.

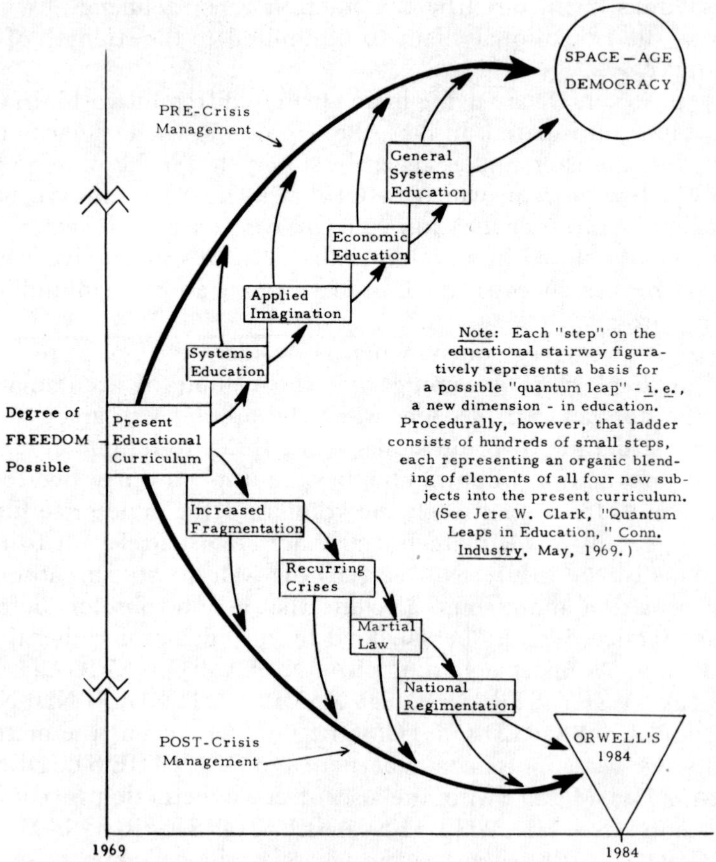

Figure 6 Alternative Stairways to 1984.

UPDATING CURRICULUM PATTERNS

Fortunately, the eco-cybernetic curriculum is flexible during the transition period as well as after. Movements in the direction of this type of program can be made slowly over a period of time or may be made completely and quickly, once the basic models have been refined and enough literature and other learning aids are available. A timely illustration of how a school system can move ahead in an intuitive and practical vein without waiting for the development and testing of the basic models is to be found in District 7, Edwardsville Community Schools, Edwardsville, Illinois, under the direction of Robert Eberle.

One of the simplest and easiest ways to begin adapting a traditional college curriculum to this approach is to provide a course or seminar in economics, one in ecology, and one in cybernetics. It is discouraging to note that less than one-third of today's college senior classes have had a course in economics, that a smaller proportion have had a course in ecology, and a still smaller proportion (perhaps less than five percent) have had a course in cybernetics.

Another fairly simple way which has been demonstrated at Southern Connecticut State College during the last three years is to incorporate an orientation to all three of these areas in the same course. This has been done successfully at the freshman level with a course in eco-cybernetics, and at the senior level with the seminar, The Control Process in the Social and Natural Sciences, and at the graduate level with the seminar, Assembly of the Sciences into a Single Discipline.

THE NEED TO LIBERATE THE ECONOMIZING PROCESS FROM ECONOMICS

With a few notable exceptions, such as Kenneth Boulding, Alfred Kuhn, John C. Schramm, and George L. Fersh, economists have been slow to recognize the relevance of the most important applications of the economizing process. Hence, one of the greatest intellectual needs of the day may be either to "liberate" this process from the economists and/or to awaken economists to the unusually fruitful opportunities that are knocking at their doors, so to speak.

This broadened interpretation of the economizing process (with built-in para-cybernetic dimensions) can be viewed as the essence of general systems theory. The similarity between the two can be seen by comparing a definition of general systems theory provided by the principal co-founder of the Society for General Systems Research and a definition of economics provided by one of the more prominent contemporary English economists. Ludwig von Bertalanffy, in 'General Systems Theory' (1969), defines general systems as follows: "A certain objective is given; to find ways and means for its realization requires the systems specialist (or team of specialists) to consider alternative solutions and to choose those promising optimization at maximum efficiency and minimal cost in a tremendously complex network of interactions." Lionel Robbins, in his classic book, 'The Nature and Significance of Economic Science,' defines economics as the study of human behavior as a relationship between scarce means which have alternative uses and given ends which can be ordered according to priority.

When we strip away the qualifying adjective "human" from the phrase, "human resources," in Robbins' statement, we get essentially the generalized statement of von Bertalanffy. This similarity between the economizing process and general systems can be strengthened further by using the definition of general systems theory often given by Buckminster Fuller. He says simply that by applying general systems theory we are able "to get more with less."

Thus we are now on the threshold of doing for economics what René Descartes did about three centuries ago for mathematics. The following excerpt from the 'Mathematics' volume of the Life Science Library series, 1963, suggests an interesting parallel:

> What he formulated (that day). . ., and subsequently elaborated, was the doctrine that all knowledge — both past and future — should be worked out in terms of mathematical reasoning. Descartes proposed that contemporary scholars cease to rely so heavily upon ancient ideas and make a frest start. (p. 82)

Against that background it may be appropriate to say that the time has come to work toward synthesizing all knowledge in economizing as well as mathematical terms. To the extent we achieve this goal, as was hinted earlier in this paper, we will be adding an explanatory dimension to the existing predominately descriptive models of general knowledge.

DEVELOPING A GLOBAL DELIVERY SYSTEM

A prerequisite to the large-scale utilization of any new idea is the development of the means whereby it can be widely disseminated, financed, and otherwise implemented. The Center for Interdisciplinary Creativity at Southern Connecticut State College, along with the collaborative ties with a few national and international organizations, already constitutes the embryo of such a "delivery system" for eco-cybernetics and related general systems approaches. Prominent in this international collaboration is the Task Force on General Systems Education of the Society for General Systems Research, which is headquartered at the Center, and the Union of International Associations, which is devoting a monthly section of INTERNATIONAL ASSOCIATIONS to general systems education throughout the U. N. International Education Year (1970).

Needless to say, one of the purposes of this conference is to encourage the cooperation of other future-oriented, interdisciplinary organizations, such as the Institute for General Semantics, The World Institute Council, The Center for Integrative Education, The World Future Society, the Joint Council on Economic Education, and the Creative Education Foundation.

An idea of the mission of and general strategy for developing this delivery system is provided in the mission profile of Figure 7. This figure is taken from a paper, "Developing a Global Network of Centers for General Systems Education," which is tentatively scheduled to be published in a forthcoming issue of 'International Associations.'

THE BEGINNING OF A DIALOGUE

This paper represents one contribution to an ongoing dialogue on the nature and necessity of updating our localized, static, impersonal, compartmentalized, inflexible education system into one that is designed to meet the demands of space-age living. The emphasis is on simplicity in organization and operation.

The suggested key to simplicity is to recognize the nature and importance of the nucleus of all knowledge, education, and action, and then to utilize the nucleus as a means of developing common denominators among all areas of knowledge and curricula. In this way, a common approach, common models, and a common language

Proposal for a

JOINT COUNCIL ON UNIVERSAL SYSTEMS EDUCATION

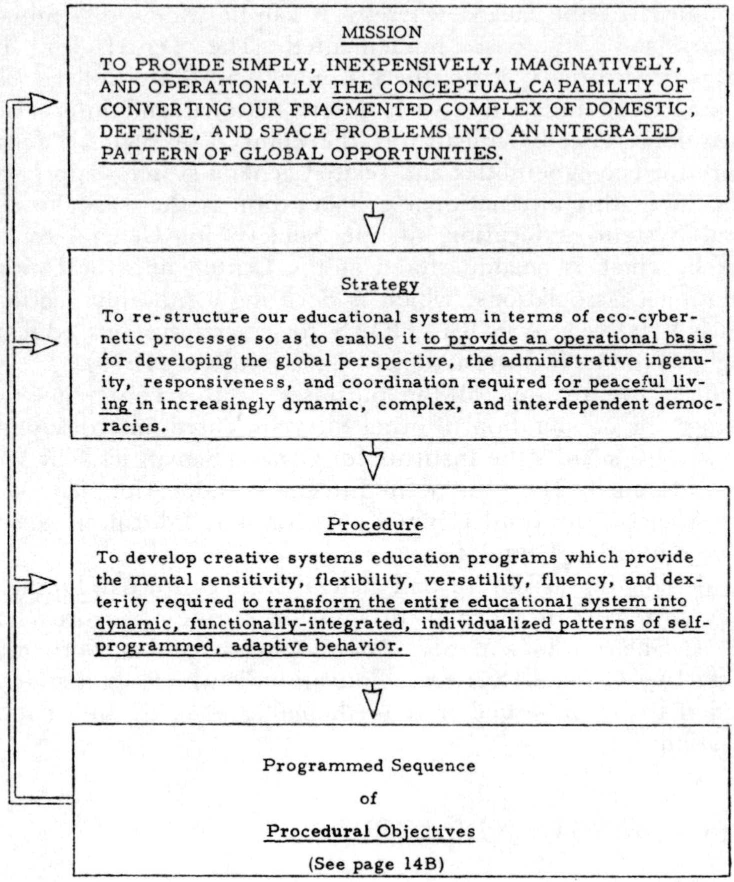

Figure 7

MISSION

Strategy

Procedure

Procedural OBJECTIVES

1. **Develop Basic Eco-Cybernetic Model and Meta-Language.**
   a. Develop an operational, organic, eco-cybernetic paradigm for functionally unifying progressively more facets of the sciences and humanities.
   b. Develop a general meta-language based on this paradigm.

2. Develop Applied Eco-Cybernetic Models:
   a. Affective Domain (Feelings)
   b. Cognitive Domain (Intellect)
   c. Psychomotor Domain (Muscles, etc.)
   d. Creative Adaptive Behavior
   e. General Curriculum

3. Blueprint Prototype of an Eco-Cybernetic Education System.

4. Encourage Development of National Network of Universal Systems Education Centers and Educational Systems.
   a. Implant prototype in a pilot community.
   b. Establish a Joint Council on U.S.E.
   c. Duplicate prototype center and education system so as to nationalize the effort.

5. Internationalize the Movement.
   a. Main export of the U.S. economy
   b. Vehicle for international exchange of ideas, products, and ideals

Figure 7 (continued)

for the basic elements of all of the areas of knowledge are being developed. As a candidate for the position of nucleus we have suggested eco-cybernetics. Eco-cybernetics is patterned after the way children learn to adapt to the world during the first months and years of their lives. It involves experimental efforts to achieve one's objectives with a minimum of means or resources, and using the results of each effort as a basis for planning the next effort. It is a self-motivating, self-directing effort to develop the whole person. At its most mature stage it is designed to help the individual achieve his own objectives in ways that will be consistent with, and ideally will reinforce, the efforts of others to achieve their goals. By concentrating over time on the fact that the public interest will be the private interest of citizens generally in the future – if there is to be a civilized future – private interests will be integrated into the global effort to continue to grow, live, and remain free.

REFERENCES

1. For this basic definition, see Jere W. Clark, "Facing the Crisis of Intellectual Poverty," 'Speech Journal,' Spring, 1970, (reprinted in 'The Journal of Creative Behavior,' Fall, 1969, and in 'International Associations,' March, 1970).
By far the most comprehensive, sophisticated, and empirically-based "para-cybernetic" model I have encountered is that developed during the last three decades by Professor Edward F. Haskell, who is Chairman of the Council on Unified Research and Education in New York City, and who teaches a graduate-faculty seminar, Assembly of the Sciences – into a Single Discipline, at the SCSC Center. For more details regarding Mr. Haskell's pioneering model and related matters, see Appendix A.
2. The graphic representation of these three domains and the nucleus should not be interpreted to mean that these relationships are mechanical or rigid. The triangle has been used here merely for graphic convenience. The same amount of space and a common shape have been given each of the domains simply as a reminder that they are equally important and hence a "balanced" education would provide for the nurture of all.
3. For the pioneer work in this great area, see Kenneth E. Boulding, 'The Image: Knowledge in Life and Society,' Ann Arbor Paperbacks, 1963. It has seemed to me since I read this book in 1958 (two years after the original hardback edition was published), that this book would prove to be one of the dozen or so outstanding books on philosophy and science of the first two-thirds of the Twentieth Century.

APPENDIX A TOWARD THE GEOMETRIC SPECIFICATION OF ALL VALUES

Edward Haskell, Chairman of the Council on Unified Research and Education, calls his para-cybernetic model (referred to in Footnote 1) "coaction theory" or "coaction cybernetics." He interprets coaction as the qualitative impact of the interaction between the strong (dominant of control) component and the relatively weak (work) component of a society with regard to any specific activity. The three major human coactions are attraction, repulsion, and indifference. The organismic coactions are: increase and decrease of life activity and neither one. For animals these might be expressed as helpful, harmful, and neutral.

When all of the possible coactions between the two elements (control and work) are considered simultaneously, the number of qualitatively different coactions are nine. Three of these are commonly used in lay circles; namely, symbiosis (mutually beneficial), parasitism (the weak helped at expense of the strong), and predation (the weak harmed while the strong benefitted). To transform these coactions into geometric terms Mr. Haskell uses the Cartesian plane coordinate system with "plus" representing beneficial effects and "minus" representing harmful effects and "zero" being neutral.

Thus we see that all of the proprieties of societies generally are said to change with this one property. To put it more pointedly, "The properties of societies are functions of their coactions." Thus coaction is proprosed as "the criterion for a natural classification of societies." (See Edward F. Haskell, "A Clarification of Social Science," 'Main Currents In Modern Thought,' Summer, 1949, especially pages 45-47.)

Several unique features of Mr. Haskell's coaction cybernetic model seem to make it superior to formal cybernetics at this early stage in the development of a near-universal model. These were enumerated in my paper, "The Role of Unified Science in Vitalizing Research and Education," at the AAAS convention in Boston, December 27, 1969, as follows:

First, coaction cybernetics is much broader in scope than is traditional cybernetics. It might be called "cybernetics," "sub-cybernetics," and "supra-cybernetics." Cybernetics is the middle link in a chain of evolution from such sub-cybernetic (closed) systems as atoms and the "cybernetic-plus" systems such as human societies which have communications capabilities and operating capabilities and operating characteristics which keep them from being considered to be cybernetic systems in traditional circles.

Second, by interpreting Mendeleyev's Periodic Table of Chemical Elements in "cybernetic" terms, and then developing the cybernetic counterpart of that table in each of several other disciplines, Mr. Haskell has been able not only to express many of the key relationships in geometric terms. And to think that the way may have been opened to express human values geometrically and in a way that can be related geometrically to other "values" is indeed remarkable.

A third unique and helpful feature of this coaction cybernetic model should be mentioned. That is the fact that it preserves the basic content and structure of each of the traditional sciences while adding the dimensions of general pattern and order, responsiveness, and coordination. It mainly adds a specially "coded," conceptual lens for viewing the various sciences and a more functional and operational way for approaching the various disciplines so as to inter-relate them meaningfully and to convert more facts into usable knowledge. This third unique feature may prove to be quite important in minimizing the amount and difficulty of changes a one-field specialist will encounter in up-dating his work. The specialist is not asked to forget what he knows and begin again, but rather to reorient his image of his field.

(The paper quoted is tentatively scheduled to be published in the proceedings of the symposium, "Extension of the Structure of Mendeleyev's Periodic Table to Physical, Biological, and Social Sciences.")

For further information on the coaction model and its implications, see the following:

1. E. F. Haskell, "A Natural Classification of Societies," 'Transactions of the New York Academy of Sciences,' March, 1947, pp. 186-196.
2. E. F. Haskell, 'Assembly of the Sciences,' I Scientia Generalis (preliminary, photo-copy edition, 700 pp.), 1969.
3. Harold G. Cassidy, 'Knowledge, Experience, and Action: An Essay on Education,' Teachers College Press, Columbia University, New York, (203 pp., paperback), 1969; especially Chapter 7, "A Scientific Theory of Value," pp. 149-178.
4. E.F. Haskell, "Unified Science: The Public Philosophy of the Space Age," 'Connecticut Review,' April, 1969.

APPENDIX B ECO-CYBERNETIC EDUCATION FOR EMPATHY

One of the greatest assets a teacher can have is empathy for his students – that is, the ability to see the world through his students' eyes and feelings, and thereby help students see the world through the eyes and values of the instructor. Likewise, one of the most valuable assets a business executive, a government official, an orchestra leader, or an artist can have is empathy with his colleagues, subordinates, supervisors, audience, customers, etc.

Important though this trait is, relatively little is done in formal education beyond the first grade to develop empathy PER SE. Given the past emphasis of our educational system on filling the mind with information, it is not surprising that the feelings of empathy tend to be lost before the child reaches high school.

On the basis of quite limited observation, it would seem to be relatively easy to develop this in almost any classroom – especially kindergarten classrooms – by combining the eco-cybernetic approach outlined in this paper with what William J. J. Gordon calls "personal analogy." In solving problems which have stumped the experts, Mr. Gordon uses the personal analogy approach to supplement three other types of analogy. If, for example, a certain type of fan belt must be fitted onto a certain kind of pulley (on an engine), Mr. Gordon may ask one of his researchers to play the role of (pretend to be) the belt and another the role of the pulley. A third researcher may even play the role of the engine. The players carry on a conversation with each other (as if each part were human and could feel and talk like humans do). In playing these roles, the players are vicariously activating (within their own minds) what are, in effect, the affective and cognitive domains of the mechanical parts. The spontaneous and imaginary dialogue – often argumentation – leads the players to observe the machines much better than they could otherwise and to formulate more meaningful hypotheses regarding the patterns of interactions among the parts.

Drawing on this general idea, suppose a kindergarten teacher were to have one hour per day devoted to this kind of personal analogy or empathy development. When animals are the objects of study, the idea might be to have each child play the role of a specific animal. The resulting conversation among the children might not be limited to oral verbal interchanges. Some parts might be written verbal; others graphic sketches of what the world (or specific things in the world) looks like to an animal. A child might be asked to compose a rhyming verse which he thinks an animal might compose to his animal lover. Dialogues might be developed between different kinds of animals.

When plants are to be studied, similar exercises might be developed, with each child playing the role of the plant in "talking" with other plants of the same and/or a different kind. Following a demonstration of a process such as osmosis, each child might play the part of one of the components of the process. In exploring genes and chromosomes, children might take the part of different units involved. In working with cells, each component of the cell structure might be represented by a child. When the local city is being studied, the usual role playing might take on more meaning.

The important thing is that the children (and the teacher) would be developing what might come to be the HABIT OF EMPATHY. This habit would enable them not only to learn more rapidly the lessons of the regular curriculum but would facilitate also exercising their imaginations. In terms of preparing for the coming global society, children would be getting practice in seeing the world through the eyes of many types of entities including the many nationalities and races of people.

[Note: The basic idea for developing this kind of exercise for this purpose has emerged indirectly over time from personal associations and from the writings of four persons. First is William J.J. Gordon, President, Synectics, Inc., Cambridge, Massachusetts. His book, 'Synectics' (Harper & Brothers, Publishers, 1961, N. Y.), specifically deals with personal analogy and the rationale for using it. Also, some of his curriculum materials which he and his associates have developed and used in a variety of elementary classrooms are highly suggestive of this type of activity.

The second person is Joseph Velez, President, Puerto Rican Human Resources Foundation in New Haven. Mr. Velez has helped me to visualize how some special projects of this kind might be developed.

General discussions with Sherman Price, President of Dynamic Perception Systems, Inc., New York City, and with Anthony J. N. Judge, Assistant Secretary-General, Union of International Associations, Brussels, have done much to emphasize the need for ways to picture any given event or situation from many points of view.]

APPENDIX C NOTES ON THE PATTERNS OF INTERACTION AMONG THE AFFECTIVE, COGNITIVE, AND PSYCHOMOTOR DOMAINS

As was pointed out in the text, the triangular representation of the affective, cognitive, and psychomotor domains in Figure 1 was used to recognize the existence of the three domains and to stress the fact that a balanced educational program must recognize and develop all three domains in some meaningful way. The question now can be raised, In what other ways might these domains and their interaction be pictured in an effort to be more realistic?

In a sense these domains might together be likened to a tree with the roots representing the affective domain; the trunk, the cognitive domain; and the branches and fruit, the psychomotor domain. This analogy at least recognizes that we are dealing with something which has more in common with an organism than with a mechanism.

Sherman Price, President of Dynamic Perception Systems, has raised the question in personal conversation, Are all three domains necessarily separate from each other? This question immediately brought to mind a classic example in economics in which a SIGNAL (interpreted by a businessman's cognitive domain) is really the same thing as an INCENTIVE (interpreted by his affective domain) looked at from a different viewpoint. That is the case of high profit potentials in a hypothetical, freely competitive market in a free enterprise economic system.

For a quick overview of some of the more traditional literature on the interplay of the cognitive, affective, and psychomotor educational objectives with some original insights and a quite selective bibliography, see Frank E. Williams, "Models for Encouraging Creativity in the Classroom by Integrating Cognitive-Affective Behaviors," 'Educational Technology,' December, 1969.

This article of Mr. Williams draws heavily on the builds upon J. P. Guilford's matrix depicting the structure of the intellect. My knowledge of the article, Mr. Guilford's model, and a similar model developed by the late engineer, John Arnold, led me initially to set up Figures 3, 4 and 5 as standard matrixes. Juanita Clark suggested the present form of these figures. In later research designed to position individual sub-components of each domain into dynamic, functional patterns of interaction, I may revert back to the standard matrix form for certain purposes.

A fertile area which needs to be explored in future research, it seems to me, is the role of positive and negative feedback in the functioning of the affective and cognitive domains. A good beginning point for such research is a section entitled, "The Autonomic Nervous System," in Douglas Stanley-Jones, 'The Kybernetics of Natural Systems: A Study in Patterns of Control,' (Pergamon Press, 1960, pp. 18-21). The discussion of a "balanced antagonism" between the parasympathetic and orthosympathetic neural networks in the human body is quite fertile. A review of the functions of the conscious and sub-conscious dimensions of the mind might be included within this context in an effort to generate new hypotheses regarding the patterns of interaction among the three domains.

The idea of using the triangular representation of Figure 1 was suggested by Juanita Clark, whom I had asked to draw a figure that was more complicated and awkward. My own attraction to the triangular form, once it was suggested, was in part due to the fact that I had been reading the thesis of Derald George Langham dealing with leadership and human behavior. (This thesis is for a second Ph.D.; the first was in genetics.) In playing around with the triangular form (as shown in Figure 1), I noticed the potential of Figure 2. It struck me that the two-dimensional hexagon shown there was similar to some of Mr. Langham's genetic building blocks of nature. Upon looking at a more accurate drawing of the hexagon (in two dimensions), I more or less accidentally noticed that the same figure could be interpreted in three dimensions as a cube. In fact, it appeared to be similar to the kind of cubical building block which Mr. Langham uses to build a model of universal phenomena. Mr. Langham's dissertation is entitled: "Genesa: An Attempt to Develop a Conceptual Model to Synthesize, Synchronize, and Vitalize Man's Interpretation of Universal Phenomena" (United States International University, San Diego, California, 1969).

Although neither time nor space will permit even a summary of Mr. Langham's model, a few more comments may be in order. In showing how nature begins simply and builds harmoniously with the resources at its command (as if to achieve a goal or set of goals) the model helps to corroborate the hypothesis that the economizing process is of the essence in nature.

The following excerpts provide some clues regarding his conceptual, 3-D space model – how it is used to integrate man's interpretations of all natural phenomena and thereby develop another simple and meaningful way to identify and utilize the indispensable minimum of knowledge:

> The method of study is to use the basic biological procreative structure of eight cells, four over four, as a form of living geometry, and build concepts around it in an orderly manner. (Thesis Abstract, p. 1)
>
> . . .I believe the octaves of sound, the octaves of the elements, the eight-fold path of Buddhism – all these and many other octaves we do not ordinarily consider – are one and the same thing. If you understand the basis of one of them, you understand ALL of them.
>
> Even in this day of the "Knowledge explosion," the basics, the vital ingredients, have not changed; and the basics are few. In my opinion we are perfectly capable of teaching every man, woman, and child in the whole world all the basics of every discipline of the Arts, the Sciences, and the Humanities. Then if they know how to expand the basics (expand the octaves) in the Genesa Space Model, they can learn by themselves from then on throughout life. (Thesis, p. 26)

Although space does not permit a description of Mr. Langham's approach to values and the cognitive, affective, and psychomotor domains, a couple of observations should be made. His model does deal with these factors in a way that appears to be scientific, unique, and unusually fertile. He incorporates one quite critical dimension in his model which Mr. Haskell's coaction model fails to consider. That is the question of what force in effect "propels" evolution to create the awe-inspiring and ingenious works of nature? His answer to the question goes back to the late Edmund Sinnott's emphasis on the hypothesis that "all living things are seekers and creators, and striving for goals is the essence of all life." (p. 209) A cursory reading of Mr. Langham's dissertation and a quite limited exposure to Mr. Haskell's model leads me to think – at least tentatively – that the genetic model does more than the coaction model to provide a feel for the way in which a little protoplasm can eventually reach out to and touch infinity in its cybernetic development. (pp. 24 and 30).

Mr. Langham's emphasis on the ultimate goal of "total awareness" is reminiscent of the exciting practical work which William Eblen and his associates, Total Education for a Total Environment, Inc., of Wilton, Connecticut, are doing in projects scattered over thirty-two states.

In any case, it is interesting to note that the coaction theory and the Genesa model appear to be fully consistent with each other and to be mutually reinforcing. In fact, these are two of eight master or meta-models which members of the Task Force on General Systems Education (of the Society for General Systems Research) have developed to interpret the entire universe. As is pointed out in my article, "Discovery of a possible Conceptual Bonanza," (in the March issue of the General Systems Bulletin), each of these eight meta-models seems to be consistent with and to fit into one other master-master or meta-meta-model. That broadest candidate for the designation, meta-meta-model, is the eco-cybernetic model outlined in this paper. Care should be taken to avoid hasty conclusions at this point, however, because the present chapter – perhaps the most exciting chapter – of the history of organized knowledge is still in the early stages of being written.

Unified Science: The Public Philosophy of the Space Age

EDWARD HASKELL

I THINK WE MUST AGREE with Walter Lippmann that "public philosophy is in large measure intellectually discredited among contemporary men;"[1] and with it, consciously or unconsciously, "the foundation of Constitutional Democracy."[2] This is a social and political catastrophe, a state so serious that many people dare not face it and therefore try to treat it as a joke. What happens when a civilization's public philosophy is discredited; when it no longer operates? What is it that, if we do not remedy it soon, is surely going to happen to us now? To convey to late 20th Century people what this means I must use terms upon which we still agree. These are the scientific and technological terms by which we have designed and operate our beautiful machines. So let me use a modern parable.

The automatic heating systems in our homes and offices and factories are based, by us, upon a built-in circuit of information; a carefully designed public philosophy. When the place gets cold, the thermostat senses it and tells the furnace, the furnace manifests inertia and says, in effect, "Why should I?" This is a question which our engineers can answer very well. They design, AND PLACE IN AUTHORITY, an instrument which knows how to anticipate this question and reply effectively. (It knows, as the hippies say, how to "turn it on.") It sends an electrical impulse to the furnace that flips a switch. The furnace, having been imbued by the engineers with the system's public philosophy, understands immediately why it should, and agrees: its motor opens its fuel and air intakes, its flames begin to roar, and the place warms up. When, presently, the thermostat senses that it has got as warm as the public philosophy requires, it informs the furnace ("turns it off") by flipping the switch the other way. The furnace understands, agrees, and complies. Is it not true that this heating system works as long as, and only as long as, all its components accept and obey its built-in public philosophy? Obviously it is. For whenever its public philosophy is discredited within any of its components, the system malfunctions or entirely breaks down.

Already, in our social system, the public philosophy has broken down so widely that the concept itself, PUBLIC PHILOSOPHY, lies outside the realm of universal public discourse. A great many people do not understand the term, and deny its validity when it is explained. Take any one of these many denials; say the one called Existentialism. "When Sartre, following Nietzsche, says that 'God is dead,' the critical point is not," in Lippmann's view, "that he refuses to believe in the existence, however, attenuated, of an anthropomorphic God. There can be, and indeed there is, great faith and deep religion without any concrete image of God. The radical unbelief lies underneath the metaphor of God's death. It is in Sartre's saying that 'if I have done away with God the Father, someone is needed to invent values . . . life has no meaning a priori . . . it is up to you to give it a meaning, and value is nothing but the meaning that you choose.' "[3]

"With this," Lippmann declares, "Sartre has done away not only with God the Father but with the recognition that beyond our private world there is a public world to which we belong. If what is good, what is right, what is true is only what the individual 'chooses' to 'invent', then we are outside the tradition of civility. . . . There is left no ground for accommodation among the varieties of men; nor is there in this proclamation of anarchy a will to find an accommodation."[4] Many who follow this view, and the related views of Herbert Marcuse, affirm that between them and the upholders of the public philosophy only complete separation or utter violence can solve the issues. And either way, socio-political disintegration is in sight.

But among us here, and our many friends, there still is a public philosophy: a fragmented and disjointed one, it is true, but one which it has now become possible to reconstruct in the natural idiom of our time. What we have is, in fact, two public philosophies and a number of smaller ones within each one of these. The man who identified and named our two principal public philosophies is C. P. Snow in his famous lecture, 'The Two Cultures';[5] the old, great, literate, artistic, religious culture on one hand, and the great new scientific, technological, industrial culture on the other. And between them the strategic point is this: up to about the 15th Century there was in the West, as in every other great civilization, a single culture. It was, and still is now, the one which we call Literate.

(Such was and is the public philosophy of each of the twenty-odd Literate civilizations discussed in Toynbee's monumental 'Study of History'.[6]) And it still operates today in Western Christendom, though with ever decreasing effectiveness. But it still works, for, as Lippmann points out, "except on the premises of this philosophy, it is impossible to reach intelligible and workable conceptions of popular elections, majority rule, representative assemblies, free speech, loyalty, property, corporations and voluntary associations. The founders of these institutions, which the recently enfranchised democracies have inherited, were all of them adherents of some one of the various schools of natural law. In our time the institutions built upon the foundations of the public philosophy still stand. But they are used by a public who are not being taught, and no longer adhere to the philosophy. Increasingly, the people are alienated from the inner principles of their institutions."[7]

Within Western Christendom's Literate culture there emerged in the 15th and 16th Centuries the at first barely perceptible scientific culture. Like all the other Literate civilizations, that of Western Christendom had consciously been built upon just one of what James B. Conant has called THE TWO MODES OF THOUGHT[8] – namely, upon the deductive-theoretical mode whose court of last resort is Authority. For though empirical observation played an important corrective role, it was applied neither consciously nor systematically. And it was heresy to appeal to experience over the voice of established authority. But Francis Bacon, the Apostle of Empiricism, reversed the position and raised the inductive-empirical mode to the place of the highest authority.[9] Thereby he set in motion what cyberneticians call a NEGATIVE feedback, which CORRECTS theory's deviations from conformity with this authority; and also a POSITIVE feedback, which actively EXTENDS theory's conformities thereto. Herewith the Scientific Revolution was set in motion, creating the ever more powerful and wide-spread Scientific Culture of the West.

The vast power of science and technology grows through the constant checking, correcting, deepening and extending of its verbal theory by the demanding process of comparing it with empirical reality; and conversely it grows through the constant improvement of observation by reviewing and improving it (and the instruments with which it is made) in the light of this evergrowing imaginative (deductive) theory. This Two-Mode progression has proved irresistible, sweeping IN ever more powerful human capabilities, and

OUT of their way, theories that disagree with them. (But with these, vast, actually valid segments of our Literate public philosophy). Lippmann points out, "A rereading of the political classics from Aristotle to Burke will not give answers to the immediate and concrete questions: to the burning issues of diplomacy, military defense, trade, taxes, prices, and wages. Nor have the classical books anything to say about repairing automobiles, treating poliomyelitis, or proceeding with nuclear fission. As handbooks for the busy man, wanting to know how to do this or that, they are now lamentably out of date. The language is archaic, the idiom is strange, the images are unfamiliar, the practical precepts are addressed to forgotten issues."[10] And what has been the result? All of the great Literate civilizations, except one, broke down and disintegrated; some of them two and even three times over. The public philosophy in each of them was, in time, discredited; the civilization went into cybernetic oscillation (Toynbee's "rhythm of disintegration") and disintegrated, its successor states transmuting downward into one or another of the lower (pre-Literate) Periods.

The sole exception was, and remains, our own Western Christendom. Because of the vast power over nature generated by its scientific revolution, instead of breaking down to a lower Period, WESTERN CHRISTENDOM TRANSMUTED UPWARD INTO THE NEXT HIGHER ONE: THE LOWER INDUSTRIAL PERIOD. And as its power spread, it transmuted Eastern Christendom (through Peter the Great) and Japan (through the Meiji). And together they are now in the process of transmuting all of the other cultures on our planet.[11]

There is, however, nothing inevitable about their succeeding in this process. The chances that they will fail, and that world civilization as a whole will disintegrate and transmute downward PREPONDERATE. And this is due to the tremendous and growing weaknesses of science. These weaknesses lie above all in the chaotic heterogeneity of its information. Since each empirical study is concrete, each concrete object or process studied has a special name; each of its presently discovered parts or characteristics has, or is soon given, another special name; and every part of these, yet still another special name, etc. The "stronger" science and technology grow (the more they explode), the feebler and narrower grows the individual scientist's view and understanding, and the smaller the proportion of the public to whom he can communicate his ever narrower and fiercer pinpoint of information. What we thus have today is an increasing host of enormously intense, minute authorities, destroying one isolated phase after another of the Literate public philosophy. But, being themselves informationally heterogeneous, their own theories are not, and cannot be assembled. So what they leave, in place of the Literate public philosophy, is what Lippmann has called The Great Vacuum.[12]

When diagrammed, however, this so-called "vacuum" turns out to have a structure; and also to stem from and imply an INSTITUTIONAL structure. This structure consists of many centrifugally directed Two Mode progressions traditionally called ACADEMIC DEPARTMENTS. The academic institution which is thus structured, Clark Kerr has called the MULTIVERSITY,[13] characterized by a number of scientific divisions, biological, chemico-physical, political, and so forth. The one-field specialists in each of these science departments stand, as Lippmann puts it, "on the hard intellectual ground where there are significant objects that are given and are not merely projected; that are compelling and not merely wished". The reason is, that for each set of scientific specialists "there is a public world sovereign above the infinite number of contradictory and competing private worlds,"[14] – namely, empirical reality, its Court of Last Resort.

And yet, the multiversity had now become the modern Tower of Babel, each of whose departmental languages grows ever less understandable to members of all other departments. As a result, the academic institution has DEVOLVED. It has already retrogressed to the condition of an echinoderm, a starfish, to the low organizational level of pre-automated systems, to systems which have no central regulator, to organisms which have no heads. This is the nature of The Crisis of the University, which Sir Walter Noberly recognized and described in 1949.[15]

Three or four centuries ago, the University did have a Rector, that is to say, an error corrector. He was a Divine; his science, theology, was uncontested queen of the sciences; for in our, as in all Literate and pre-Literate civilizations, religion is the public philosophy. The Rector therefore could and did correct deviations by any and all departments from the public philosophy. But with the cybernetic hook-up of the Two Modes of Thought by Francis Bacon and the Royal Society, scientific communication exploded into ever further exploding and self-developing communication bits and pieces. Unable any more to understand or communicate with its proliferating departments, the Rector was reduced to the enormous but self-defeating work of nourishing (by raising funds) and nurturing (by keeping house) for what, to use plain English, had become an uncontrolled and therewith cancerous kind of growth. The University, having lost both its public philosophy and its academic head, DEVOLVED into the modern multiversity.

What happens when the heating system's thermostat gets disconnected? What comes about when the central and generating institution within the information system of Lower Industrial civilization has lost its controller and has gone out of phase, gone haywire? The physical and biological sciences and technologies are roaring ahead exponentially while the psycho-social sciences are endlessly spinning out of gear; together they are knocking the arts, music and literature into corkscrew gyrations, philosophy into Logical Positivism and Existentialism, theology into atheism, democracy into egalitarianism, ethics into promiscuity, private enterprise into anarchy, crime into a science, and thus the environment into incipient collapse.[16] Students around the world are rioting, armies are occupying each others' territories and girding for general holocaust, while the public is watching commercial television and increasing its consumption of alcohol, tobacco and narcotics. The universal questions, "Why should I?" and "Why shouldn't I?" have become widely rhetorical; for the commercial editor, the permissive parent, the atheistic minister, and the fund-raising university president do not have compelling answers. Our civilization's Minority has lost all but gadget creativity. For our public philosophy has, as Walter Lippmann puts it, been eclipsed.[17]

Some people try to believe, with the rather grim Prussians, that "the situation is serious but not hopeless". That may be. Yet one may hold with the kidding Austrians that while "the situation might well be hopeless, it is not serious". And for the following reasons. Our terrifying disordered academic-social system has been EXAMINED thoroughly; the nature and source of its malfunction has been successfully DIAGNOSED; the PRESCRIPTION for its correction has been drawn up and successfully pilot-tested; and will be outlined presently. The prognosis, which has been written up realistically is (granted certain conditions) positive; and the concrete conditions for EXECUTING THE PRESCRIPTION have been created. At the heart of the matter is the need for bringing a multiplicity of disciplines into unity.

Everyone can see today, when it is pointed out, what only a

Leibniz grasped centuries ago, that all empirical scientists are IN FACT working at one great discipline, and that what keeps them from building the necessarily new public philosophy is just that they are unaware that they are doing so. They are quite capable of grasping that what they are doing is discovering, and describing ever more accurately PARTS OF THE UNIVERSE, a single and actually brilliantly working system. As more and more parts of this description emerged, somebody was bound one day (as Leibniz foresaw) to realize that what these apparently diverse sciences have created, being a description of a single system, could be made to form a single discipline. We may postulate that all the superficially or empirically diverse scientific disciplines have, in fact, a common structure that may be represented geometrically as the Periodic co-ordinate system. Leibniz, in fact, had predicted this geometry and had named it the Universal Characteristic.[18] Essential to The Renewal, of the Public Philosophy[19] is the now clearly apparent fact (predicted by Leibniz) that this model of the Universal Characteristic encompasses law, medicine, political science, and even metaphysics. Also, that this geometry tends to free the diverse disciplines' term of communication-noise, disclose gaps in their vocabularies, and helps to fill them in. This is to be the conceptual framework on which the public philosophy of the next higher human Period, the Space Age, can be constructed. Theoretical unification of the sciences could be, and had to be, brought about by an Inter-Divisional Committee. Once such a synthesis has been brought to the point of a working model, as it now has, institutional academic unifications can be put in train. But it requires, even to begin, an Inter-Disciplinary Center, located within a college or university, whose development it can engage and then transform. What, then, is the PRESCRIPTION for correcting this communications breakdown, and then moving forward? Is it not the reversal of the centrifugal directions in which the languages of the several sciences have long been developing, so that they presently become CENTRIPETAL? Our aim is Leibniz's, "If we could find characters or signs appropriate for expressing all our thoughts . . . . as exactly as arithmetic expresses numbers or geometric analysis expresses lines [then] if someone were to doubt my results, I should say to him: 'Let us calculate, Sir,' and thus by tak g to pen and ink, we should settle the question."[18] This result has been repeatedly obtained under test conditions. A Cartesian coordinate system was used to define and code the terms of discourse used in political controversies at a large, well known American college.[20] It was obtained by executing (in an elementary way and on a very small scale) the PRESCRIPTION for the crisis-ridden multiversity as a whole. In each discipline there would be formulated specific concepts and notations. These would be synthesized in an accommodation with other disciplines. The ultimate synthesis would be fed into (and initiated by) the Administration itself. That PRESCRIPTION is the converse of the multiversity's diagnosis. Thus it consists of the same number of Two–Mode developments, but instead of increasingly specialized languages, which diverge farther and farther from each other, these languages converge, becoming more and more understandable to its speaker.

The PROGNOSIS IS, that the members of all science departments will understand each other better and better, and therefore be able to benefit from each other's discoveries more and more. Each specialist would thus become better able to grasp the side-effects which manifest themselves outside his own field. With his colleagues' cooperation, he would find ways of reducing or preventing his destructive, and enhancing his constructive side-effects as Bacon said, "to the glory of God and the relief of man's estate".

Another result would be emergence of the University's long lost ACADEMIC head. As specialized languages converge, and the various kinds of scientists understand each other with ever greater ease, a point must presently be reached at which the university president finds himself understanding all his departments. When that occurs, a University Council made up of representatives from all major departments, and charged with coordinating and planning the institution's academic affort, becomes feasible, as feasible as a government's cabinet or the executive board of any corporation. And when that occurs, our habitat's confused destruction will probably decelerate and presently be reversed into construction. (The other expectable benefits of this convergence of academic languages must be omitted here, but are discussed elsewhere.)

The question now is this: how can the Center for Interdisciplinary Creativity make this convergence possible? And the answer is as follows. By teaching representatives from every discipline how to program, encode, or map their specialized data into the Periodic coordinate system (our model of the Universal Characteristic), thereby transforming such persons into what really amounts to PRACTITIONERS OF a single discipline with a common language.[21] This discipline with its geometric framework, its Periodic coordinate system, was abstracted from, and represents, the structure which all these kinds of data have in common. When these data are organized cybernetically, they constitute a Systems-hierarchy of nested Periodic Tables. Beginning with the Periodic Table of Stable Particles, it proceeds with the Periodic Table of Chemical Elements, Periodic Table of Biopoetic Molecules, the Periodic Table of Plant Ecosystems, the Periodic Table of Animal Ecosystems, to the Periodic Table of Human Cultures.

All these tables display, geometrically speaking, the same Sub-Strata, Strata and Periods, so that people in all fields of science will eventually be using the same basic terms. And what is probably even more important, all these Periodic tables display the same Groups, the same basic relations which philosophers over the Ages have called VALUES. This being the case, the Existentialists' anarchic contention that "value is nothing but the meaning that you choose" can be tested unequivocally in every discipline, and that, in both of the Two Modes of Thought, deductive and inductive.

The Center for Interdisciplinary Creativity is the means by which we will train participants from diverse disciplines to reason together in a common language. Is it not obvious that those who learn this language will have the capacity either to represent their departments on the University Council or, other things being equal, to chair such councils, thereby becoming academic heads of Universities? They should be equipped to rewrite the textbooks in their several fields so that science students no longer have to work against the Tower of Babel's bewildering cacophony. This seems to be the kind of thing which C. P. Snow saw emerging here in the United States, as he recorded in his 1963 revision, 'The Two Cultures – And a Second Look'.[22]

> At Yale and Princeton and Michigan and California ... scientists of world standing are talking to non-specialized classes: at M.I.T. and Cal. Tech. were students of the sciences receiving a serious humane education. In the last few years, all over the country, a visitor cannot help being astonished by the resilience and inventiveness of American higher education. ... It is probably too early to speak of a third culture already in existence, but I am convinced that it is coming."

It was too early in 1963 to speak of a third culture as already in existence. Our model of Leibniz's SCIENTIFICA GENERALIS had not yet become fully operational; no courses, let alone a Center for Interdisciplinary Creativity, had been organized; and no copies of even a preliminary textbook had been offset printed. Nor, for that matter, had protests and demonstrations closed famous universities around the world, preparing administrators and faculties to consider the well-coordinated change by which the disorganized crew of our Space-Ship Planet Earth will come to think in terms of the public philosophy of the Space Age: Unified Science. Today, however, these things have come about. Now it is possible, as Walter Lippmann hoped, "to alter the terms of discourse." For:

> "A convincing demonstration can be made that the principles of the good society are not, in Sartre's phrase, invented and chosen – that the conditions which are to be met if there is to be a good society are there, outside our wishes, where they can be discovered by rational inquiry, and developed and adapted and refined by rational discussion.
>
> If eventually this were demonstrated successfully (as I affirm it can) it would . . . rearm all those who are concerned with the anomy of our society, with its progressive barbarization, and with its descent into violence and tyranny. Amidst the quagmire of moral impressionism they would stand again on hard intellectual ground where there are significant objects that are given and not merely wished. Their hope would be re-established that there is a public world sovereign above the infinite number of contradictory and competing private worlds. Without this certainty, their struggle must be unavailing. . . . For political ideas require operative force in human affairs when, as we have seen, they acquire legitimacy, when they have the title of being right which binds men's consciences. That they possess, as the Confucian doctrine has it, 'the mandate of Heaven.'
>
> In the crisis within Western Society, there is at issue now the mandate of heaven."[23]

In our Scientific-Industrial civilization, the title of being right, which binds men's consciences, bears the sign of logical consistency and the seal of empirical validation. Some of the sciences' traditional terms of discourse have now, I believe, been altered in the manner which Lippmann advocates, producing the result he intuited and expressed more movingly than any scientist. However, belonging to the Literate Culture as he does, he could point out but could not himself lead the way into the Promised Land. The recreation of public philosophy in our society must be led by the keepers of its signs and seals of legitimacy – by scientists. They will achieve it by converging the sciences into a single discipline. For this is the public philosophy of the Space Age. This they can achieve because, and only because, with all its novelty and power, it is in basic harmony with the old Literate public philosophy which the fragmented sciences had brought into decline. The major forces of disintegration will, under this mandate of Heaven, be turned around and organized constructively.[24]

REFERENCES

1. Walter Lippmann: 'The Public Philosophy – On the Decline and Revival of Western Democracy'. Little, Brown & Co., Boston, 1955. (pp. 99-101, 179). Quoted by permission of Little, Brown, Publishers.
2. Idem.
3. Jean Paul Sartre, 'Existentialism', translated by Bernard Frechtman, (1947), p. 58.
4. Lippmann, ibid, p. 176.
5. C. P. Snow, 'The Two Cultures and the Scientific Evolution'. Cambridge University Press, New York, 1959.
6. Arnold J. Toynbee, 'A Study of History' (Somervell abridgement of Vols. I-VI) Oxford University Press, New York, 1949.
7. Lippmann, ibid., pp. 101-2.
8. James B. Conant, 'The Two Modes of Thought', Trident Press, New York, 1964.
9. M. E. Prior, "Francis Bacon" in 'Roots of Scientific Thought', P. P. Weiner and A. Noland, Editors, Basic Books, New York, 1958.
10. Lippmann, ibid., p. 103.
11. Edward F. Haskell, "Assembly of the Sciences – Into a Single Discipline". (With a chapter by Harold G. Cassidy) Preliminary version offset-printed, New York, 1968.
12. Ibid., pp. 96-104.
13. Clark Kerr, 'Uses of the University', Harvard Press, Cambridge, Mass., 1963.
14. Lippmann, ibid., p. 180.
15. Walter Noberly, 'The Crises in the University', S.C.M. Press, London, 1949.
16. Harold G. Cassidy, "Incipient Environmental Collapse".
17. Lippmann, ibid., Chapter VIII, "Eclipse of the Public Philosophy".
18. Robert McRae: "The Unity of Science: Bacon, Descartes, Leibniz" in 'Roots of Scientific Thought', P. P. Weiner and A. Noland, Editors, Basic Books, New York, 1958.
19. Lippmann, ibid., Chapter IX, "Renewal".
20. Edward F. Haskell, "Assembly of the Sciences – Into a Single Discipline". Vol. II. Res. Cognitans, m.s.
21. 'Calculemus', Volume III of "Assembly of the Sciences – Into a Single Discipline" by 20 scientists, E. F. Haskell, Editor, m.s.

22. C. P. Snow: 'The Two Cultures: And a Second Look'. New American Library, New York; New English Library, London, 1963.
23. Lippmann, ibid., pp. 180-1.
24. Edward F. Haskell, "The Religious Force of Unified Science", 'The Scientific Monthly', LIV, June, 1942, 545-551.

LIST OF CONTRIBUTORS

BELA H. BANATHY is a Program Director at the Far West Laboratory for Educational Research and Development, Berkeley, California. Prior to joining the Far West Laboratory he was Director of the East Europe-Middle East Division of the Defense Language Institute and has served as consultant in systems design and evaluation to school systems, to research, publishing and industrial concerns. He has conducted professional and graduate level courses at San Jose State College and he has authored language textbooks, standardized tests, teacher training materials, research studies and numerous articles in professional journals. He is a contributing author to 'Trends in Language Teaching' (McGraw-Hill, 1966), to the 'Britannica Review of Foreign Language Education' (Encyclopedia Britannica, Inc., 1969), author of 'Instructional Systems' (Fearon, 1968), and co-author of 'Design of a Functional Competence Training Program for Development, Dissemination and Evaluation Personnel at Professional and Paraprofessional Levels in Education' (Far West Laboratory, 1970).

He is a member of the National Task Force on Systems Education of the Society of General Systems Research and Chairman of the Western Division of the Task Force. He has been conference leader on Applied Linguistics at recent annual meetings of the Modern Language Association.

ALTON BARBOUR is Associate Professor and Director of Graduate Studies in Speech Communication at the University of Denver. He describes himself as a generalist rather than a specialist with interests ranging from poetry to humanistic psychology. He devotes his time to the study and teaching of communication because, as he says, "Communication is germane to practically everything".

J. SAMUEL BOIS lives the very active life of a retired industrial psychologist and management consultant. A former Canadian (Ph.D. from McGill University, Montreal; president of the Canadian Psychological Association, 1948-49), he is now a U.S. citizen, extension lecturer at U.C.L.A., instructor at the School of Public Administration, U.S.C., co-founder and director of Viewpoints Institute, a 'center for the study, practice, and development of general semantics,' in Los Angeles. Some general semanticists see him as 'the leading theoretician' (ETC, Dec. 1969); some find him a disturbing innovator. His books, 'Explorations in Awareness', (Harper, 1957); 'The Art of Awareness', (Wm. Brown Co., 1966); 'Communication as Creative Experience', (Viewpoints, 1968); 'Breeds of Men', (Harper & Row, 1970); cover a broad field which he has renamed 'Epistemics, A Time-binding Emergent from General Semantics', (ETC, June 1971).

D. DAVID BOURLAND, JR., recently founded Semantics Research Corporation with his wife, K. L. R. Bourland who also contributed a paper to this volume (see below). In addition to serving as President of that consulting operations research firm, he also participates in general semantics activities as a member of the Board of Trustees of the Institute of General Semantics and has acted in various editorial capacities with the 'General Semantics Bulletin' since 1962. A long-time student in the field (five seminars with Korzybski, holder of a Korzybski Fellowship during 1949-1950), he has written a number of papers inquiring into the epistemological foundations of Korzybski's methodology. He holds A.B. (Mathematics) and M.B.A. degrees from Harvard.

KAREN L. R. BOURLAND recently founded Semantics Research Corporation with her husband, D. D. Bourland, Jr., who also contributed to this volume. In addition to serving as Vice President of that Washington, D.C.-based operations research firm, she has assisted in producing and editing the 'General Semantics Bulletin' since 1968. She holds an A.B. degree from the University of California at Riverside, and currently pursues a doctoral program at the George Washington University.

JAMES M. BROADUS teaches and learns with his fellow students at Trannsylvania University in Lexington, Kentucky. He has been permitted to do this for nine years as Chairman of the Department of Education and Physical Education. Some of his other experiences include: (1) Growing up in rural Alabama, where conformity may have fostered an awareness of the desirability of differences, (2) Study at the University of Alabama, where he discovered that the emotional evaluation of education (and everything else?) is a function of man, (3) Teaching chemistry and band where it became evident that without children, people, there would be no chemists nor musicians, nor even anyone who missed them, (4) Time in W.W. II Air Force, where he concluded people ought to look for better ways of doing things, e.g., resolving differences, (5) Work at Johns Hopkins, where he was awarded credentials which are considered more important by some people than by others, (6) Election to the local School Board, where idealism and practicality seem less and less disparate, (7) Study of Korzybski in search for more good things and more good people.

JERE W. CLARK is professor of Economics and Director, Center of Interdisciplinary Creativity at SCSC. For the last three years he has served as the international Chairman of the Task Force on General Systems Education, Society for General Systems Research. A charter member of the Leadership Council, Creative Education Foundation, Dr. Clark has served as Program Chairman for three national, and one international, conferences on the future of education.

During the U.N. International Education Year (1970), he served as Systems Creativity Editor for the monthly journal, 'International Associations'. In 1963 he received the first-place award in the National Awards Program of the Calvin K. Kazanjian Economics Foundation for the best course in Economics at the college level.

Dr. Clark received the B.B.A. and M.A. degrees at the University of Georgia and the PH.D in Economics at the University of Virginia.

He served as editor for the first three volumes of 'Economic Education Experiences of Enterprising Teachers', 1963-1965. He also edited 'Systems Education Patterns on the Drawing Boards for the Future,' 1969, and 'Practical Action Programs in Education', 1971. Dr. Clark is Consulting Editor for 'The Journal of Creative Behavior' and Associate Editor of 'General Systems Bulletin'. He is now a member of the Coordinating Committee, Connecticut Chapter, World Future Society.

STUART C. DODD has been a research professor in Sociology at the University of Washington in Seattle for the last twenty-four years, and the Director of the Washington Public Opinion Laboratory during its legislature-financed lifetime. For a score of years before that he directed the Social Science Research Section at the American University of Beirut and organized the popular Village Welfare movement for which the Lebanese government decorated him.

His life interest in synthesizing science developed through publishing 'Dimensions of Society' (1942), 'Systematic Social Science' (1947) and 'The Epicosm Model for the Material and Mental Universes' (1969). His 150 research articles, pursuing this trend, are in press for republication by Gordon and Breach in a four volume series of 'Systemed Studies on Human Transactions.'

ALLEN FLAGG has the position of Executive Director, New York Society for General Semantics. He is Secretary of the General Semantics Foundation and a Trustee of the Institute of General Semantics. He has taught over 75 courses, seminars and workshops on communication and behavior, for such organizations as Fairfield University's Graduate School for Corporate and Political Communication, International Business Machines, Inc., Institute of General Semantics, New York Society for General Semantics, and Queens College. He received his B.A. degree from New York University, majoring in mathematics. He read his paper 'Dream Education as a Neglected Part of the Education Process' at the 1968 International Conference on General Semantics at Denver University. His present concern is the development of his Epistemic Differential, to measure cognitive or denotative meanings of concepts.

ALVIN A. GOLDBERG is professor and chairman of the Speech Communication Department at the University of Denver. Professor Goldberg has served on the faculty of the University of Hawaii, Northern Illinois University and Northwestern University. He received his Bachelors degree from Wayne State University, his Masters from the University of Hawaii and his Doctorate from Northwestern. Professor Goldberg is a past chairman of the Freedom of Speech Committee of the Speech Communication Association. In addition to freedom of speech, his scholarly research interests include interpersonal and group communication, communication theory and general semantics. Professor Goldberg is author of more than fifty journal articles and papers in these areas. He is a co-editor of 'Language Behavior: A Book of Readings' published by Mouton Press, The Hague.

GLYNN HARMON is an Associate Professor of Information Science, Graduate School of Library Science, University of Texas at Austin. He holds B.A. and M.A. degrees from the University of California at

Berkeley, and M.S. and Ph.D. degrees from Case Western Reserve University. His publications include "On Integration of the Integrative Disciplines", 'Research Designs in General Semantics' (Gordon and Breach), "Mathematical Approach to Scientific Discovery," 'Nature' (January 8, 1971) and 'Human Memory and Knowledge: A Systems Approach,' (Greenwood Publishing Corporation, 1971). His hobbies are running, swimming, real estate and private flying.

EDWARD HASKELL'S father's family – American missionaries to the Balkans (1860-1935) – profoundly imprinted in him the public philosophies of the pan-English and pan-Slavic societies; and Switzerland, his mother's country, imprinted those of the pan-German and pan-Romance societies.

His experience of the shocking breakdown of public philosophy throughout the Atlantic Community began at Oberlin College and increased through ten years of graduate studies at Columbia, Harvard, and the University of Chicago until – in his seventh graduate year (1940) – he revolted. He refused to participate in the public philosophy's further breakdown by preparing for a doctorate in any one-field discipline.

Thereupon the University of Chicago established an Interdivisional Committee to supervise the work of Edward Haskell and awarded him a three year fellowship, initiating what may prove to be the trend's reversal, from what Toynbee calls SCHIZM to TRANSFIGURATION. In 1948 Haskell organized the Council for Unified Research and Education (C. U. R. E., Inc.), which he still chairs; and in 1953 began his three-volume 'Unified Science – Assembly of the Sciences into a Single Discipline'. In 1968 he began teaching unified science at the New School for Social Research, and in 1969 moved to the Center for Interdisciplinary Creativity at Southern Connecticut State College in New Haven. The National Institute of Health and the IBM Systems Research Institute xeroxed enough copies of 'Unified Science, Volume I' to supply his experimental courses' textbook needs.

The first book publication in Unified Science is scheduled for 1972 by Gordon and Breach, Science Publishers: 'Full Circle – The Moral Force of Unified Science'.

C. ANDREW HILGARTNER first became interested in general semantics in 1950. In 1959 he began his career as a research worker, and while attending the 1960 Seminar/Workshop of the Institute of General Semantics embarked on his current project. Initially his efforts went to 'make his own life more to his own liking,' by using the methods of general semantics, etc., to 'make explicit, test and if necessary revise' his own life-assumptions. Eventually he became curious about the structure of these experiences of 'self-correcting,' and wrote two papers (1963, 1965) exploring the topic of 'major behavioral change.' In presenting the latter of these two papers, he made logical claims for his viewpoint; and then became interested in the topic of how to subject these claims to test. In a collaborative setting, he succeeded in translating his doctrine into the form of an axiomatic system stated in a mathematical (set theory) notation, which has turned out as a formal non-Aristotelian system, a 'general theory' of unprecedented scope. Besides its implications for the logical foundations of mathematics, this 'general theory' has already delivered results of fundamental import on a number of other 'levels' of human interest: for example, in applications to the human context of classical physics; to the level of biological theory (where it has provided a rigorous and formally satisfactory definition of the term 'living system'); to the topic of the behavior-and-experience of individuals; and to the 'social' levels of 'small group' phenomena and of 'large social institutions'.

Dr. Hilgartner attended Exeter, Amherst College and Johns Hopkins Medical School. He worked in biochemical research from 1959 to 1968 at the University of Rochester. From 1968 to 1969 he held a fellowship at the Center for Theoretical Biology, State University of New York at Buffalo, and since that time has pursued his interest in general non-Aristotelian theories as a free-lance investigator.

With his wife and five children, Dr. Hilgartner lives at present in Rochester, New York.

B.R. HUNT is a systems analyst on appointment with the University of California, Berkeley, at the Los Alamos Scientific Laboratories, Los Alamos, New Mexico. He received his Ph.D. from the University of Arizona in 1967. Papers by him have been published in 'Mathematical Biosciences', 'IEEE Transactions on Computing', 'IEEE Transactions in Audio', and numerous proceedings of conferences on systems, engineering, and medical sciences.

RUSSELL W. JENNINGS, Ph.D. is an Associate Professor in Speech at Southern Illinois University, Carbondale Campus, and is associated with the SIU Center for the Study of Communication. He received his B.S. Degree from Western Montana College in 1959, the M.A. Degree from the University of Montana in 1966, and the Ph.D. Degree from Southern Illinois University in 1968. For the past three years, he has conducted research on Interpersonal Communication in urban tension confrontation and cross-cultural encounter situations. The focus of his research has been on the functional role of communication in formal organizations as they attempt to affect decision-making and coping behavior in social-political action.

KENNETH G. JOHNSON is an associate professor of mass communication at the University of Wisconsin-Milwaukee. He is also on the staff of the University of Wisconsin Extention. For seven years he was chairman of the department of journalism at UWM, before it became the department of mass communication.

Dr. Johnson has been concerned with problems of communication for more than 20 years, based on a background of newspaper reporting, science writing, technical writing and editing, and teaching of journalism and semantics.

The study of the relationships between language and human behavior has been his specialty. He received a bachelor's degree in chemistry, a master's in journalism, and Ph.D. in mass communication from the University of Wisconsin. His doctoral dissertation concerned problems in the communication of science to a general public.

Dr. Johnson is author of several articles on communication, science, and semantics and a booklet entitled 'General Semantics: An Outline Survey', He has edited a book, 'Research Designs in General Semantics', and is senior author and editor of a book of communication activities for groups (in process).

His is a member of Sigma Delta Chi, professional journalism society, the Association for Education in Journalism, American Psychological Association, The Institute of General Semantics, the International Society for General Semantics, and the Creative Education Foundation. He serves as an associate editor of the General Semantics Bulletin. During the summer of 1963, he was visiting scholar at the Institute of General Semantics.

JAMES C. McFARLAND is currently an Instructor in the Department of Speech-Communication at Northern Illinois University where he teaches courses in Communication Skills and Language and Thought in Society. His B.S. Degree is from the University of

Utah and his M.A. is from San Francisco State College. He is currently completing his Ph.D. at Northwestern University. In addition to Communication Theory his academic interests include interpersonal and small group communication. He has done dissertation research and has conducted workshops on Interpersonal Self-Actualization and the concept of Dialogue in Communication. He is also active in conducting Encounter groups, Self Discovery Groups and other types of Growth Groups in the Chicago area.

ELWOOD MURRAY is Emeritus Director, Institute of General Semantics and Emeritus Director, School of Speech, University of Denver, where he has his office. He was program chairman of the International Conference on General Semantics held in Denver, 1968, papers of which have been edited by Lee Thayer and published by Spartan Books, Inc., N.Y.C. 1970. He was also program chairman of a Conference at Pennsylvania State University, on Research Methods in General Semantics, now ready to be distributed by Gordon and Breach, N.Y.C., 1972. At present he is engaged in carrying through several other joint research conferences with learned societies related to applied epistemology and communication.

THOMAS J. PACE, JR. is Professor of Speech at Southern Illinois University where he teaches the Laboratory in Interpersonal Communication. He received his Ph.D. at the University of Denver, did postdoctoral work in Speech and in Philosophy at Northwestern University, and was a member of an Institute of General Semantics Seminar. His research interest is in the Philosophy of Communication.

ROBERT PULA has taught communication courses in the public schools and at various colleges, universities and government agencies. The agencies include the National Institutes of Health, Federal Trade Commission, Atomic Energy Commission, NASA, Federal Communication Commission and others. He is currently teaching reading improvement courses at Forst Meade and Holabird and for the Adult Education Division, Baltimore Public Schools. He is also part of an instructional team presenting a pilot program (college preparatory) in communications and mathematics to military personnel at Fort Meade. The communications part of the program is built around a core of general semantics formulations. The program is presented by the University of Maryland and funded by the Veterans Administration.

Mr. Pula is a lecturer at the Baltimore College of Commerce (general semantics) and a member of the seminar staff of the Institute of General Semantics. He has lectured on general semantics at Pennsylvania State University, New York State University, Bard College and the University of Denver. In October, 1970, he was a featured speaker at a communications seminar at the University of Missouri. He conducted a unique one-day seminar on critiques of general semantics in New York (December 5, 1970) and was the main lecturer at the Institute's winter seminar-workshop at Rye, N.Y. He has been invited to give a paper for a conference on creativity at New York State University (Buffalo) and will lecture at two general semantics seminars at Colorado Academy in July, 1971.

Mr. Pula serves on the Editorial Board of the 'General Semantics Bulletin', has published several papers in his field, and has a book-length study in general semantics scheduled for publication in late 1971.

CHARLOTTE S. READ (Mrs. Allen Walker Read) is Literary Executor and Trustee of the Alfred Korzybski Estate, and is also associated with the Institute of General Semantics, Lakeville, Connecticut, in a number of ways. She was Acting Director 1966-67 and 1970-71, is Co-Editor of the 'General Semantics Bulletin,' Secretary of the Board of Trustees and a staff member for many years at Institute Seminars. She was Korzybski's assistant at the Institute 1939-50. She now resides in New York City.

Mrs. Read received her B.A. from the University of Wisconsin in Zoology and M.A. from the University of Chicago in Education. She studied and taught physical education and dance, and is particularly interested in her present teaching in the development of kinesthetic awareness in relation to personal growth and integration. She has written numerous articles for the 'General Semantics Bulletin,' and has contributed to 'Communication; General Semantics Perspectives' (Lee Thayer, ed.).

EDWARD S. SAFFORD is President of the Denver Division of the Cahners Publishing Company, Inc. He serves on a number of business organization and business association boards, and is presently President of The National Center of Communication Arts and Sciences. Mr. Safford is listed in eight different versions of Who's

Who. He lectures regularly before business and academic audiences and has published numerous articles in the field of communication.

FRANKLIN H. SILVERMAN is an Associate Professor of Speech at Marquette University and has been on the faculty of the University of Illinois at Urbana-Champaign and of the University of Iowa. He is an Associate Editor of the 'Journal of Speech and Hearing Research.' He received his B.S. in Speech (1960) from Emerson College, his M.A. (1961) from Northwestern University, and his Ph.D. (1966) from the University of Iowa. He is a member of the American Association for the Advancement of Science, the American Speech and Hearing Association, the American Statistical Association, the Institute of General Semantics, the Psychometric Society, and Sigma Xi. His publications include numerous articles in the areas of research methodology and stuttering.

DENNIS R. SMITH is an Assistant Professor in Speech Communication at Temple University, Philadelphia, Pennsylvania. He received his Ph.D. from the University of Kansas in 1967. He has directed research under the auspices of the U.S. Office of Economic Opportunity and the U.S. Office of Education. Articles by him have appeared in the 'Quarterly Journal of Speech' and 'Today's Speech'. Along with Donald Washburn he is an editor of the present volume.

LEE THAYER is Gallup Professor of Communication Research and Director, Center for the Advanced Study of Communication, the University of Iowa. He was previously Professor of Administration and Director of Research, School of Administration, University of Missouri at Kansas City. He has served as consultant on organizational and communication problems to government and industry for fifteen years. His papers have appeared widely in professional and scholarly journals. He is Past President of the International Communication Association, and serves on several boards of directors, including the Institute of General Semantics. He has recently edited 'Communication: General Semantics Perspectives', and 'Communication: Ethical and Moral Issues', in addition to his widely-used text, 'Communication and Communication Systems'. He will be Editor-in-Chief of the new international journal, 'Communication. . .' to appear in 1972.

JOSEPH VINOVICH is a Graduate Fellow in Speech at Southern

Illinois University, Carbondale Campus, and is completing his doctoral program in association with the SIU Center for the Study of Communication. He received his B.A. and M.A. Degrees from the University of New Mexico in 1967 and 1969. For the past two years he has conducted research on Interpersonal and Organizational Communication in communication systems and design. The focus of his research has been the use of participant simulation experience for human relations situations.

DONALD WASHBURN is chairman of the English Department at North Adams State College in the Berkshire mountains of Massachusetts. His interest in General Semantics dates from his undergraduate years at Yale University, where he became acquainted with the work of Korzybski and attended one of the IGS seminars held at Lakeville, Connecticut. Later, at the University of Denver he combined a wide range of studies, including General Semantics, communication methodology, literature, speech, and linguistics. His doctoral dissertation was an analysis of the work of C. G. Jung from the point of view of communication theory.

For many years he was a member of the English Department at Edinboro State College in northwestern Pennsylvania, where he taught World Literature over closed-circuit television and wrote two textbooks for the course. He also conducted graduate courses in General Semantics, linguistics, and contemporary poetry. During the summer of 1969, he was a staff member of the IGS Seminar held at Edinboro. He has published in the 'General Semantics Bulletin,' the 'Journal of Communication', and in collections such as 'Communication: General Semantics Perspectives', Lee Thayer, ed. and 'Research Designs in General Semantics', Kenneth G. Johnson, ed. Along with Dennis Smith, he has served as editor of this volume.

ANTHONY WILDEN (University of California, San Diego) received his Ph.D. from Johns Hopkins in 1968 for his study of structuralism in French psychoanalysis, centered on the translation and analysis of a work by Jacques Lacan ('The Language of the Self', Baltimore: Johns Hopkins Press, 1968).

The inadequacy of the linguistic and mechanistic models proposed by structuralism led him to the study of cybernetic, communications, and systems theory, and eventually to an analysis of the epistmological and ideological basis of current French and Anglo-American approaches to modeling the social ecosystem. Wilden has published a number of journal articles on Freud, Lacan, Marcuse, Bateson, Lévi-Strauss, and Piaget, as well as a theoretical study of analog and digital communication and an outline of a communicational model of social and evolutionary change. In 1971-72, he was a visiting professor in sociology at the Sorbonne (Ecole Pratique des Hautes Etudes, Center for the Study of Mass Communications). His 'System and Structure: Essays in Communication and Exchange' was published in 1972 by Tavistock Publications, who will also publish his latest book, 'Ecology and Ideology.' He is presently working on an analysis of Marx from an ecosystemic and communicational perspective.

Printed in Great Britain by Biddles Ltd, Guildford, Surrey